Optimizing STEM Education With Advanced ICTs and Simulations

Ilya Levin
Tel Aviv University, Israel

Dina Tsybulsky
Tel Aviv University, Israel

A volume in the Advances in Educational Technologies and Instructional Design (AETID) Book Series

IGI GLOBAL
DISSEMINATOR OF KNOWLEDGE

www.igi-global.com

Published in the United States of America by
IGI Global
Information Science Reference (an imprint of IGI Global)
701 E. Chocolate Avenue
Hershey PA, USA 17033
Tel: 717-533-8845
Fax: 717-533-8661
E-mail: cust@igi-global.com
Web site: http://www.igi-global.com

Library of Congress Cataloging-in-Publication Data

Names: Levin, Ilya, editor. | Tsybulsky, Dina, editor.
Title: Optimizing STEM education with advanced ICTs and simulations / Ilya Levin and Dina Tsybulsky, Editors.
Description: Hershey PA : Information Science Reference, [2017]
Identifiers: LCCN 2017005138| ISBN 9781522525288 (hardcover) | ISBN 9781522525295 (ebook)
Subjects: LCSH: Science--Study and teaching--Computer-assisted instruction. | Science--Study and teaching--Technological innovations. | Educational technology. | Information technology.
Classification: LCC LB1585 .O68 2017 | DDC 371.33--dc23 LC record available at https://lccn.loc.gov/2017005138

This book is published in the IGI Global book series Advances in Educational Technologies and Instructional Design (AETID) (ISSN: 2326-8905; eISSN: 2326-8913)

British Cataloguing in Publication Data
A Cataloguing in Publication record for this book is available from the British Library.

All work contributed to this book is new, previously-unpublished material.
The views expressed in this book are those of the authors, but not necessarily of the publisher.

For electronic access to this publication, please contact: eresources@igi-global.com.

Advances in Educational Technologies and Instructional Design (AETID) Book Series

ISSN:2326-8905
EISSN:2326-8913

Editor-in-Chief: Lawrence A. Tomei, Robert Morris University, USA

MISSION

Education has undergone, and continues to undergo, immense changes in the way it is enacted and distributed to both child and adult learners. From distance education, Massive-Open-Online-Courses (MOOCs), and electronic tablets in the classroom, technology is now an integral part of the educational experience and is also affecting the way educators communicate information to students.

The **Advances in Educational Technologies & Instructional Design (AETID) Book Series** is a resource where researchers, students, administrators, and educators alike can find the most updated research and theories regarding technology's integration within education and its effect on teaching as a practice.

COVERAGE

- Curriculum Development
- Digital Divide in Education
- Instructional Design
- E-Learning
- K-12 Educational Technologies
- Virtual School Environments
- Web 2.0 and Education
- Adaptive Learning
- Collaboration Tools
- Online Media in Classrooms

IGI Global is currently accepting manuscripts for publication within this series. To submit a proposal for a volume in this series, please contact our Acquisition Editors at Acquisitions@igi-global.com or visit: http://www.igi-global.com/publish/.

The Advances in Educational Technologies and Instructional Design (AETID) Book Series (ISSN 2326-8905) is published by IGI Global, 701 E. Chocolate Avenue, Hershey, PA 17033-1240, USA, www.igi-global.com. This series is composed of titles available for purchase individually; each title is edited to be contextually exclusive from any other title within the series. For pricing and ordering information please visit http://www.igi-global.com/book-series/advances-educational-technologies-instructional-design/73678. Postmaster: Send all address changes to above address.

Titles in this Series

For a list of additional titles in this series, please visit:
http://www.igi-global.com/book-series/advances-educational-technologies-instructional-design/73678

Exploration of Textual Interactions in CALL Learning Communities Emerging Research...
Jonathan R. White (Högskolan Dalarna, Sweden)
Information Science Reference • ©2017 • 195pp • H/C (ISBN: 9781522521426) • US $120.00

Mobile Technologies and Augmented Reality in Open Education
Gulsun Kurubacak (Anadolu University, Turkey) and Hakan Altinpulluk (Anadolu University, Turkey)
Information Science Reference • ©2017 • 366pp • H/C (ISBN: 9781522521105) • US $185.00

Cases on STEAM Education in Practice
Judith Bazler (Monmouth University, USA) and Meta Van Sickle (College of Charleston, USA)
Information Science Reference • ©2017 • 375pp • H/C (ISBN: 9781522523345) • US $195.00

Handbook of Research on Driving STEM Learning With Educational Technologies
María-Soledad Ramírez-Montoya (Tecnologico de Monterrey, Mexico)
Information Science Reference • ©2017 • 590pp • H/C (ISBN: 9781522520269) • US $300.00

Integrating an Awareness of Selfhood and Society into Virtual Learning
Andrew Stricker (The Air University, USA) Cynthia Calongne (Colorado Technical University, USA) Barbara Truman (University of Central Florida, USA) and Fil Arenas (The Air University, USA)
Information Science Reference • ©2017 • 383pp • H/C (ISBN: 9781522521822) • US $200.00

Intercultural Responsiveness in the Second Language Learning Classroom
Kathryn Jones (Lamar University, USA) and Jason R. Mixon (Lamar University, USA)
Information Science Reference • ©2017 • 327pp • H/C (ISBN: 9781522520696) • US $175.00

For an enitre list of titles in this series, please visit:
http://www.igi-global.com/book-series/advances-educational-technologies-instructional-design/73678

IGI GLOBAL
DISSEMINATOR OF KNOWLEDGE

www.igi-global.com

701 East Chocolate Avenue, Hershey, PA 17033, USA
Tel: 717-533-8845 x100 • Fax: 717-533-8661
E-Mail: cust@igi-global.com • www.igi-global.com

Editorial Advisory Board

Table of Contents

Foreword xv

Preface xvii

Chapter 1
From Wearing to Wondering: Treating Wearable Activity Trackers as Objects of Inquiry 1
Joel R. Drake, Utah State University, USA
Ryan Cain, Utah State University, USA
Victor R. Lee, Utah State University, USA

Chapter 2
The Use of Complementary Virtual and Real Scientific Models to Engage Students in Inquiry: Teaching and Learning Climate Change Science 30
Allan Feldman, University of South Florida, USA
Molly Nation, University of South Florida, USA
Glenn Gordon Smith, University of South Florida, USA
Metin Besalti, University of South Florida, USA

Chapter 3
Using Virtual Labs in an Inquiry Context: The Effect of a Hypothesis Formulation Tool and an Experiment Design Tool on Students' Learning 58
Tasos Hovardas, University of Cyprus, Cyprus
Nikoletta A. Xenofontos, University of Cyprus, Cyprus
Zacharias C. Zacharia, University of Cyprus, Cyprus

Chapter 4
Collaborative Learning and Assessment of Science: Structuring Effective Groups 84
Yigal Rosen, Harvard University, USA

Chapter 5
Learning About Genetics in an Elementary Classroom Using a Web-Based Inquiry Science Environment (WISE) Unit 107
Amal Ibourk, Florida State University, USA
Michelle Williams, Williams Learning Solutions, Inc., USA
Merle Heidemann, Michigan State University, USA

Chapter 6
STEM Inquiry Through INPLACE Mobile Games 134
Denise M. Bressler, Rutgers University, USA

Chapter 7
Learning about Complex Systems from the Bottom Up: Role-Playing Together in a Participatory Simulation 158
Sharona T. Levy, University of Haifa, Israel

Chapter 8
Simulations in Chemistry for Conceptual Understanding and Assessment of Student Knowledge 186
Tanya Gupta, South Dakota State University, USA
Zachary P. Ziolkowski, South Dakota State University, USA
Gregory Albing, South Dakota State University, USA
Akash Mehta, Independent Researcher, USA

Chapter 9
Science Educator Professional Development: Big Data and Inquiry Learning 219
Anna Lewis, University of South Florida – St. Petersburg, USA
George Matsumoto, Monterey Bay Aquarium Research Institute, USA

Chapter 10
A Case Study of Infusing Collaborative STEM Inquiry Learning With Available Technology Into Undergraduate Student Learning: Perceptions, Benefits, and Challenges 245
Katarin Alinta MacLeod, St. Francis Xavier University, Canada

Chapter 11
The Use of Tablet Technology to Support Inquiry Science for Students Incarcerated in Juvenile Justice Settings....267
Michael Krezmien, University of Massachusetts – Amherst, USA
Wardell Powell, University of Massachusetts – Amherst, USA
Christina Bosch, University of Massachusetts – Amherst, USA
Tracey Hall, Center for Applied Special Technology, USA
Martina Nieswandt, University of Massachusetts – Amherst, USA

Compilation of References....296

About the Contributors....335

Index....341

Detailed Table of Contents

Foreword xv

Preface xvii

Chapter 1
From Wearing to Wondering: Treating Wearable Activity Trackers as Objects of Inquiry 1
Joel R. Drake, Utah State University, USA
Ryan Cain, Utah State University, USA
Victor R. Lee, Utah State University, USA

Wearable technologies represent a rapidly expanding category of consumer information and communications technologies. From smartwatches to activity tracking devices, wearables are finding their way into many aspects of our lives, changing the way we think about ourselves and the world around us. The rapid adoption of these tools in everyday life hints at the possibilities these devices may hold in school and other educational settings. Drawing on examples taken from a five-year study using wearable fitness tracking devices in elementary and middle school classrooms, this paper presents two examples of how wearable devices can be appropriated for use in school settings. These examples focus on instances where students turned activity trackers into objects of inquiry using data from familiar activities.

Chapter 2
The Use of Complementary Virtual and Real Scientific Models to Engage Students in Inquiry: Teaching and Learning Climate Change Science................30
Allan Feldman, University of South Florida, USA
Molly Nation, University of South Florida, USA
Glenn Gordon Smith, University of South Florida, USA
Metin Besalti, University of South Florida, USA

This chapter reports on a four-year study to change how climate change science is taught and learned in schools. The goal of the Climate Change Narrative Game Education (CHANGE) project is to take what is known about reform-based practices, incorporating students' lived experiences into the curriculum, and the integration of Information and Communications Technology (ICT) into the classroom. CHANGE uses the following: scientifically realistic text narratives (text stories with local characters, 50-100 years in the future, a local, place-based approach, a focus on the built environment, the use of simulations and games based on scientific data, and a web-based "intermedia" eBook narrative where sections of narrative text alternate with simulations and computer games. The chapter reports on the ways that we have used the above principles to connect classrooms and communities and school science with academic science to facilitate student inquiry into climate science by combining virtual serious educational games with in class, hands-on inquiry using scientific models.

Chapter 3
Using Virtual Labs in an Inquiry Context: The Effect of a Hypothesis Formulation Tool and an Experiment Design Tool on Students' Learning.........58
Tasos Hovardas, University of Cyprus, Cyprus
Nikoletta A. Xenofontos, University of Cyprus, Cyprus
Zacharias C. Zacharia, University of Cyprus, Cyprus

The present study employed two different Go-Lab scaffolds/tools. These tools were used by primary school students to carry out successive learning tasks during experimentation. The first tool assisted learners in formulating hypotheses, while the second tool guided students in designing experiments (EDT). Both tools were designed to take into account the trade-offs between structuring and problematizing student inquiry. Our aim was to investigate the effect of each tool separately, as well as the combined effect of the tools in supporting student work. Participants were 41 fifth graders from two classes of a public primary school in Cyprus. They were randomly assigned to four conditions: Condition 1 involved use of both tools, Condition 2 included the hypothesis tool only, Condition 3 included the EDT only, and Condition 4 had no tools provided. Conditions including one of the two tools outperformed the condition with no tools in the corresponding skill scaffolded by the tool. The cumulative effect of both tools seems to have been greater than the effect of each tool separately.

Chapter 4
Collaborative Learning and Assessment of Science: Structuring Effective Groups 84
Yigal Rosen, Harvard University, USA

In order to understand potential applications of collaborative problem solving (CPS) assessment tasks, it is necessary to examine empirically the multi-faceted student performance that may be distributed across collaboration methods and purposes of the assessment. Ideally, each student should be matched with various types of group members and must apply the skills in varied contexts and tasks. One solution to these assessment demands is to use computer-based (virtual) agents to serve as the collaborators in the interactions with students. This paper proposes human-to-agent (H-A) approach for formative CPS assessment and describes an international pilot study aimed to provide preliminary empirical findings on the use of H-A CPS assessment to inform collaborative learning. Overall, the findings showed promise in terms of using H-A CPS assessment task as a formative tool for structuring effective groups in the context of CPS online learning.

Chapter 5
Learning About Genetics in an Elementary Classroom Using a Web-Based Inquiry Science Environment (WISE) Unit 107
Amal Ibourk, Florida State University, USA
Michelle Williams, Williams Learning Solutions, Inc., USA
Merle Heidemann, Michigan State University, USA

Genetics is an increasingly important topic in today's society, yet continues to be an ongoing challenge to the science education community and an important aspect of school student learning. This chapter examines how STEMGenetics, an inquiry-based technology-enhanced online unit, engages upper elementary students in building a better understanding of the concepts of inheritance of traits, variation of traits, and the life cycle of plants. Using pre/post measures, supplemented with qualitative analyses of students' responses, the authors show how upper elementary students' understanding of inheritance and variation of traits, as well as the life cycle of plants, resulted in greater learning gains. In addition, engaging in learning technologies, such as dynamic visualizations, provided the students opportunities to interact with the scientific phenomena and enhanced their reasoning about inheritance of traits and its relationship to the life cycle of plants.

Chapter 6
STEM Inquiry Through INPLACE Mobile Games 134
Denise M. Bressler, Rutgers University, USA

To make STEM learning meaningful, students need to feel the relevance and authenticity of the learning activity. Games—particularly mobile games—offer a unique opportunity for students to be immersed in collaborative STEM inquiry. INPLACE mobile games combine the best practices of what the games and learning field knows about using the affordances of mobile devices to engage students, to support collaboration, and to promote authentic practice in a discipline. INPLACE is an acronym that stands for Interdependent, Networked, Participatory Learning, Augmented, Collaborative Experience. School Scene Investigators is a game series designed according to the INPLACE framework; students playing it demonstrated higher engagement and scientific inquiry than students in a control activity. Ultimately, INPLACE provides a design framework that teachers and researchers can use for building mobile games that heightens engagement and increases inquiry-based learning.

Chapter 7
Learning about Complex Systems from the Bottom Up: Role-Playing Together in a Participatory Simulation 158
Sharona T. Levy, University of Haifa, Israel

The topic of learning through collaborative role-playing in computer-based participatory simulations of complex systems in STEM is presented. Participatory simulations are networked classroom activities aimed at learning about complex systems. In the process of learning, students query its underlying structure and explore its spatial, temporal and mathematical patterns in various conditions. The importance of understanding complex systems is highlighted, driving the main question in this chapter: How can we design learning experiences that support students' deep learning of emergent systems? The motivations behind using participatory simulations and their various designs are described as well as some of the more central learning research, cumulating with five studies into designs for such activities in science. Based on this research, eight design principles are introduced and future research directions are proposed.

Chapter 8
Simulations in Chemistry for Conceptual Understanding and Assessment of Student Knowledge 186
Tanya Gupta, South Dakota State University, USA
Zachary P. Ziolkowski, South Dakota State University, USA
Gregory Albing, South Dakota State University, USA
Akash Mehta, Independent Researcher, USA

Simulations are dynamic resources that have been found useful for communicating abstract fundamental ideas such as stoichiometry and several other concepts. In this chapter the authors present their recent work on designing and implementing an interactive simulation called Combustion Lab based on reaction stoichiometry - a topic that has continually been a challenge for chemistry learners. Several researchers have reported persistent student misconceptions in stoichiometry. In order to address this challenge, a novel computer simulation was developed to assess student understandings of stoichiometry based on student problem solving performance, and also to promote student conceptual understanding. The Combustion lab was particularly focused on the stoichiometry of these reactions, problem solving, and the relevance of stoichiometry for its everyday applications. Results of this sequential exploratory study show that the simulation was effective in revealing student understanding and student treatment of stoichiometry problems based on analysis of various data collected.

Chapter 9
Science Educator Professional Development: Big Data and Inquiry Learning 219
Anna Lewis, University of South Florida – St. Petersburg, USA
George Matsumoto, Monterey Bay Aquarium Research Institute, USA

Scientific discovery, problem-solving, and hypothesis testing requires observation, data analysis and synthesis of new knowledge. In today's world, this process is highly dependent on computer-based data exploration of high volume, high velocity, and high variety data streams (3HV) However, though the power of 3HV surpasses the amount of information gathered from more familiar lab experiences, data-intensive science has not yet achieved the same impact or prominence in public education. This chapter provides an examination of the Education And Research: Testing Hypotheses (EARTH) Science Educator Professional Development (PD) which was developed to bridge this gap, bringing data-intensive science into the classroom while supporting inquiry learning practices.

Chapter 10
A Case Study of Infusing Collaborative STEM Inquiry Learning With Available Technology Into Undergraduate Student Learning: Perceptions, Benefits, and Challenges 245
Katarin Alinta MacLeod, St. Francis Xavier University, Canada

This chapter describes the development, launch, and findings of an undergraduate course entitled "Teaching and Learning in the Physical Sciences". It was a second-year course that introduced students to content commonly taught within a Bachelor of Education program and connected this content to the literature found in the areas of Physics Educational Research and Chemistry Educational Research. It infused inquiry-based STEM learning allowing students to re-examine content and connect to context, while also examining the pedagogical significance and implications. This chapter provides a qualitatively based insight into students' experiences of the course content and students' voices of the transferability of content from this course to others citing lessons learned, implications for physical science students and recommendations for the future offering.

Chapter 11
The Use of Tablet Technology to Support Inquiry Science for Students Incarcerated in Juvenile Justice Settings 267
Michael Krezmien, University of Massachusetts – Amherst, USA
Wardell Powell, University of Massachusetts – Amherst, USA
Christina Bosch, University of Massachusetts – Amherst, USA
Tracey Hall, Center for Applied Special Technology, USA
Martina Nieswandt, University of Massachusetts – Amherst, USA

This chapter describes the challenges in implementing science instruction in juvenile corrections settings and present a tablet-based model for meeting the complex challenges. Project RAISE is a Project-based Inquiry Science (PBIS) curriculum designed in the Universal Design for Learning framework. It is developed in a tablet platform, and is designed to meet the unique needs of incarcerated learners. The chapter describes the juvenile justice educational setting, the characteristics of the classrooms, the learners, and the teachers. It provides an overview of one iBook that has been co-designed and tested with incarcerated learners.

Compilation of References 296

About the Contributors 335

Index 341

Foreword

Since the early decades of the previous century, substantial technological transformations in the field of Information and Communication technologies have transformed irreversibly the ways individuals learn, work, communicate, spend leisure time and function in everyday life. As well, these transformations have had profound social, economic, cultural and political impact – we live in what has been already acknowledged as post-industrial, digital, knowledge society.

Foundational knowledge underlying these developments relates to the STEM disciplines. STEM related knowledge is the main engine driving innovation entrepreneurship and development, which in turn affect all venues of life. Thus, the understandable social demand for encouraging high quality STEM education at all levels in the educational system, starting from the early stages.

In order to answer this demand and reach the goal of optimizing STEM education, ICTs play a substantial role. Thus, an interesting recursive cycle has been generated in education: STEM knowledge is at the foundation of ICTs development, which in turn have the potential to support the learning of STEM knowledge – fostering the goal to continue to enhance ICTs development.

This book is extraordinary evidence of the multiple ways ICTs might be harnessed to support STEM education. First, they describe the use of a wide range of technologies, e.g., mobile devices, the Web, wearable trackers, simulation software, big data analysis or virtual labs. As well, they present a variety of pedagogical modalities both for individual and collaborative learning addressing a range of cognitive skills. They also cover a sample of subjects showing the potential of ICTs to support learning in different content areas.

As ICT itself, the use of the technologies for teaching and learning continues to be an "under construction" field. New and intriguing technological/pedagogical solutions are being created, studied and implemented continuously. STEM education, by its nature, has proven to be a highly rich field for ICT-based pedagogical innovation. The works presented by the authors of the chapters in this book are evidence of this richness, of the gains already obtained, and of the numerous questions still to be explored as the field advances.

David Mioduser
Tel Aviv University, Israel

David Mioduser *is a Head of the School of Education at Tel Aviv University. His research concerns the cognitive aspects and learning processes in the encounter between learners and technology, and focuses on two main fields: one – young children's development of conceptions, skills and technological thinking, and the second – current technology-integrated learning. Prof. Mioduser participates in projects taking place in educational institutions in the country's center and the periphery. During recent years he has been a research partner in international studies conducted under the auspices of the European Union, the OECD and the IEA on integrating up-to-date technology in teaching and learning.*

Preface

The change our society experienced as we entered the digital age, and especially the corresponding transformation of science education, gave rise to research activities at the intersection of inquiry-based STEM learning, emerging technologies, and digital society. This type of research corresponds to the field of study known as Science, Technology and Society (STS). However, given the digital character of the contemporary STS transformations, we call this extended field of studies Science, Technology and Digital Society (STDS). The research presented in our book, *Optimizing STEM Education With Advanced ICTs and Simulations*, belongs to this new field.

Thematically, the book presents a continuation of our previous book, titled *Digital Tools and Solutions for Inquiry-Based STEM Learning*. In comparison with the first book, which mainly includes theoretical chapters that discuss emerging research directions and approaches for integrating ICT in science education, the present book reports the results of educational research that examined the implementation and optimization of STEM educational practices in which digital technologies are a prominent component.

A brief description of each of the chapters follows.

One of the most promising emerging technologies in education, wearable technology, is the subject of Chapter 1, written by Joel R. Drake and his coauthors. The chapter presents real examples of using wearable devices in school settings. These examples are cases in which students study themselves as objects of inquiry, tracking their own activities, and then gathering and analyzing the data. The chapter demonstrates that using wearable devices as tools of inquiry constitutes a highly valuable method for gaining a comprehensive understanding of a variety of STEM topics.

Chapter 2 reports on a study about learning and teaching climate change in school science lessons. The authors of the chapter, Allan Feldman and his colleagues, describe the Climate Change Narrative Game Education (CHANGE) project, which integrates ICT in the classroom in numerous ways. The project incorporates students' lived experiences into the curriculum, and takes advantage of the great opportunities that ICT offers, to facilitate students' inquiry into climate science. It combines specific educational games with hands-on activities, to engage students in inquiry-based learning of climate-change science. The authors mention ways to connect school science with academic science, in order to attract students to study the topic even further.

The focus of the study presented in Chapter 3 by the authors Tasos Hovardas et al. is inquiry-based learning conducted in virtual labs. These are science labs based on emerging digital technologies. In the study, the virtual labs are used by primary school students to carry out successive learning tasks involving experimentation. The virtual lab tools assist students in two ways: in formulating scientific hypotheses and in designing the experiments. Both of these are metacognitive, inquiry-based student activities that did not exist in the traditional, pre-digital classroom. This is a brilliant example of integrating emerging technologies in STEM learning in a way that enriches the science lesson significantly.

Chapter 4 deals with Collaborative Learning Science and its assessment, focusing specifically on measuring students' performance in collaborative problem solving (CPS) tasks. One way to standardize the assessment of collaborative skills is to use computer-based agents to serve as the collaborators in the interactions with students. The author of the Chapter, Yigal Rosen, posed a dual goal: to empirically examine students' multi-faceted performance and the distribution of collaborative skills across collaboration methods and settings, and to propose a computer-based approach for CPS assessment. The author describes a pilot study that was conducted to address these goals.

Chapter 5, written by Amal Ibourk and her coauthors, deals with integrating technology in middle school. The authors analyze an inquiry-based, technology-enhanced online unit for learning genetics. The focus on highly complex topics in genetics such as heredity and variations of traits provides a great example of the advantages of integrating technology in the discipline; indeed, integrating ICT simply makes the learning of genetics feasible. Dynamic visualizations offer the students opportunities to interact with the scientific phenomena, performing various experiments. In genetics education in the digital age, The STEMGenetics software plays the same role as did the legendary drosophila fly.

One of the important directions in the development of future education is educational gaming. The well-known dilemma of "*Homo faber* vs. *Homo ludens*" is representative of the type of deliberations that preoccupy our changing society. The notion of Homo faber, "Human as maker," stresses our ability to create. This is perhaps one of the most important and transformational elements of the networked world, which provides a unique set of affordances for understanding the relationship between new media and learning. *Homo ludens*, "human as player," is perhaps the most important—albeit overlooked—element for understanding our relationship to new media. Chapter 6 deals with a specific and promising kind of educational game, which utilizes the emerging mobile technologies. A specific kind of gaming, mobile gaming, is in the focus of the chapter. The author of the chapter, Dennis M. Bressler, argues that this type of game presents a unique opportunity for students to be immersed in collaborative STEM inquiry. Students playing the game INPLACE demonstrated higher levels of engagement and scientific inquiry than did students in a conventionally controlled activity.

The unique affordances of role-playing in networked participatory simulations for learning to make sense of complex systems in STEM is described in Chapter 7. Participatory simulations engage both individual participants' action and involvement and collaborative experimentation on the explored system. In this chapter, Sharona T. Levy maps out the existing applications through several dimensions, and demonstrates the scarce research that has been conducted into this form of learning. The chapter introduces a number of new studies that show the diversity of applications in STEM and the inclusive nature of these learning activities; and accents the mechanistic conceptual learning that takes place as well as the learning patterns.

Chapter 8 deals with inquiry-based learning of chemistry through the use simulations. Tanya Gupta and her colleagues, the authors of this chapter, present the design and implementation of interactive simulations of topics in organic chemistry that are typically difficult for chemistry learners. Moreover, the proposed simulation provides an option for assessing student understanding based on problem solving abilities. Results of the study show that simulation-based learning is highly effective, both for advancing students' understanding of the topics and for the purpose of student assessment.

Professional development of science educators in light of the new demands of the digital age is the topic of Chapter 9. As the digital age is changing the ways scientists and citizens view and understand the world, both students and educators have to be prepared for the new reality. Anna Lewis and George Matsumoto, the authors of the chapter, deals with the remarkable phenomenon of the digital age, namely, the shift from scarcity to abundance of information, which, in turn, gave birth to the emerging science research paradigm - Data Intensive Science. Along with traditional inquiry-based STEM learning, educational activities oriented to big data, represent a new, intensively developing and highly-facilitating inquiry-based educational activity, which should be an integral part of STEM teachers' professional development.

Chapter 10 is a case study of pedagogical practices in the realm of teacher education. The author, Katarin Alinta MacLeod, describes a pedagogical approach and the corresponding BEd course titled "Teaching and Learning in the Physical Sciences". The author is currently a professor in the Faculty of Education and a former graduate of the Physics Department of the same institute. Her personal experience and awareness afford an insider's view of the problems that teacher educators must contend with in our changing world. Methodologically, the course she describes combines three main components of STEM education: content, pedagogy and technology, with an emphasis on inquiry-based orientation and intensive ICT integration.

Teaching special categories of students, as well as teaching in non-standard conditions, are traditionally topics of particular interest, as such teaching requires both dedicated methods and special means. Addressing the problems and challenges of teaching incarcerated students is the focus of Chapter 11, written by Michael Krezmien and his coauthors. Developing scientific ways of thinking is considered highly desirable for such students, as it increases their chances of successfully reintegrating into society and becoming active and productive citizens.

We hope that educational researchers, educators, and teacher educators will find in this book a comprehensive reference, not only to STEM studies that focus on the role of digital technologies, but also to a new field of studies: Science, Technology, and Digital Society (STDS).

Ilya Levin
Tel Aviv University, Israel

Dina Tsybulsky
Tel Aviv University, Israel

Chapter 1
From Wearing to Wondering:
Treating Wearable Activity Trackers as Objects of Inquiry

Joel R. Drake
Utah State University, USA

Ryan Cain
Utah State University, USA

Victor R. Lee
Utah State University, USA

ABSTRACT

Wearable technologies represent a rapidly expanding category of consumer information and communications technologies. From smartwatches to activity tracking devices, wearables are finding their way into many aspects of our lives, changing the way we think about ourselves and the world around us. The rapid adoption of these tools in everyday life hints at the possibilities these devices may hold in school and other educational settings. Drawing on examples taken from a five-year study using wearable fitness tracking devices in elementary and middle school classrooms, this paper presents two examples of how wearable devices can be appropriated for use in school settings. These examples focus on instances where students turned activity trackers into objects of inquiry using data from familiar activities.

DOI: 10.4018/978-1-5225-2528-8.ch001

INTRODUCTION

Wearables represent a rapidly growing category of information and communications technologies (CCS Insight, 2015). From smartwatches to activity tracking devices, wearables unobtrusively capture and collect large amounts of data relating to aspects of wearers' experiences that were previously unavailable. Using sensors, like accelerometers, wearables can quantify a user's activity (e.g., steps, sleep, breathing) and make it available for inspection. The subsequent analysis of these data can change the user's sense of self and their relation to the world around them (Lee & Drake, 2013a). Prior to the introduction of wearable technologies, these data required active intervention on the part of the individual to capture and track relevant data—manually measuring distances traveled, logging places visited, etc.

Given the potential wearables have for producing personally-relevant data and their increasing social recognition, it is only a matter of time before these devices find their way into school and classroom settings. To use these devices to their full potential, teachers and researchers must work to understand the opportunities and challenges presented by using these devices in schools. For multiple years, the authors (along with a team of researchers, teachers, and designers) have worked with 5th and 6th grade classrooms in the United States to understand how the use of personally-relevant data from wearable activity trackers affects students' engagement with and appropriation of statistical content and practices. Over the course of the study, the authors developed, tested, and refined a statistics curriculum and video recorded and analyzed classroom interactions in order to examine how students leveraged their familiarity with activity in making sense of data from wearables. We have seen familiar activities inspire students to pursue lines of inquiry, develop inclusion criteria, and provide context for their interpretation of data (Lee, Drake, Thayne, & Cain, 2015).

Other publications by the authors have focused on using students' own activities as objects of inquiry—how using activity tracker data can help students better understand these activities (e.g., Lee & DuMont, 2010). The aim of this chapter is to examine how students' use of activity tracker data to better understand the tracker itself—how well activity trackers capture and quantify activity—can be a productive strategy for fostering evidence-based discourse in classrooms. Through presentation of the classroom examples below, the authors argue that students' knowledge of familiar activities can be used to foster skepticism towards wearable devices that leads to productive inquiry in a statistics unit, including the eventual resolution of that skepticism through evidence-based discourse. Each of the examples shows a different way that students may use activity data to critically examine the functionality of the wearables and how the students resolve their questions. In the first example, students test whether the trackers they are using are accurate enough for use in other inquiry activities. The second example involves students investigating whether and how well their devices can track a particular activity and how that ability affects their interpretation of previously collected data.

BACKGROUND

In an effort to reverse US students' underperformance in science, technology, engineering, and mathematics (STEM) topics (OECD, 2014) and increase interest in STEM careers (President's Council of Advisors on Science and Technology, 2010), researchers and educators developed the Next Generation Science Standards in part to encourage students' engagement in scientific practices rather than simply memorize facts (Achieve, Inc., 2013). Among the practices encouraged by the NGSS are: 1) Asking questions and defining problems, 3) Planning and carrying out investigations, 4) Analyzing and interpreting data, 6) Constructing explanations, and 7) Engaging in argument from evidence (Achieve, Inc., 2013, Appendix F). Each of these practices is important to the accomplishment of productive inquiry. However, students can struggle with inquiry curricula for a variety of reasons, including difficulties asking appropriate questions, inexperience with school-based inquiry learning, and underdeveloped representational and interpretive skills (Elby, 2000; Kanari & Millar, 2004; Krajcik et al., 1998; Kuhn, 2007). Research in statistics education has also noted that students' ability to engage in meaningful inquiry requires an understanding of variability as a natural part of measurement (Petrosino, Lehrer, & Schauble, 2003). Some researchers even recommend that variability be the central concept in any statistics or data-based inquiry curriculum (Konold & Pollatsek, 2002).

Familiarity with the data collection processes can be a powerful tool for interpreting data representations (Hug & McNeill, 2008; Nemirovsky, Tierney, & Wright, 1998).

This idea of collecting data first-hand is central to both the commercial popularity wearables and the increasing interest they are receiving from researchers. When researchers look to examine the application of wearable technologies to educational settings, their gaze seems to focus on devices like Google Glass and the Oculus Rift (e.g., Bower & Sturman, 2015). The augmented and virtual reality aspects of these devices offer tantalizing glimpses of what education and life may be like in the future. However, this focus on the technological "deep end" overlooks a variety of device capabilities and applications that are currently realizable (e.g., Sandall, 2016). Over the past several years, researchers used wearable devices to help students and educators examine various aspects of their own experiences. Wearing custom temperature sensors, students collected and later analyzed changes in temperature across various spaces throughout the day (Resnick, Berg, & Eisenberg, 2000).

Point-of-view cameras (e.g., GoPro) have been used to help educators reflect on their teaching practices (Sherin, Russ, Sherin, & Colestock, 2011). While wearing custom "thinking tags," students were able to study the spread of disease through interpersonal contact (Klopfer, Yoon, & Rivas, 2004). Researchers have used wearables and students' movements to understand broader scientific phenomena including animal foraging systems (Moher et al., 2014; Peppler et al., 2010) and the effects of global warming (Lyons, Silva, Moher, Pazmino, & Slattery, 2013). Capturing larger-scale body movements has been used in teaching geometry (Hall, Ma, & Nemirovsky, 2014) and GIS-related topics (Taylor & Hall, 2013).

With their more limited feature set, wearable activity trackers are often overlooked when the future of wearable devices in education. However, their relatively lower price point coupled with their ability to provide personally-relevant, health-related data has led to their rapid adoption by adults desiring to track their health (Choe, Lee, Lee, Pratt, & Kientz, 2014; Lee & Drake, 2013a). Noting the popularity of activity trackers and the obesity epidemic, researchers have begun to look into ways that youth can use wearables to better understand their own health (Ching & Schaefer, 2014; Norooz, Mauriello, Jorgensen, McNally, & Froehlich, 2015).

Wearable Activity Trackers in This Study

For this study, the authors used wearables created by Fitbit, which are essentially internet-connected pedometers; they count steps and use that information to estimate other activity-related quantities (e.g, distance traveled, calories burned). Unlike standard pedometers, these activity trackers can provide step counts in minute-level increments. The Fitbit Ultra (Figure 1, left), used in early implementations, was designed to be worn at the waist. An on-device LCD display showed a real-time step count. The Fitbit Flex (Figure 1, center), used in the last year of the study, was worn on the wrist and had a five-LED display to indicate progress toward the 10,000-step goal. Like other activity trackers, these devices passively sync activity data to a web service via wireless connection to a smartphone or personal computer. Users can interact with their data via a "dashboard" at fitbit.com (Figure 1, right) or via mobile app.

The authors focused on steps as the primary unit of measurement for the students. Steps are familiar, countable events and are already tracked by a wide number of wearable and mobile devices. Steps can be easily translated into a measure of length or a rate (e.g., steps per minute). However, steps are impermanent and cannot be reexamined unless tracked.

Figure 1. Fitbit Ultra (l), Fitbit Flex (c), and activity display and aggregate steps tiles from the Fitbit Dashboard, fitbit.com (r)

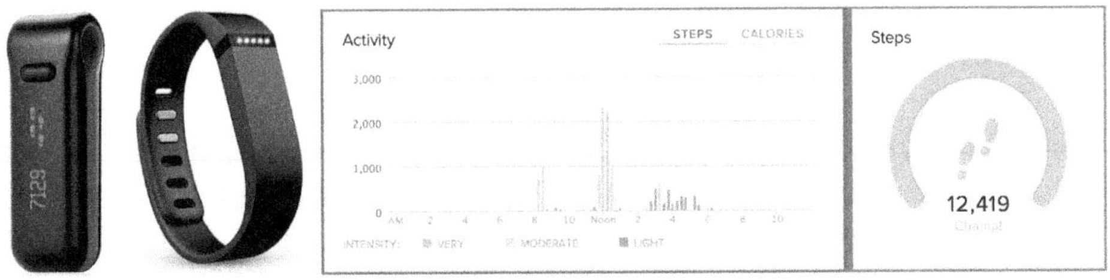

Modern activity trackers use three-axis accelerometers or gyroscopes to sense movement. Every movement is tracked as "activity." Proprietary algorithms compare the motion signature detected to the shock profile of a step (Figure 2). Movements that match this profile are counted as steps. The accuracy of the devices is based on how reliably the algorithm can differentiate between steps and non-step movements.

Figure 2. Acceleration profile of three steps measured by a waist-mounted accelerometer
Adapted from Khandelwal & Wickstrom, 2016.

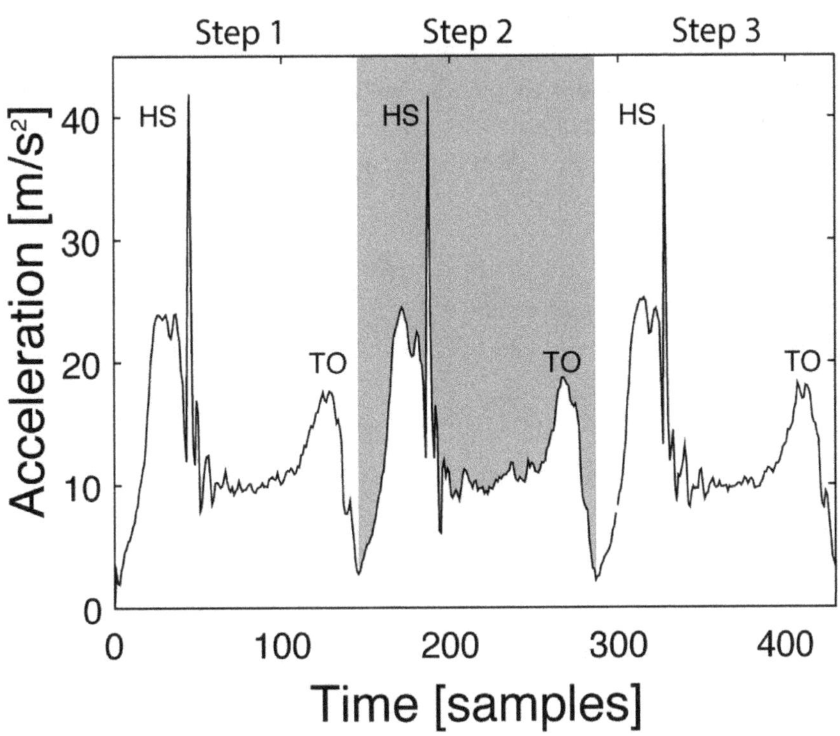

Fitbit devices have been shown to track steps accurately in laboratory tests (Diaz et al., 2015; Takacs et al., 2014).

Wearable Curriculum Implementation

The curriculum included frequent opportunities for students to explore data from their routine activities (Lee et al., 2015). These data explorations typically included students viewing data representations, discussing as a class the meaning of the data represented (e.g, typicality, variation, and shape) and comparing features of the representation to other datasets. Because standard dashboard for these trackers is not intended for such analysis, the authors developed a PHP web form that retrieved data from the Fitbit servers in one-minute increments. These data were returned to the user as comma-separated values (CSV). The CSV was then imported into Tinkerplots (Konold & Miller, 2005) software for data visualization and analysis. Tinkerplots was designed specifically for elementary and middle grade students to create and interpret data visualizations. Using a drag-and-drop interface, students can quickly organize and reorganize data in ways that make sense to them. For this study, the authors focused students' attention on understanding and interpreting time-ordered and frequency distributions.

These data explorations were supplemented with instruction on statistical concepts (e.g., measures of center, variability) to help students concretize their comparisons. Through these discussions and the supplementary instruction, the students came to question their assumptions about the nature of their daily activities. In some instances, students' investigations led them to explore the functions of the devices themselves. It is these latter instances that the authors wish to explore in this chapter.

Figure 3. Time-ordered (l) and frequency distribution (r) for one student's recess data; each dot represents one minute of recess activity.

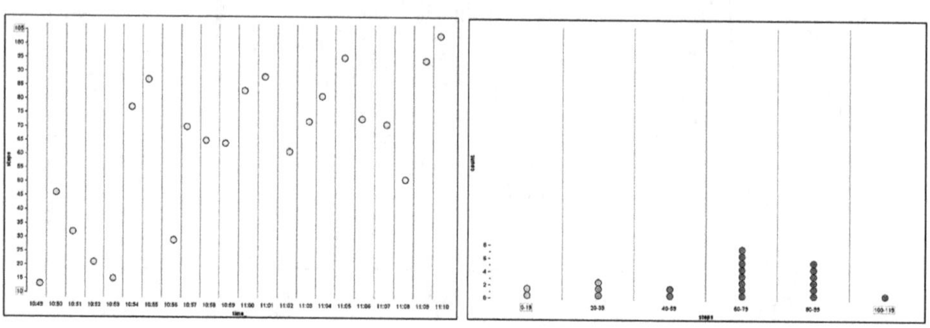

METHODS

The data for this study comes from three years of a design research study (Brown, 1992) examining how students could leverage their familiarity with their own activities while learning and applying statistical techniques to interpret data collected during those activities. Over these three years, the curriculum was iteratively refined between enactments in six classrooms. These enactments averaged 24, approximately hour-long sessions. Classroom interactions during each session were captured using two video cameras resulting in over 145 hours of video.

To analyze these videos, the research team collectively viewed each video and timestamped major segments, identified as changes in conversational focus. During this initial pass, team members flagged for further analysis those segments where the students' conversation appeared to lead to important insights. For the current study, the authors identified instances where the students used their developing statistical knowledge in an effort to understand the output of the devices. These instances were then examined for structural similarities. Through repeated examination and discussion, these similarities were refined into the characteristics presented in this chapter.

CHARACTERISTICS

As noted previously, other publications by the authors have examined the use of wearable devices in an inquiry curriculum (Lee & Drake, 2013b; Lee & DuMont, 2010) and have presented quantitative results demonstrating the effectiveness of this approach (Lee, Drake, & Thayne, 2016; Lee et al., 2015). The characteristics presented in this chapter are intended to provide guidance for helping students view the data from these devices critically and focus their inquiry on the devices themselves.

In examining instances in the video data where students turned their attention to the devices themselves, the authors identified four characteristics in the students' inquiry: 1) Noticing Discrepancies, 2) Questioning the Devices, 3) Hypothesizing Conditions, and 4) Investigating. Whether due to natural variability or a misunderstanding of what is being measured, discrepant data is a part of measurement. The key for having students critically examine a measuring device (wearable sensor or otherwise) is to provide a means for them to *notice* that discrepancy. With activity trackers, that noticing may be facilitated by the children's familiarity with the activities being measured. Inconsistency between a recalled experience and a representation of the resulting data can lead students to *question the device* and their assumptions about it. In an effort to resolve their questions, students may *hypothesize conditions* under which the device might (not) work. Small-group and whole-class discussions are useful for helping students refine these hypotheses into a testable form. Once the students have a testable hypothesis, they can begin an *investigation*. This investigation includes defining the data collection methods, performing the collection, and then analyzing a visualization of the resulting data. Recollection of the data collection activities can play as important a role as statistical methods in students' analysis of this data. This analysis may result in further refinement of the conditions and another round of investigation or a resolution of the students' initial questions.

It should be noted that the goal of these investigations in the current study is an acceptable *resolution* to the students' questions based on the data rather than achieving a specific *solution*. The analysis is often "quick and dirty," as the students may be learning the processes of inquiry and data analysis at the same time.

EXAMPLES

The examples that follow seek to show how these characteristics operate in practice and explicate their ties to the NGSS Scientific Practices. These examples were chosen because they demonstrate different approaches to data collection and visualization. The first example comes from the first year of the project. During this year, the students used aggregated data across the entire class to examine the accuracy of the trackers. So that students could see their individual contributions to the aggregate display, the authors chose to use a decidedly low-tech solution to data visualization—sticky notes on butcher paper. In the second example, data from the students' investigations are represented and displayed to the class using the Tinkerplots data visualization software (Konold & Miller, 2005).

Classroom Example 1: Are Activity Trackers Accurate?

At the beginning of this project, activity tracking was a new phenomenon. The public was just becoming aware of activity trackers and possibilities they offered. Prior to this, physical activity tracking was limited to specific episodes (e.g., "going for a run"). While tracking in this manner was embraced by the athletic community (Lee & Drake, 2013a; Lee & DuMont, 2010), the "always on" tracking of routine activity represented a fundamental shift in what activities were trackable and who would be doing the tracking.

In the year prior to this case, the authors had used activity trackers for small-scale studies (e.g., Lee & Drake, 2013b) to identify avenues for their use in classroom activities. At the time of this case, the researchers were working to develop and implement a unit with a fifth-grade class twice a week from September through December. The curriculum used was designed to promote student-led inquiry, with teachers and researchers encouraging the students to focus on questions that could be answered statistically using data from the trackers.

Introduction to Activity Tracking

The unit began with a device exploration period during which the students each wore an activity tracker and moved around the school to see how they worked. After this free-exploration time, the class assembled in small groups to share what they noticed about the devices and come up with a shared idea of how well they worked. The discussion proceeded with the students offering their assessment of how (well) the trackers worked and why they came to that conclusion. The students immediately began questioning how accurately the devices could count steps.

Jocelyn: *Mine felt inaccurate because when I just took 90 steps, it took me up to about 200 steps and I was very confused.*
Houston: *Mine was perfectly accurate. When I took 10 steps, it added 10 steps...*
Mischa: *Like Jocelyn said, when I walked to the library, it said I took 200 steps, but I actually took a lot less than that.*
Students: *[Chorus of agreement]*

In this excerpt, the authors note evidence of both *noticing discrepancies* and *questioning the device*. Each of the students cited their own performance ("I just took 90 steps") and the on-device count ("it took me up to about 200 steps") as evidence. The Fitbit Ultras used by these students had an on-device LED display that showed the current daily step count. The students frequently checked this display against their own step counts. Jocelyn and Mischa both recognized discrepancies in these comparisons. Although Houston claimed perfect agreement in his exploration, this claim was largely ignored. Conversely, the chorus of agreement following Mischa's statement indicated that many of the other students had noticed discrepancies during their trials as well.

Beyond noting discrepant data, the students started questioning the devices using these discrepancies as evidence. In positing that the devices were inaccurate, the students were stating what would become the driving problem for the remainder of the inquiry activity (NGSS Practice #1: Asking questions and stating problems).

Making Hypotheses Public

Following a second free-investigation period focused on the accuracy of the devices, the class reconvened to discuss their findings. The goal of this discussion was for the students to create a shared understanding of observations they made or questions that could be answered using the trackers. These observations and questions were captured on a large sheet of poster paper hung at the front of the room. This list remained visible throughout the remainder of the unit. The segment that follows shows how the language for each item was discussed and refined.

Clara: *There could be a little person who takes really small steps and it's hard for the Fitbit to know if you just took a step or not or there could be a really big tall person who takes big obvious steps and the Fitbit can process it faster.*

Erin: *I think what she basically said is it depends on the person and the place it's put and the Fitbit.*

While the students could still advance any hypotheses they chose, the class seemed to focus on *hypothesizing conditions* that might affect the accuracy of the devices. Clara's suggestion that accuracy depends on the size of the user made sense within the context of the class. It allowed for the devices to be accurate for use by adults (the students knew that their teacher was using an activity tracker) while allowing for perceived inaccuracy in student measurements.

Erin's "clarification" advanced another condition—that accuracy depended on "the place it's put," referring to where the device was worn (i.e., trackers may be more accurate when worn at the waist than on a lapel).

The conditions offered by students can serve to further define the main problem and focus the inquiry. These conditions may also identify new problems. In either case, giving students the opportunity to express conditions under which the device might (not) work, gives the students experience defining investigable problems (NGSS Practice #1). In the case of Clara's hypothesis, the class took up a version of it in a subsequent investigation.

Through this and the previous class discussions, the students began to flesh out a problem definition—that the activity trackers were inaccurate. In identifying and shaping this problem, the students leveraged their familiarity with the activities about which they were collecting data (i.e., walking, running, and jumping).

Planning the Investigation

After completing the class hypothesis list, the students elected to continue trying to determine whether the activity trackers were accurate enough to continue using them for further inquiry investigations. To do this, the students decided to use a sample of seven trackers to try out their test methods. During a whole-class discussion, Erin and James proposed methods that would eventually shape the structure for the test.

James: *I think there should be one person with seven Fitbits... When they're walking, they count their steps and then they look at all of the Fitbits and see if they all have the same answer.*

Erin: *I think we should have one person focusing on how many steps they take... and then another person just thinking about other things while they're walking.*

Planning this investigation (NGSS Practice #3: Planning and carrying out investigations) was a collaborative effort. The suggestions continued to draw on students' familiarity—direct experience with the devices, earlier discussions, and conditions on the class list created earlier. Investigation planning continued the pattern of refinement begun in previous share-outs. The final-form method that the class agreed to had three roles: a Walker, Counters, and Monitors. The Walker would wear all of the devices. Counters would follow the Walker and silently count the Walker's steps. Monitors would be assigned to one device each; they would record data on a sticky note—a device identification number, the step count from the Counters, and the starting and final step counts from their device. The Walker and Counters would walk a set distance then return to the classroom. Each Monitor calculated the number of steps recorded by the device (final count minus starting count) and the residual between the device and Counter step counts.

Figure 4. Students testing accuracy of Fitbit Ultra devices by having a single student wear multiple devices, manually counting the student's steps, and comparing the counts

In planning this investigation, the students collaboratively developed a method that parallels methods used by professional studies looking into the accuracy of these devices (Diaz et al., 2015; Takacs et al., 2014). In these studies, multiple devices are tested per trial, and their output is compared to observer-counted steps. Building off each other's statements, the students were able to devise a reliable and repeatable method for testing the functionality of the devices. This method resulted in a corpus of directly comparable data points that could then be used as the investigation turned toward representing the data as a means for answering their question.

Conducting the Investigation

The students conducted 11 separate trials with varying numbers of activity trackers per trial, resulting in 56 individual sticky-note data points. Following each trial, the trackers placed their sticky notes on a poster with two categories marked: "Not very accurate" and "Pretty accurate." The difference between the categories was kept intentionally vague to allow the students to decide what constituted "accurate enough" individually. Because this approach would result in a less conclusive representation, the authors and the teacher hoped it would spur the students to desire more formal categorization of the data. Although the "pretty accurate" bin contained a few more sticky notes (see Figure 5), the students did not see this initial

Figure 5. Detail of a video frame capture showing the first representation of data from the Fitbit accuracy trials. The left column is "Not very accurate;" the right column is "Pretty accurate".

representation as providing the conclusive evidence they needed to decide whether the Fitbits were accurate.

To help the students decide for themselves whether the devices were accurate, the authors prepared four posters on which the students could create frequency plots. Each of the four posters divided the data up differently to show the students that different representations can be made of the same data and that the right representation can help answer questions.

Working in small groups, the class created four frequency plots using sticky notes. The finished plots were hung in front of the class for inspection.

In presenting their plots to the class, students from each group reported how the plot convinced them that the trackers were accurate or inaccurate. Each of the presenters said they believed that the devices were accurate, offering as evidence the larger clumps of data in the bins representing lower residual step counts (NGSS Practice #6: Constructing explanations). After each of the groups had presented, the teacher invited Sabine to share her interpretation of the plots. Sabine was invited to share because she had been reluctant to accept anything short of perfect agreement before agreeing that the devices could be accurate. As she stood in front of the plot (Figure 7), Sabine noted the tall peak in the first column and the rapid drop from there to the bins with greater differences. Using similar reasoning, the rest of the class accepted that the devices were accurate enough for continued use.

Before creating the plots (NGSS Practice #4: Analyzing and interpreting data), the students relied on individual trial results when making their claims. After seeing the data in aggregate, the students were able to use a single number (i.e., the mode) to summarize a data distribution, which has been identified as an early stage in learning to statistically analyze and interpret data (Lehrer, Kim, Ayers, & Wilson, 2014).

Figure 6. Students presenting sticky note frequency plots

Figure 7. Sabine describing how this plot helps her see the Fitbits are "mostly accurate"

Example 2: Do Jumps Count as Steps?

The second classroom example comes from an enactment in a sixth-grade classroom. Students in this enactment each wore a Fitbit Flex (Figure 1) throughout the school day. Because the Fitbit Flex does not provide immediate feedback, students had to wait until the next day to see their data. By using the devices in this way, the students could not directly compare step counts from particular activities to output from the devices. As a result, the students generally viewed the trackers as authoritative, a view encouraged by the teacher throughout the unit. However, as in the previous example, the students come to question the abilities of the devices when their output is viewed in light of the students' own experiences with familiar activities.

This example begins with the students examining three unlabeled sets of data from the previous day's recess (Figure 8); the activity's objective was for the class to identify the student represented by each dataset and justify the identification. This activity was intended to encourage students to attend to various features of the data and make connections between distinct features and the students' recalled activities. With various visual landmarks specified (Moher et al., 2014), students could then consider how less distinct features mapped onto the data-generating experiences. This data examination activity spurred discussion of what exactly it means to "jump rope" at recess and whether or not this is a worthwhile approximation of what it means to "jump." This in turn led the students to wonder whether or not the activity tracker recorded jumps as steps, the tracker's primary data product. To answer this question, the students devised several data collection and analysis activities as they refined their understanding of both the activity of jump rope and how to test the devices.

Interrogating the Device: Initial Talk About Jumping Rope

At the beginning of math class each day, data from a handful of students was projected on the whiteboard in front of the class. The students analyzed the data using a combination of statistical tools and their knowledge of the activities represented in the data (NGSS Practice #4). The analysis involved the students whose data were displayed introducing the activities that produced the data. The whole class would then engage in a discussion to identify features in the representation, compare overlaid data sets, et cetera. The nature of the analysis task changed from day to day in an effort to broaden the students' analytical repertoire.

The data investigation in this example began with unlabeled recess data from three students (Kelly, Chad, and Jason) projected in front of the class (Figure 8). To introduce the data, these students were asked to share what they did during the morning recess and which data points corresponded to their activities. Kelly claimed the yellow points (white dots in Figure 8) represented her recess because she walked around at the beginning of recess then joined a game of jump rope and did not move much after that. After the other students shared their recess, Nick began the exchange that follows by challenging Kelly's selection of the yellow points with an alternative interpretation of the data.

Nick: *I think Kelly is red 'cause she said she was just walking around in the beginning, and it's kind of low. And at the end she said she was jump roping, and jump roping takes a lot of steps, so I think she is red.*

Mrs. Bryson: *I wonder if she was walking around while she was waiting. I wonder about these zeros though.*

Nick: *I said red not yellow.*

Mrs. Bryson: *She said yellow. Right, Kelly? So, you're thinking she is red instead. Does anyone have anything to add to that?*

Researcher: *I have another question. Why do you think that jump roping wouldn't get you very many steps?*

Kelly: *I was swinging most of the time, but I don't know*

Figure 8. TinkerPlots time-ordered graph of three students recess activity by minute, organized horizontally. Each student's recess is represented by one of the three sets of colored dots—yellow, red, and blue. Yellow dots are shown as white, red dots are gray, and blue dots are black. The data were encountered in the TinkerPlots data visualization software.
Konold & Miller, 2005.

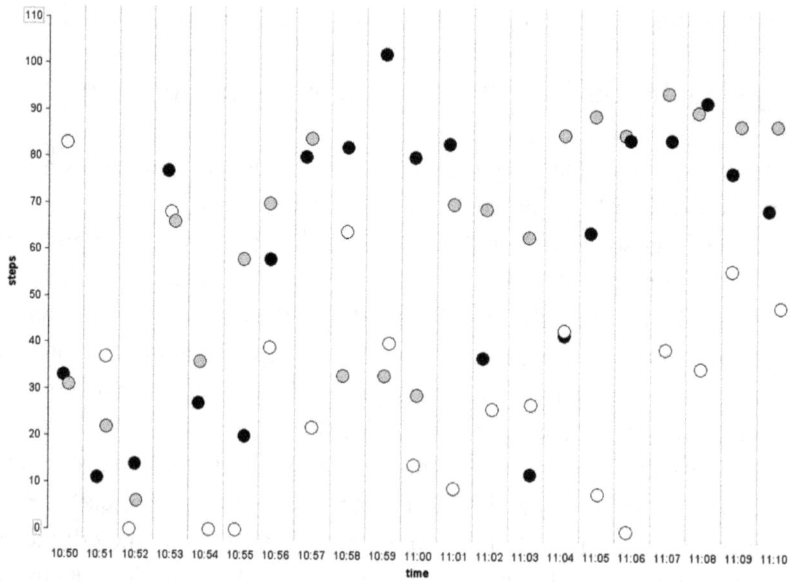

Nick's interpretation accepted Kelly's general recess narrative—walking around followed by playing jump rope. However, Nick asserted that "jump roping takes a lot of steps," in opposition to Kelly's claim of having not moved much after joining the game. This assertion seems to align well with the red data, which have higher step counts at later times. Mrs. Bryson returned to Kelly's claim and raised a concern that it did not fully account for the zero-valued yellow dots. This statement made explicit an assumption that Kelly had made but not yet stated aloud: playing jump rope involves times when one does not move and thus does not get any steps in a given minute. To clarify Kelly's reasoning, a researcher asked why she thought that "jump roping wouldn't get…very many steps?" Kelly responded that she was swinging the jump rope, which does not involve taking steps, although she was unsure if that made a difference.

In this excerpt the *discrepancy* the students noticed was between the data and students' expectations. Two views of the nature of jump roping were advanced—one involving stationary periods of time and another that involved continuous bodily movement. With datasets that could potentially be linked with either of these views, there was no easy resolution to this question. The discussion made jump rope a more complicated activity. Further, the assumption that jump rope might produce fewer steps, relative to other activities, was opened for examination by the researcher. In the next excerpt, this assumption was taken up again by the teacher, who helped maintain an ongoing focus for this discussion on the question about whether jumps would be counted as steps.

Asking About the Tracker's Step Counting

Mrs. Bryson: *I have a question about the about the jump rope, this is a Fitbit question, if you are just jumping up and down in the same place...*
Kaylie: *...does it count?*
Mrs. Bryson: *Does it count steps?*
Alondra: *I guess so.*
Adam: *I would say so because you using, doing exercise. You're exercising*
Carson: *I think it thinks that your movement is that you are running around*
Mrs. Bryson: *Yeah, but the Fitbit is smart, but is it smart enough to do that? Can we find out?*

Out of genuine uncertainty, Mrs. Bryson more formally asked the question about whether jumps were counted as steps by the wearable devices. This *question* changed the focus of the discussion to the device while maintaining jumping as the *condition* under which the device's function was being questioned. While Mrs. Bryson first advanced the question, Kaylie's interjection hints that the question was likely one the students were already anticipating at that moment. Adam volunteered the view that because jump rope was exercise, a *fitness-tracking* device should track it and make jumping count. Other students seconded the idea that the tracker would interpret jumping as steps. These statements suggested students already suspected that jumps counted, but Mrs. Bryson made an important move next by changing the question from "Does the Fitbit count jumps?" to "Can we find out?" The change of phrasing implied that simply assuming how the devices worked would not be enough and encouraged the students to find a way to answer this question.

Following this, several more suggestions were made about how the class could devise a way to address the initial problem of finding a way to determine if the trackers counted jumps as steps (NGSS Practice #3). This ultimately led to a suggestion from a student that received class support: three students would spend all of that day's recess playing jump rope, and three other students would be sure to walk for the entire recess. According to their plan, having a walking group and a jumping group would let the class see if the numbers were equal and could tell if jumping was registered as walking (taking steps).

While the teacher initially voiced the question, the students seemed to take it up as their own question to answer. Initially, they offered their assumptions that the device (which they took as presenting the "truth" about an activity) would work. After Mrs. Bryson changed the question to "Can we find out?" the students took it on themselves to plan an investigation (NGSS Practice #3) to use the device's data to probe a potential limitation of the device.

Reviewing the Data From Walkers and Jumpers

When the class convened the next morning, Mrs. Bryson recapped the class's plan to determine if jumping was counted as steps by the devices then asked the participating "walkers" and "jumpers" to share what they had done. All six "walker" and "jumper" students reported attempting to do the activities that they were assigned. However, with the exception of Adam, each of these students was eventually absorbed into other existing activities. Some students got "bored" and changed activities, and some played regular games (such as four square) and tried to incorporate walking. For those who did jump rope, the standard play rules had continued to apply. Two of the students stated they had limited time to jump, and the third did not get to jump at all.

Based on how students tried to express that they had at least attempted to walk or jump amidst their recess and how animated students were during the previous day's discussion, it did not seem that the students were being dismissive of the data collection activity. However, other aspects of recess, such as play, socializing, and

Figure 9. TinkerPlots graph comparing "jumpers" (black dots) and "walkers" (white dots)

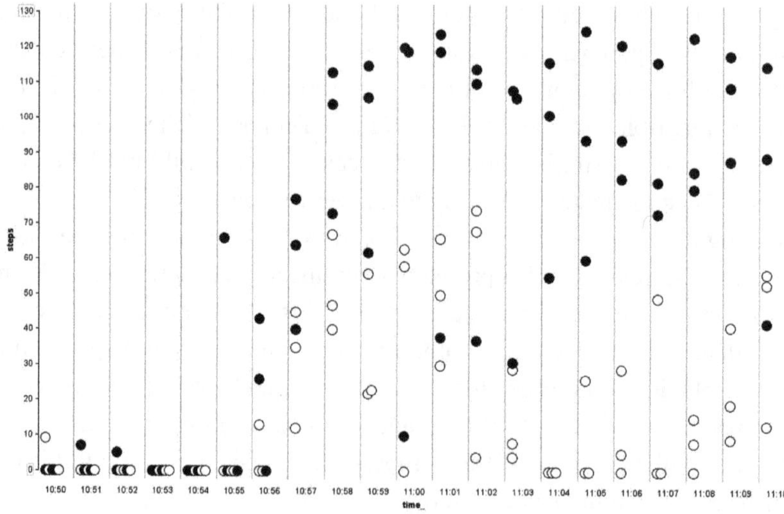

following established playground rules, took precedence. Because the conditions for this test were set explicitly within established recess activities—jumping rope and walking—the influence of established recess order in the data collection should come as no surprise. These conditions harkened back to the initial data analysis task (i.e., identifying data corresponding to a student who jumped rope) that led to the question of whether the devices counted jumps as steps.

When the data were projected (Figure 9), the students could see how these conditions were problematic for answering the question, leading them to discuss the validity of their experiment's design.

Evaluating the Design of Their Investigation

After being presented with the students' recollections of recess and the accompanying data plot, Adam began the following exchange by questioning how data should be interpreted in light of his own observation of a jumper, Melissa.

Adam: *I watched Melissa, and I saw her, and she was standing in line for the jump rope a lot.*
Mrs. Bryson: *She said that when she talked about that.*
Kate: *We have to stand in line, because you're not really...*
Adam: *I know, I'm just saying, I am not blaming you or anything. I am just saying, just that maybe it is not very accurate because the walkers walked the entire break, but the jumpers just jumped some of the break and part of it they were standing in line*

Adam recalled seeing Melissa "standing in line for the jump rope a lot" during recess. Due to the layout of the playground and his own role as a walker, Adam was able to watch what others were doing on the playground when they were supposed to be walking or jumping. That position of observing what happened when data were collected made the regular noticing of what was happening on the school playground a relevant resource for the conversation.

His main point was that there was a *discrepancy* between the conduct of the investigation (i.e., jumpers spent too much time standing in line) and conditions sufficient to answer the question ("Do activity trackers count jumps as steps?"). This point echoed the conditions that served as the impetus for the investigation, but in the context of Adam's statement it was important because the nature of the test itself was being questioned again. Adam recognized that the experiment as run did not meet the conditions described in the question—the recess data included periods of standing, jumping, and several other activities. In order to draw any conclusions relating to the main question, the investigation needed to be refined.

Proposing a New Investigation

This concern was recognized by the class and Mrs. Bryson. For several minutes, students offered a number of possible ideas for new data collection. These ideas were vetted for feasibility and utility. In the segment below, Casey suggested the new data collection activity for the class that would then be implemented.

Casey: *We could do it (jump) for a minute and count how much we do and then we can look at it (data) tomorrow. And we could look at someone who remembers how many they have and they could write it down or something, and we could see how much steps it counted.*

Mrs. Bryson: *Can we do that? We can do that right now, couldn't we?...*

Jimmy: *Just do it for a minute... Wait until the next minute and then we stop right where the time is and then we just look at [the data].*

The students' plan for this new *investigation* (NGSS Practice #3) involved all the students jumping for one minute during math class. After getting similarly inconclusive results from looking at data from two recesses, the students decided to perform data collection during class. This choice freed them from the constraints of recess and shortened the time in which the students needed to focus on data collection. Having all the students jump allowed everyone to participate and resulted in a similar number of points to the previous data sets the students had analyzed. The students' decision to count their own jumps for comparison against the tracker counts parallels the investigation in the previous example, with the students' self-counted jumps being the authoritative count.

On a signal from Mrs. Bryson, the students all proceeded to jump, with many students counting aloud. When the minute ended, the students excitedly began calling out their numbers of jumps; Mrs. Bryson told the students to quickly write the numbers down so they could compare the next day after the data from the Fitbit devices had been synched and the minute of jumping had been made visible.

Reviewing Data From the New Investigation

The next and final day of this unexpected jumping investigation began with Mrs. Bryson saying that she had access to all of the students' counted steps during the minute of jumping from the previous day. She called on various students to share what number they had counted and recorded themselves and then reported back the number that the Fitbit devices had counted. The numbers that were obtained by the Fitbit devices generally within about 15 jumps, with a few students having numbers that differed by several dozen. Following this, Mrs. Bryson asked how many students had their own count as higher than what was recorded on their Fitbit device (10 students), how many had an exact match (1 student), and how many had counts below the Fitbits (8 students). The class then examined the distribution of steps recorded during the minute of jumping and discussed what were the various measures of center for it (NGSS Practice #4).

Eventually, Mrs. Bryson suggested to the class that even though the numbers were not exact, it did appear that many of the Fitbit observations of steps were close to human counts. For the purposes of the class, that meant the Fitbit would usually count jumps as steps. Even though that was established as understood and accepted knowledge for the class, some students still pushed for other possible variations on their test in an attempt to mitigate sources of error. Mrs. Bryson noted the cleverness of these suggestions, but in the interest of getting back onto the unit as planned, steered conversation to other topics related to data and distribution.

SOLUTIONS AND RECOMMENDATIONS

The examples above show the students from two elementary school classes participating in an inquiry-based statistics unit that featured wearable activity trackers to produce data for analysis. The authors' goal for this unit was to see how students could leverage their familiarity with their own activities as a tool for analyzing and interpreting displays of activity data. The thought being that students would develop a deeper understanding of both their own activities and the statistics content. On occasion, however, the students turned the tracker's data back on itself, making it the object of inquiry. In these instances, the authors noted certain characteristics to the students' inquiry. The students 1) noticed a discrepancy in the data (either relative to their personal experience or to their expectations), 2) questioned the device, 3) hypothesized conditions which might explain the discrepancy, and 4) planned and carried out an investigation.

Students' ability to notice and call out discrepancies in the data is key in this process. Recognizing irregularities requires a degree of familiarity with both the data and the processes by which those data were captured. Because wearables capture data related to the wearer's mundane experience, familiarity with the data collection activities is assumed to be built in. This familiarity brings with it expectations of how the activity is supposed to proceed, which students do indeed bring to bear on data interpretation tasks (Lee et al., 2015). However, students' familiarity with representations of the resulting data should not be taken as given (Ching & Schaefer, 2014). To help students develop the same degree of familiarity with data that they have with their own activities, they need to be provided frequent opportunities to inspect and interact with that data (Lee et al., 2016). In the curriculum used in this study, the students viewed and publically interpreted their data on a daily basis.

Throughout the students' analysis activities in this study, the students' recollections of their experiences were treated as valuable sources of information. The teachers encouraged students to share this knowledge, particularly as it related to the data being examined. Establishing classroom norms requiring students to ground their explanations in data is a powerful tool in inquiry (Ryu & Sandoval, 2012).

The students in the examples above were able to resolve their questions about the devices to their satisfaction. Along the way, they were able to gain valuable experience with many of the scientific practices targeted by the Next Generation Science Standards (Achieve, Inc., 2013). It should be noted that the investigations and analyses the students engaged in and their eventual resolutions are all of the "rough and ready" variety. The driving questions were binary, yes-or-no questions—Are the trackers accurate? and Do they count jumps as steps? The resolutions really only hold within the context of the classroom and for the students involved. That is okay. In each of these examples, the investigation was the students' first experience with inquiry learning, and they were only just beginning to learn much of the statistical content for their grade level (National Governors Association Center for Best Practices, Council of Chief State School Officers, 2010). These early, small-scale experiences with inquiry can give students the confidence necessary to continue to more advanced investigations and analysis techniques.

FUTURE RESEARCH DIRECTIONS

The research that has been done with wearable activity tracking technologies has shown the promise of these devices in inquiry-based STEM education. However, given their relative newness, much may still be done to better understand what these devices may afford for education. The authors note several avenues that they feel remain to be explored. Among these are the following:

- **Expanding the Devices' Uses With Students of Different Ages:** The current study worked with fifth and sixth grade students. Small scale studies (e.g., Lee & Briggs, 2014; Lee & DuMont, 2010) and afterschool clubs (e.g., Ching & Schaefer, 2014) have begun to examine how high school students and undergraduates (Thayne & Lee, 2016) might use the data from these devices. Further work remains to be done to understand how other grades might use these devices and how those uses might be leveraged for productive STEM inquiry.

- **Expanding the Breadth of STEM Subjects:** As noted above, existing research has incorporated wearables into a broad range of STEM content areas. It is not hard to imagine these devices contributing to even more stem topics. The expanding capabilities of these devices immediately lend themselves to uses in Physics and Biology. With a little creativity, even broader applications are possible.
- **Examining Equity:** Wearable technologies may have the potential to democratize access to a broad range of health-related data and the greater personal insight that comes through its analysis. However, use of wearables "in the wild" is dominated by wealthy, middle-aged white males. The authors have done some initial research to examine how underrepresented populations (do not) use these devices (Lee & Briggs, 2014), but more remains to be done. Only by understanding the motivations behind their (dis)use can they be used in a culturally-responsive, and truly productive, manner.

CONCLUSION

Wearable technologies are becoming more prevalent in everyday life. Using these devices as tools for inquiry holds many possibilities for understanding a variety of STEM topics (e.g., Moher et al., 2014; Resnick et al., 2000). The current study used wearable activity trackers as part of an elementary school statistics curriculum. While the goal of this curriculum is for students to use the data from these trackers as a tool for learning statistical content while examining their own activities, the students occasionally turned those data back on the devices in an effort to answer questions they had about the devices. This chapter presented characteristics of these instances that resulted in productive engagement in inquiry. These characteristics are that the students 1) noticed a discrepancy, 2) used that discrepancy to question the device, 3) hypothesized conditions under which the device might (not) work, and 4) planned and carried out an investigation to test their hypotheses and resolve their questions. These characteristics were then situated in two examples from different years of the current study, each using different activity trackers and means of representing the data.

In the above examples, the students were familiar with the physical activities that had generated the data, both as active participants and as observers monitoring the activities. Discrepancies between students' expectations for these activities and the data provided by the trackers provided the impetus for the students to question the devices. Through whole-class discussion and with the teachers' support, the students were able to develop tests that they could carry out using the trackers and would resolve their questions. Students leveraged their familiarity with their activities to even question the conditions of their own investigations.

Through these inquiry experiences, the students were able to demonstrate many of the science practices in the Next Generation Science Standards (Achieve, Inc., 2013). They asked questions that could be answered using scientific methods (NGSS Practice #1). They collaboratively planned and carried out investigations (NGSS Practice #3). Using statistical methods and their own experience conducting the investigations, they were able to visualize, analyze, and interpret the resulting data (NGSS Practice #4). In the second example, they noticed that their data collection activity would not answer their question and iterated through these practices again.

As researchers and educators work to integrate advancing information and communications technologies into STEM classrooms and reignite interest in STEM topics among students, wearable activity tracking devices represent one option. By enabling students to investigate data from familiar and routine activities, activity trackers can be used to foster inquiry focusing not only on those activities but also on device itself.

REFERENCES

Achieve, Inc. (2013). *Next generation science standards*. Washington, DC: Author.

Bower, M., & Sturman, D. (2015). What are the educational affordances of wearable technologies? *Computers & Education*, *88*, 343–353. doi:10.1016/j.compedu.2015.07.013

Brown, A. L. (1992). Design experiments: Theoretical and methodological challenges in creating complex interventions in classroom settings. *Journal of the Learning Sciences*, *2*(2), 141–178. doi:10.1207/s15327809jls0202_2

CCS Insight. (2015, September 1). *Wearables Market to Be Worth $25 Billion by 2019*. Retrieved July 21, 2016, from http://www.ccsinsight.com/press/company-news/2332-wearables-market-to-be-worth-25-billion-by-2019-reveals-ccs-insight

Ching, C. C., & Schaefer, S. (2014). Identities in motion, identities at rest: Engaging bodies and minds in fitness gaming research and design. In V. R. Lee (Ed.), Learning technologies and the body: Integration and implementation in formal and informal learning environments (pp. 201–219). Routledge.

Choe, E. K., Lee, N. B., Lee, B., Pratt, W., & Kientz, J. A. (2014). Understanding Quantified-selfers' Practices in Collecting and Exploring Personal Data. In *Proceedings of the 32nd Annual ACM Conference on Human Factors in Computing Systems* (pp. 1143–1152). New York, NY: ACM. doi:10.1145/2556288.2557372

Diaz, K. M., Krupka, D. J., Chang, M. J., Ma, Y., Goldsmith, J., Schwartz, J. E., & Davidson, K. W. (2015). Fitbit: An Accurate and Reliable Device for Wireless Physical Activity Tracking. *Circulation*, *131*(Suppl 1), AMP11–AMP11. PMID:25795203

Elby, A. (2000). What students learning of representations tells us about constructivism. *The Journal of Mathematical Behavior*, *19*(4), 481–502. doi:10.1016/S0732-3123(01)00054-2

Hall, R., Ma, J. Y., & Nemirovsky, R. (2014). Rescaling bodies in/as representational instruments in GPS drawing. In V. R. Lee (Ed.), Learning technologies and the body: Integration and implementation in formal and informal learning environments (pp. 112–131). Routledge.

Hug, B., & McNeill, K. L. (2008). Use of First-hand and Second-hand Data in Science: Does data type influence classroom conversations? *International Journal of Science Education*, *30*(13), 1725–1751. doi:10.1080/09500690701506945

Kanari, Z., & Millar, R. (2004). Reasoning from data: How students collect and interpret data in science investigations. *Journal of Research in Science Teaching*, *41*(7), 748–769. doi:10.1002/tea.20020

Khandelwal, S., & Wickstrom, N. (2016). Gait Event Detection in Real-World Environment for Long-Term Applications: Incorporating Domain Knowledge into Time-Frequency Analysis. *IEEE Transactions on Neural Systems and Rehabilitation Engineering*. https://doi.org/10.1109/TNSRE.2016.2536278

Klopfer, E., Yoon, S., & Rivas, L. (2004). Comparative analysis of Palm and wearable computers for Participatory Simulations. *Journal of Computer Assisted Learning*, *20*(5), 347–359. doi:10.1111/j.1365-2729.2004.00094.x

Konold, C., & Miller, C. D. (2005). TinkerPlots. Emeryville, CA: Key Curriculum Press. Retrieved from http://scholar.google.com/scholar?cluster=5929212600541009408&hl=en&oi=scholarr

Konold, C., & Pollatsek, A. (2002). Data Analysis as the Search for Signals in Noisy Processes. *Journal for Research in Mathematics Education*, *33*(4), 259–289. doi:10.2307/749741

Krajcik, J., Blumenfeld, P. C., Marx, R. W., Bass, K. M., Fredricks, J., & Soloway, E. (1998). Inquiry in Project-Based Science Classrooms: Initial Attempts by Middle School Students. *Journal of the Learning Sciences, 7*(3–4), 313–350. doi:10.1080/10508406.1998.9672057

Kuhn, D. (2007). Reasoning about multiple variables: Control of variables is not the only challenge. *Science Education, 91*(5), 710–726. doi:10.1002/sce.20214

Lee, V. R., & Briggs, M. (2014). Lessons Learned from an Initial Effort to Bring a Quantified Self "Meetup" Experience to a New Demographic. In *Proceedings of the 2014 ACM International Joint Conference on Pervasive and Ubiquitous Computing: Adjunct Publication* (pp. 707–710). New York, NY: ACM. doi:10.1145/2638728.2641321

Lee, V. R., & Drake, J. (2013a). Digital physical activity data collection and use by endurance runners and distance cyclists. *Technology, Knowledge and Learning, 18*(1–2), 39–63. https://doi.org/10.1007/s10758-013-9203-3

Lee, V. R., & Drake, J. (2013b). Quantified recess: Design of an activity for elementary students involving analysis of their own movement data. In *Proceedings of the 12th International Conference on Interaction Design and Children* (pp. 273–276). New York, NY: ACM. doi:10.1145/2485760.2485822

Lee, V. R., Drake, J., Thayne, J., & Cain, R. (2015). Opportunistic Uses of the Traditional School Day Through Student Examination of Fitbit Activity Tracker Data. *Proceedings of IDC 2015*. Retrieved from http://works.bepress.com/victor_lee/31

Lee, V. R., Drake, J. R., & Thayne, J. (2016). Appropriating Quantified Self Technologies to Support Elementary Statistical Teaching and Learning. *IEEE Transactions on Learning Technologies*. https://doi.org/10.1109/TLT.2016.2597142

Lee, V. R., & DuMont, M. (2010). An Exploration into How Physical Activity Data-Recording Devices Could be Used in Computer-Supported Data Investigations. *International Journal of Computers for Mathematical Learning, 15*(3), 167–189. doi:10.1007/s10758-010-9172-8

Lehrer, R., Kim, M.-J., Ayers, E., & Wilson, M. (2014). Toward establishing a learning progression to support the development of statistical reasoning. In *Learning over Time: Learning Trajectories in Mathematics Education*. Charlotte, NC: Information Age Publishers.

Lyons, L., Silva, B. L., Moher, T., Pazmino, P. J., & Slattery, B. (2013). Feel the Burn: Exploring Design Parameters for Effortful Interaction for Educational Games. In *Proceedings of the 12th International Conference on Interaction Design and Children* (pp. 400–403). New York, NY: ACM. doi:10.1145/2485760.2485791

Moher, T., Ching, C. C., Schaefer, S., Lee, V. R., Enyedy, N., Danish, J., … Fong, C. (2014). Becoming Reflective: Designing for Reflection on Physical Performances. *ICLS 2014 Proceedings*, 1273–1282.

National Governors Association Center for Best Practices, Council of Chief State School Officers. (2010). *Common Core State Standards*. Washington, DC: Author.

Nemirovsky, R., Tierney, C., & Wright, T. (1998). Body Motion and Graphing. *Cognition and Instruction*, *16*(2), 119–172. doi:10.1207/s1532690xci1602_1

Norooz, L., Mauriello, M. L., Jorgensen, A., McNally, B., & Froehlich, J. E. (2015). BodyVis: A New Approach to Body Learning Through Wearable Sensing and Visualization. In *Proceedings of the 33rd Annual ACM Conference on Human Factors in Computing Systems* (pp. 1025–1034). ACM. Retrieved from http://dl.acm.org/citation.cfm?id=2702299

OECD. (2014). PISA 2012 Results: What Students Know and Can Do – Student Performance in Mathematics, Reading and Science (Volume I, Revised edition, February 2014). Pisa: OECD Publishing. doi:10.1787/9789264201118-en

Peppler, K., Danish, J., Zaitlen, B., Glosson, D., Jacobs, A., & Phelps, D. (2010). BeeSim: Leveraging Wearable Computers in Participatory Simulations with Young Children. In *Proceedings of the 9th International Conference on Interaction Design and Children* (pp. 246–249). New York, NY: ACM. doi:10.1145/1810543.1810582

Petrosino, A. J., Lehrer, R., & Schauble, L. (2003). Structuring Error and Experimental Variation as Distribution in the Fourth Grade. *Mathematical Thinking and Learning*, *5*(2-3), 131–156. doi:10.1080/10986065.2003.9679997

President's Council of Advisors on Science and Technology. (2010). *Prepare and inspire: K-12 educatioin in science, technology, engineering, and math (STEM) for America's future*. Retrieved from http://www.whitehouse.gov/administration/eop/ostp/pcast/docsreports

Resnick, M., Berg, R., & Eisenberg, M. (2000). Beyond Black Boxes: Bringing Transparency and Aesthetics Back to Scientific Investigation. *Journal of the Learning Sciences*, *9*(1), 7–30. doi:10.1207/s15327809jls0901_3

Ryu, S., & Sandoval, W. A. (2012). Improvements to elementary childrens epistemic understanding from sustained argumentation. *Science Education*, *96*(3), 488–526. doi:10.1002/sce.21006

Sandall, B. K. (2016). Wearable Technology and Schools: Where are We and Where Do We Go From Here? *Journal of Curriculum, Teaching, Learning, and Leadership in Education*, *1*(1). Retrieved from http://digitalcommons.unomaha.edu/ctlle/vol1/iss1/9

Sherin, M. G., Russ, R. S., Sherin, B. L., & Colestock, A. A. (2011). Accessing mathematics teachers' in-the-moment noticing. In M. G. Sherin, V. R. Jacobs, & R. A. Philipp (Eds.), *Mathematics teacher noticing: Seeing through teachers' eyes*. Routledge.

Takacs, J., Pollock, C. L., Guenther, J. R., Bahar, M., Napier, C., & Hunt, M. A. (2014). Validation of the Fitbit One activity monitor device during treadmill walking. *Journal of Science and Medicine in Sport / Sports Medicine Australia*, *17*(5), 496–500. https://doi.org/10.1016/j.jsams.2013.10.241

Taylor, K. H., & Hall, R. (2013). Counter-mapping the neighborhood on bicycles: Mobilizing youth to reimagine the city. *Technology, Knowledge and Learning*, *18*(1–2), 1–29. https://doi.org/10.1007/s10758-013-9201-5

Thayne, J., & Lee, V. R. (2016). *Making statistics matter: Connecting statistical inquiry to the life of the students.* Presented at the Association for Educational Communications and Technology Conference, Indianapolis, IN.

KEY TERMS AND DEFINITIONS

Activity Tracker: A subcategory of wearable technologies with a feature set that emphasizes collecting of activity data such as steps taken, calories burned, and distance traveled.

Fitbit: A company specializing in the production and marketing of a line of wearable activity trackers.

Personally-Relevant Data: Data collected about activities or topics of particular importance to the individual(s) interpreting the data.

Tinkerplots: Software for supporting data visualization and analysis in elementary and middle grades, developed by a team from University of Massachusetts lead by Konold and Miller.

Typicality: A rough approximation of where most of the data are in a given distribution. Typicality was used as an entry point to more canonical measures of center such as mean and median.

Variability: A measure of how much the points in a dataset differ from each other and from what is typical or average, can be quantified as standard deviation, range, variance, etc.

Wearable Technology: A category of technologies that are worn as part of the technology's standard use case. These devices often include motion sensors for capturing data relating to the wearer's activities.

Chapter 2
The Use of Complementary Virtual and Real Scientific Models to Engage Students in Inquiry:
Teaching and Learning Climate Change Science

Allan Feldman
University of South Florida, USA

Glenn Gordon Smith
University of South Florida, USA

Molly Nation
University of South Florida, USA

Metin Besalti
University of South Florida, USA

ABSTRACT

This chapter reports on a four-year study to change how climate change science is taught and learned in schools. The goal of the Climate Change Narrative Game Education (CHANGE) project is to take what is known about reform-based practices, incorporating students' lived experiences into the curriculum, and the integration of Information and Communications Technology (ICT) into the classroom. CHANGE uses the following: scientifically realistic text narratives (text stories with local characters, 50-100 years in the future, a local, place-based approach, a focus on the built environment, the use of simulations and games based on scientific data, and a web-based "intermedia" eBook narrative where sections of narrative text alternate with simulations and computer games. The chapter reports on the ways that we have used the above principles to connect classrooms and communities and school science with academic science to facilitate student inquiry into climate science by combining virtual serious educational games with in class, hands-on inquiry using scientific models.

DOI: 10.4018/978-1-5225-2528-8.ch002

INTRODUCTION

It is an interesting time for science education. We are faced with global problems associated with advances in science and technology, while at the same time we seek solutions to them in new scientific knowledge and technological innovations. It is also a time in which there are massive changes in the ways that we share and use information as print and wired media are replaced by digital and wireless communications in the social sphere. All that said, classrooms in schools remain the familiar places in which the teaching and learning of science has changed little over the years. In particular, there are few visible effects of innovative uses of Information and Communications Technology (ICT) or the use of inquiry in the teaching and learning of science in schools.

The use of ICT has become ubiquitous in our contemporary lives. We see this in schools where students use it to write assignments, look up information, and communicate with teachers. Similarly, teachers use ICT to write lesson plans, handouts, quizzes and tests; to look up information; to present lessons; to record and calculate student grades; and to communicate with students, parents and administrators. For the most part these are examples of how ICT has replaced older technology, like printed books, paper and pencil, overhead projectors, calculators, and telephones. While there are teachers and students who engage with ICT in more innovative ways including the use of probeware, access to and analysis of real scientific data, and models and simulations, in the US, where we are located, these innovative uses of ICT are few and far between (e.g., Bang & Luft, 2013).

Similarly, the effects of a more than 50-year effort to transform the teaching of science to be more inquiry oriented are not common. In the mid 20th century governmental agencies around the world funded the development of courses and curriculum in which inquiry was central (e.g., Donahue, 1993; Nuffield Foundation, 2016; Tamir & Jungwirth, 1975). However, Stake and Easley (1978) found that in elementary schools there were only occasional efforts to do more than read about science topics, and that in secondary schools science was seldom taught as scientific inquiry and subjects were presented as what experts had found to be true. Twenty years later the National Science Education Standards (National Research Council, 1996) called again for the teaching of science as inquiry, and more efforts were made to transform teachers' practice (National Research Council, 2000). However, as with ICT, there is little visible evidence that teachers use inquiry oriented pedagogy.

There are many possible reasons why what happens in classrooms changes slowly if at all, including the negative effects of high stakes examinations, ill-prepared teachers, and the overall difficulty of enacting changes in large bureaucracies like school systems. There are also practical problems such as limited access to computers and other devices, and broadband access; and lack of scientific supplies and equipment. Another possible reason is that often the use of ICT and inquiry are presented as add-ons, rather than being incorporated into existing courses and curricula. In this chapter, we report on our efforts to change how climate change science is taught in schools within the context and structure of an existing high school course – Marine Science. We chose this course because many of the topics in its curriculum leant themselves to climate change science.

Our project -- the NSF-funded Climate Change Narrative Game Education (CHANGE)[1] project -- takes what we know about reform-based practices, the need to incorporate students' lived experiences into the curriculum, and the integration of Information and Communications Technology (ICT) to facilitate students' inquiry into climate science. Its aims are for students to gain a better understanding of the impacts of climate change on the natural environment and human-built world particularly within the place-based setting of the southeast US. It attempts to connect climate change and sea level rise to students' everyday lives so that they become motivated to learn about issues related to their region (Hallar, McCubbin, & Wright, 2011; Theobald, 2006). CHANGE does so by employing the following principles: (a) the use of scientifically realistic text narratives about future southeast US residents, (b) a local, place-based approach grounded in scientific data, (c) a focus on the built environment, (d) the use of simulations and games to help students learn principles of Global Climate Change (GCC), and (e) a web-based "intermedia" approach where sections of narrative text alternate with simulations and computer games. All CHANGE materials are available on our website (http://climatechange.usf.edu).

BACKGROUND

In this chapter we report on the how CHANGE uses serious educational games (SEGs) and hands-on activities in a complementary manner to engage students in inquiry through the use of models to learn climate change science. In the next sections we provide some background on global climate change, the teaching of climate change science, scientific inquiry, and serious educational games. We then provide examples of the CHANGE materials after which we present data from our pilot testing of them in classrooms.

Global Climate Change

The Earth's climate is rapidly changing as the concentration of atmospheric greenhouse gases increases. These heat-trapping gases, data show, raise global temperatures, which causes sea level to rise and can lead to an increase in the frequency of extreme weather events (IPCC, 2013; World Bank, 2012; UCAR, 2015). All this has a great impact on the built and natural environments (Jollands, Bernier, & Golubiewski, 2007). With more than 100 million Americans living within the areas directly affected by climate change and coastal flooding, the impact is being seen environmentally and economically in the destruction of coastal infrastructure and communities; the incursion of salt water into fresh water aquifers as the seas rise; the disruption and even destruction of beaches and natural coastal habitats, which are the breeding sites for large numbers of marine organisms; and coral bleaching.

Given the above, there are still significant numbers of people in the US who are ill informed and skeptical about climate change. Politicians, policy institutes, and television or radio commentators often misrepresent climate change as a highly controversial and debatable issue. This may be due in part to misinformation campaigns by energy companies like ExxonMobil (Brulle, 2014; Kaiser & Wasserman, 2016), which questions whether climate change is real and the authenticity of anthropogenically induced climate change (McCright & Dunlap, 2011). In addition, many people consider climate change to be a significant threat globally, but do not feel threatened personally (Leiserowitz, 2005). However, agreement among over 97% of scientists worldwide suggests that the globe is warming at an alarming rate, and that humans are largely to blame (IPCC, 2013).

Difficulty of Teaching and Learning Climate Science

An understanding of science, including its processes and methods for warranting knowledge, is fundamental to understanding the forces that produce climate change and its effects on coastal regions. Clearly students need to be better educated about climate change science if they are to make intelligent and reasonable decisions about what to do in response to its effects on their communities (Kohut, Doherty, Dimock, & Keeter, 2009; Maibach, Roser-Renouf, & Leiserowitz, 2009). This is particularly important for those living in coastal areas, such as those in Florida, who need to be prepared to make decisions about the future of coasts and infrastructure.

Putting climate change science into national science standards (NGSS Lead States, 2013a) is not enough. Teachers need to be able to teach it, and students need to be able to learn it. However, climate change is a difficult topic to teach because: (1) the science involved is complex and multidisciplinary, (2) people do not notice the changes in climate on a daily basis, and (3) the media portrays it in misleading ways.

Its multidisciplinary nature requires knowledge from different fields such as geography, geology, chemistry, physics, ecology, and marine science. Typically, these topics are taught in a disjointed way, for example, a unit on the carbon cycle, another on weather and climate, and others on environmental effects. Therefore, for students to gain a comprehensive understanding of climate change, researchers and educators agree that an interdisciplinary approach is necessary (Summers, Childs, & Corney, 2005).

The global nature of the problem makes climate change difficult to observe at the local level (Pruneau, Liboiron, Vrain, Gravel, Bourque, & Langis, 2001). While GCC is happening faster than ever before, its effects are incremental and noticeable over decades. Therefore, it is necessary to find ways to zoom in, from the global to the local level, and to somehow reduce the perception that it is a problem of the future, not of the present.

Finally, when an issue is controversial, it can affect teachers' instructional decisions due to, among other things, concern that instruction of the topic will lead to objection from stakeholders within the community (Boykoff & Boykoff, 2004). When this controversy is combined with teachers' lack of understanding of the complexities of climate science (Herman, Feldman, & Vernaza-Hernandez, 2015), teachers may opt out of including it in their enacted curriculum (Gehrke, Knapp, & Sirotnik, 1992). While many teachers believe that climate change is an important topic, discrepancies exist about which aspects of climate change should be taught and how much time to allocate to it (Liu, Roehrig, Bhattacharya, & Varma, 2015). There are clear motives as to why teachers avoid controversial topics in their classrooms, including the complexity of the science, opposition from stakeholders, and curricular constraints. However, teachers' report that the greatest challenge when teaching controversial issues is that students are misinformed, lack knowledge regarding the topic, and are therefore unable to actively engage in classroom discussions and activities (Kuş, 2015). This suggests that successful teaching and learning of climate change science in schools requires new and innovative approaches.

Inquiry, Models, and ICT

In this section we examine what it means to engage in inquiry, the various types of scientific models, and the use of serious educational games in school settings.

The Role of Inquiry in Science Education

The use of inquiry has been promoted as a way to engage students in science so that they develop conceptual understanding of science content and of the nature of science (National Research Council, 2000). While the use of inquiry should not be seen as a panacea, many research studies have shown that it can meet the stated goals of its use (e.g., Anderson, 2002; Crawford, 2007; Hofstein, Navon, Kipnis, & Mamlok-Naaman, 2005).

What is meant by inquiry in the science education literature has been somewhat ambiguous. In this chapter we use it to mean what scientists actually do when they engage in science (Anderson, 2002; Bybee, 2006; National Research Council, 1996). For example, in their examination of the use of various forms of guidance to support virtual science laboratories, Zacharia et al. (2015) state that when students engage in inquiry, they do so in "an investigative process that resembles what scientists do" (p. 258), in order to develop scientific knowledge and skills as well as gaining an understanding of the nature of inquiry. In defining inquiry, they provide a list of skills and practices that is echoed by other researchers (e.g., Donnelly, O'Reilly, & McGarr, 2013; Linn, Clark, & Slotta, 2003; National Research Council, 2000), and have been included as the practices of science that all students are expected to be able to engage in according the Next Generation Science Standards (NGSS Lead States, 2013a) and the 2013 PISA draft framework (OECD, 2013). The CHANGE materials are designed so that students engage in these types of science practices, both with hands-on and virtual activities as part of SEGs.

Scientific inquiry in schools can also be enacted at different levels of complexity, depending upon how skilled students, and teachers, are in its practices. Level 1, Confirmation Inquiry is when students engage in an activity that confirms a scientific principle for which the results are known in advance. Level 2, Structured Inquiry, is when the teacher originates the question and students follow a prescribed procedure. Level 3, Guided Inquiry is when the question still comes from the teacher but the students design the procedures. The most complex is Level 4, Open Inquiry, in which students investigate questions that they formulate and devise ways to find answers to their questions (Bell, Smetana, & Binns, 2005).

Scientific Models

Climate change is difficult to teach and learn because it is global in scope and takes years to observe its effects. Therefore, climate change education materials need to rely on models to provide students with inquiry opportunities. As with inquiry, the term scientific model has been defined in multiple ways in the literature (e.g., NGSS Lead States, 2013b; Schwarz & Passmore, 2012). In this chapter, we rely on the list of attributes of scientific models laid out by Cartier, Rudolph, and Stewart (2001, pp. 2-5):

- Models are constituted by empirical or theoretical objects and the processes in which they participate.
- Models can be used to explain and predict natural phenomena.
- Models are consistently assessed on the basis of empirical and conceptual criteria.

- Models are useful as guides to future research.

Given the complexity of how we think about and use models, Harrison and Treagust (2000) developed a typology of models and how they are used in science teaching and learning. They divided models into four main categories: (1) *Scientific and teaching models* are used by teachers to explain entities not easily observed in the classroom. (2) *Pedagogical analogical models for conceptual knowledge* include symbolic models, like chemical formulae, mathematical models, and theoretical models designed to help students understand concepts. (3) *Models depicting multiple concepts and/or processes* include maps, diagrams and tables, concept-process models, and simulations. Concept-process models emphasize that scientific concepts, when often represented by nouns, are complex processes. Simulations are a type of model that can help students to understand the underlying processes through manipulation of the various factors that interact in the process. (4) *Personal models* include mental models and students' alternative conceptions that combine their intuitive mental models with accepted scientific models.

Feldman, Cooke, and Ellsworth (2010) provided a list of types of physical models based on how they can be used in scientific inquiry. A physical model 1 (PM1) can be used simply to represent a physical phenomenon in the form of a three-dimensional model, drawing, videos, or animation. PM1s can be used to teach or explain the phenomenon to students, but are not generally useful for inquiry. PM2s allow students to manipulate the model to explain and predict natural phenomena so that they can understand the processes inherent in the phenomena. A PM3 is a representation of the phenomenon that can be manipulated and used to test hypotheses, assess empirical and conceptual criteria, or guide future research (Cartier, Rudolph, & Stewart, 2001).

Feldman et al.'s (2010) typology of physical models can be applied to virtual models such as computer simulations or games. Simulations or games could simply represent the phenomenon (e.g., demonstrating how sea level rises when glaciers melt); students could manipulate them to show, for example, the rising of the seas due to thermal expansion; or they could include ways for students to manipulate them to test ideas that describe the natural phenomena that underlie climate change. We will refer to these as DM1-DM3 digital models (see Table 1). As we will show later in this chapter, when students engage in inquiry with CHANGE materials they are interacting with concept process models (Harrison & Treagust, 2000) that can be manipulated to test hypotheses (Feldman et al., 2010).

The Use of Digital Tools for Inquiry in STEM Learning

In this chapter we examine the ways in which students can use different types of ICT to help them engage in science inquiry activities. Rather than review the uses of ICT in a decontextualized manner, we do so in the context of the CHANGE project, with appropriate references to the literature. The CHANGE project uses a wide variety of forms of educational technology including computer simulations and models, serious educational games (SEGs) (Annetta et al., 2013), electronic measuring technologies, eBooks (G. G. Smith et al., 2013), and webquests (Subramaniam, 2012), among others.

These different forms of educational technology are used for different purposes in the CHANGE project. For example, CHANGE uses a web-based intermedia[2] eBook narrative where sections of narrative text alternate with simulations and computer games to tie together the units that are distributed through the marine science course. The narrative uses a place-based story-telling approach, situating climate change firmly in the west-central Florida Gulf Coast communities, where our high school students live, to make climate change feel more immediate to them (G. A. Smith, 2007). It consists of scientifically realistic, short stories about high school students in the future who grapple both with how climate change plays out in Florida (hurricanes, sea level rise, etc.) and typical issues high school students encounter such as concerns about grades, graduation and job prospects, friendships, and family issues. Each story includes at least one serious educational game (SEG), embodying some of the core climate change science content from the current unit.

SEGs are games "in which education (in its various forms) is the primary goal, rather than entertainment" (Michael & Chen, 2006, p. 17) and "use the artistic medium of games to deliver a message, teach a lesson, or provide an experience" (p. 23). Annetta, Cook, & Schultz (2007) argue that when students are immersed in well-designed SEGs, they engage in metacognition, learn new content, and revise existing schemata through the help of cognitive scaffolds built into the games. In addition, SEGs can provide students with lifelike experiences that are not possible to replicate in traditional classrooms or field trips and can be the site of problem-based learning activities.

CHANGE uses SEGs for several purposes including providing students with science content needed for classroom inquiry, introducing them to the use of actual instruments used by scientists in their research, and to engage in inquiry through a virtual field trip. It is important to note that the CHANGE SEGs are in fact games, and not simply simulations. We can see this for example, in a game called Water Gauge Warrior in which students engage in various activities to construct a tide gauge that can withstand a severe hurricane. Water Gauge Warrior has the key elements of computer games as, described by Malone (1981), Crawford (1982), and Gee (2003), i.e., (1) *rules based on the game play structure*: the player has to stay within their budget as they select parts for their water level gauge, (2) *a start state*: starting with no water level gauge built and no budget, (3) *a goal for winning*: with your budget, earned by answering questions on the topic, create a water level gauge that will withstand the ravages of a tropical storm, (3) *immediate feedback on progress towards the goal*: selecting a set of parts to build your water gauge that stays within your budget, (4) *a game play space with multiple options*: many possible choices to be made in building your water level gauge, (5) *competition*: against yourself, and implicit competition with classmates, and (6) *fantasy:* the storyline that you have been hired by the city of Tarpon Springs to build a water level gauge.

CHANGE also incorporates the use of ICT in its curriculum materials and its pedagogical approach. The materials are provided to teachers via the Internet (http://climatechange.usf.edu), and web-based surveys and other instruments, distributed using sites like SurveyMonkey, facilitate data collection for formative evaluation and research. Most of the background information is provided to teachers in PowerPoint files or MS Word documents, so that teachers can modify them to suit their educational situations.

EXAMPLES OF THE COMPLEMENTARY USE OF VIRTUAL AND REAL SCIENTIFIC MODELS TO PROMOTE LEARNING OF CLIMATE CHANGE SCIENCE IN CHANGE

The primary way that CHANGE engages students in inquiry is by combining SEGs and hands-on classroom laboratory activities. As the school year progresses the amounts and levels of inquiry in the SEGs and activities increases. In this section of the chapter we provide descriptions of the SEGs and hands-on activities for three CHANGE topics: sea level rise, coastal sediments, and harmful algal blooms. Each of these topics are part of a larger unit of the marine science course: ocean exploration, marine geology, and populations: producers, respectively.

Sea Level Rise

To see how global warming affects sea level, one needs to be able to measure it, for example with a tide gauge. Tide gauges are complex pieces of equipment, distributed throughout the world, that continuously measure and record the height of the sea at a fixed point. Modern ones have on-site computers, self-contained power supplies, and can transmit data remotely. To teach students how a tide gauge works and the importance of its different components, CHANGE developed an SEG called "Water Gauge Warrior" (Figure 1). In this game students are challenged to design a tide gauge that would withstand a category 5 hurricane. In the first level of the game, students answer questions about the purpose of the different components to earn money to build their own tide gauge. As they design their gauge they need to make decisions about the cost and properties of the components that they assemble

Figure 1. Screenshots from the Water Gauge Warrior game

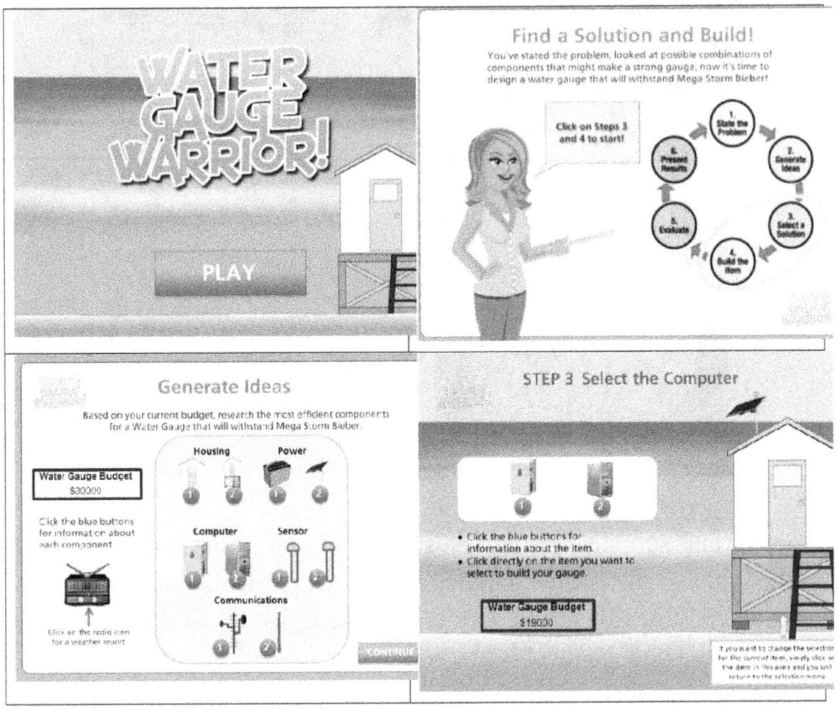

virtually for their gauge. Once completed, their gauge is "tested" to see whether it can withstand a large storm.

We see the Water Gauge Warrior game as an example of a type 2 digital model (DM2) because it does more than simply demonstrate how the tide gauge works, but does not allow for the testing of hypotheses (see Table 1). It is also what Harrison and Treagust (2000) call a concept-process model because it helps students to understand the complex processes that underlie the term sea level. We also see the game as an opportunity for students to engage in Level 2 inquiry because the students are given the problem and the methods with which to solve it.

While the Water Gauge Warrior game provides a way for students to learn about the workings of an actual piece of scientific equipment, we also want them to be able to inquire into how different factors affect sea level using a hands-on laboratory activity. It became clear to us that a physical model that mimics the operation of a tide gauge would either be too expensive, or too complicated to build using low-cost or found materials. We solved the problem by using the ubiquitous (in the U.S.) one-gallon plastic jug. The jug model exemplifies CHANGE's use of low cost of found materials in recognition of the limited budgets that teachers have for the purchase

Figure 2. Plastic jug that is used as a model for measuring sea level

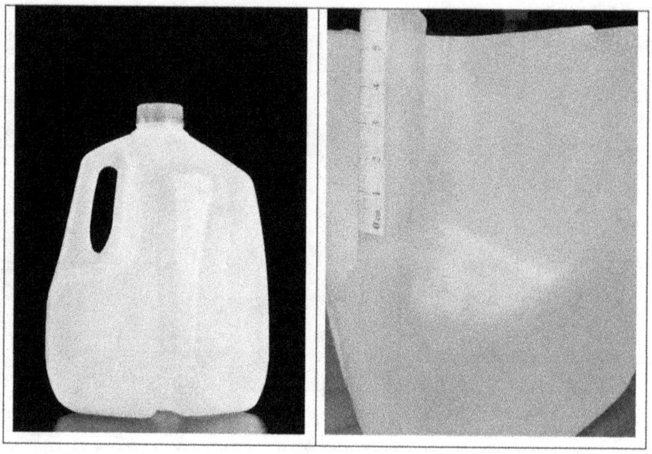

of equipment and supplies. As can be seen in Figure 2, the main part of the jug models the sea, and the hollow handle with a ruler, which is connected to the rest of the jug, models the tide gauge. In this way, the model distinguishes between the body of water and how its level is measured.

The students use the jug model to investigate the effects of the melting of sea and continental ice, and of temperature changes on sea level. In this activity the inquiry questions are teacher generated, and students are provided with the tools and materials to investigate them. Hence, it an example of Level 2 inquiry. We see it as a physical model 3 (PM3) because students can manipulate it to the test their ideas about the relationships between the effects of global warming – the melting of sea and continental ice and warming of the oceans – on sea level (see Table 1). It is also what Harrison and Treagust (2000) call a concept-process model because, like the Water Gauge Warrior game, it helps to unpack the complex processes hidden in the noun phrase, *sea level rise*. In addition, this activity incorporates the CHANGE place-based approach. After the students investigate how GCC affects sea level, they then go to a website of the US National Oceanic and Atmospheric Administration (NOAA) (http://www.coast.noaa.gov/slr/) in which they can see predictions for sea level rise would affect their community where they live, as well as those of relatives and friends.

The combination of SEG and hands-on inquiry activity helps to solve a problem identified by Waight and Abd-El-Khalick (2011) in their article about the attempt to use the Biology Student Workbench (BSW), a student version of a web-based tool that enables scientists to model protein and nucleic acid sequencing. They found that the BSW was not enacted in schools for two main reasons: 1) The difference between

Table 1. Types of physical and digital models

Type of Physical Model	Type of Digital Model	Properties	Examples
PM1	DM1	Representation that can be used to teach or explain the phenomenon.	Three-dimensional models, drawings, videos, animations,
PM2	DM2	Representation that can be manipulated to explain and predict nature phenomena.	Physical models with moving parts, simulations, serious educational games
PM3	DM3	Representation of the phenomenon that can be manipulated and used to test hypotheses, assess empirical and conceptual criteria, or guide future research.	Bench-scale laboratory models, simulations, serious educational games

Based on Feldman, Cooke and Ellsworth (2010) and Cartier et al. (2001).

the ways in which the scientists who developed and used the Biology Workbench did science, and the teachers' beliefs about what constitutes school science. 2) While the BSW was significantly modified from the Biology Workbench, it still was not appropriate technology for the doing of science in school. In the CHANGE sea level activities, we dealt with these issues by having a simplified but accurate model of a tide gauge in the Water Gauge Warrior game, and used school appropriate materials for the hands-on activity. The structure of the hands-on activity helps to scaffold a change in teachers' beliefs, from lab activities that were purely conformational towards ones in which students can begin to inquire into natural phenomena.

Sediments

Students investigate the connection between coastal sediments and historical changes in sea level in the Beachrock Decoder SEG (Figure 3) and a classroom activity in which they investigate a teacher-presented question while using student designed and selected procedures.

In the game students participate in a virtual field trip to Hutchinson Island, Florida. Hutchinson Island has large beachrock outcroppings showing evidence of past sea level, estimated to be six to eight meters higher than it is today. The students travel virtually to the island with Dr. Ping Wang, the geologist on the CHANGE team, to investigate the properties of an outcropping by observing it through a magnifying glass and by testing it with acid to determine its composition. Later in the game, students examine an ancient sandbar and a sub-tidal worm reef, both of which were at one time located below sea level, but are now a few meters above it. These investigations

Figure 3. Screenshots from the Beachrock Decoder game

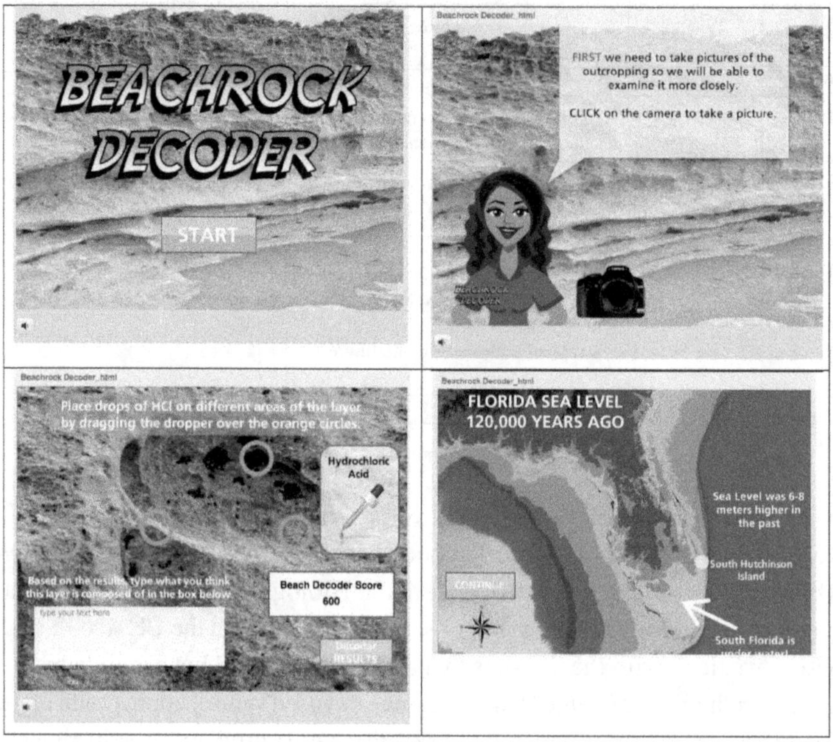

lead the students to uncover clues related to the past sea level, concluding that the area's sea level was once six to eight meters higher than it currently is.

The Beachrock Decoder game engages the students in Level 2 structured inquiry because the students are presented with a question to investigate – "What do the observations of the beach rock outcroppings and the other formations tell us about Florida's past sea level?" – and they use the investigative methods built into the game. We also see it as a DM3 (see Table 1) because the students engage it to test hypotheses about the relationships of the characteristics of the swash zones, sandbar and worm reef, with sea level.

After students have completed the virtual field trip, they engage in a hands-on inquiry activity in which they examine the properties of sediment samples, including color, size, Wentworth classification, texture, sorting, and composition, that were collected from different sites around Florida. Using their observations and measurements they deduce the location of the collection site for each of the samples. The characteristics of the samples and where they were collected are then used to formulate hypotheses about Florida's past sea level and coastline. As with the Beachrock Decoder game, this is an example of Level 2 inquiry because students are given both the question and the methods to investigate them.

In this topic students engage in the science practices that Dr. Wang and his students use when they investigate the formations at Hutchinson Island and coastal sediments. In the game they use those practices in a virtual setting, while in the classroom they can work directly with the sediments to engage in structured inquiry. This combination provides an important opportunity for teachers to engage their students in the types of science practices described in the NGSS (NGSS Lead States, 2013) and by PISA (OECD, 2013).

Harmful Algal Blooms (HABs)

A third example from the CHANGE curriculum focuses on the study of Harmful Algal Blooms (HABs), which are often referred to as red tide. HABs are rapid growths of algae in marine or freshwater environments due to influxes of nutrients and warm water temperatures. They have major negative effects on the environment because they deplete the dissolved oxygen in water needed by fish and most other organisms. They can also be toxic to humans and other animals. It is believed that algal blooms will increase in frequency, and possibly in size, as a result of GCC because of both increases in water temperature and increased washing of nutrients in fertilizers used by farmers and homeowners, from the land to the water by larger storm events. Because the study of algae was already a part of the Marine Science curriculum, the CHANGE materials on HABs and climate change fit easily into what the teachers were already expected to cover as part of the course.

The SEG for this topic is Whack Algae (Figure 4) in which students gain information about HABs and the factors that contribute to their growth. It is a DM2 because students can manipulate the factors that cause HABs, but would find it difficult to use it to test hypotheses (see Table 1). We believe that it is generous to call it an inquiry activity because its primary purpose is for students to gain the knowledge of what factors can cause an HAB in Tampa Bay and how they are related to climate change. They then can use this knowledge when they engage in inquiry in the hands-on activity.

The HAB hands-on activity is CHANGE's most authentic use of science practices, modified to be appropriate for schools. The students are given instruction in how to set up a bioreactor — a vessel in which to grow algae. In the research laboratory at our university, bioreactors are usually custom-built containers constructed of glass or Plexiglas, costing hundreds of dollars. The students' bioreactors are clear plastic 20-ounce (~0.6L) water bottles. Instead of toxic marine algae, they grow a freshwater species provided to them by the university laboratory. At the university, the concentration of algae is measured using a method called fluorospectroscopy

Figure 4. Screenshot from Whack Algae game showing the factors that students can manipulate to cause a harmful algal bloom

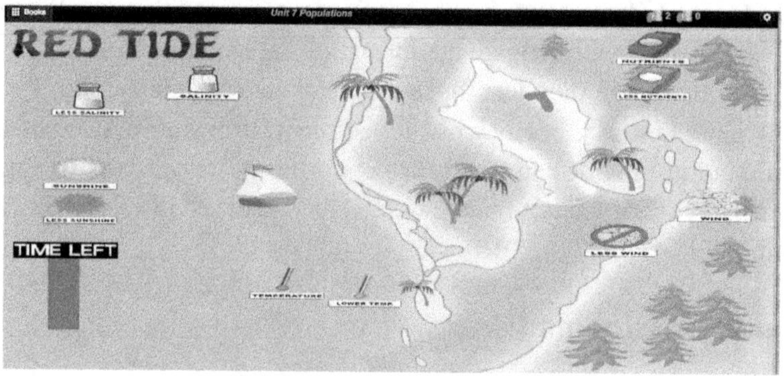

(Halfhide, Wilkie, & Ergas, 2016). In the classroom we provide the students with a color chart, which is a set of test tubes, each containing a different concentration of algae (Figure 5). The students match the color of their bioreactor against the colors in the photograph to estimate the concentration algae.

In the activity students make a control by adding fertilizer to a bottle containing the non-toxic algae species and distilled water. They then choose a variable to alter from the control to examine whether it will increase or decrease HAB growth over time. While the teacher provides students with supplies that can be used in their bioreactor, students are free to choose any variable they like that could be related to an effect of climate change. In the piloting of the activity, students have changed

Figure 5. Color chart for measuring the concentration of algae in the bioreactors

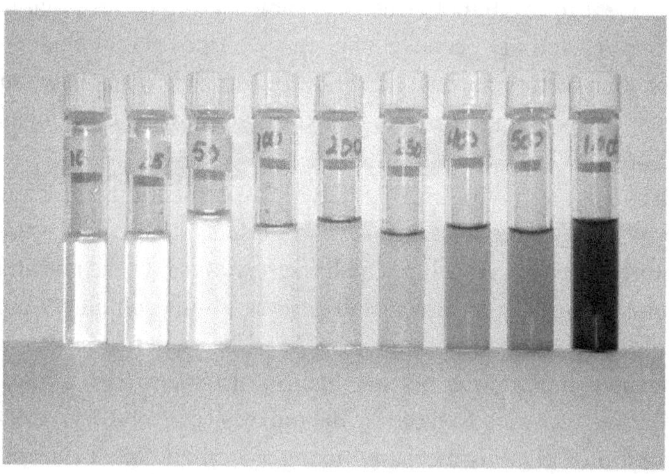

the amount of light, fertilizer and salinity. They have also looked at the effect of temperature and even the presence of bacteria. For five days the students measure the concentration of the algae in the bioreactor using the color chart. They then use their data to draw conclusions about the relationship between their variable and the growth of algae. We see this as an example of Level 4: Open inquiry in which the students investigate topic-related questions that are student formulated through student designed and selected procedures.

Finally, we ask whether the HAB activity is an example of the use of a model. While the water bottle is a physical model of an actual bioreactor, we prefer to think of the whole activity as a scientific model that fits the criteria suggested by Cartier et al. (2001), and what we would call a DM3. In the HAB activity the students construct ideas that describe the process by which different factors associated with climate change can affect the growth of algae. It is an empirical model in which they make measurement of algae concentrations; they use it to explain and predict the phenomenon of HABs; they test their model against their empirical evidence; and although they are limited by the time constraints of the Marine Science curriculum, they could be used as guides for future research.

THE EFFECTS OF STUDENTS' ENGAGEMENT WITH VIRTUAL AND REAL SCIENTIFIC MODELS IN CHANGE

The CHANGE curriculum materials were developed and pilot-tested during the 2013-14, 2014-15, and 2015-16 academic years. In this section we describe the data collection and analysis methods that we used, and the findings from the pilot testing.

Data Collection

During the three pilot years we collected data to 1) provide formative information for the improvement of the CHANGE materials; and 2) test instruments that we would use in the full implementation year to determine the efficacy of the materials. We collected the data for the former through online surveys that were administered at the end of each unit of the marine science course (End of Unit Surveys). For the latter, we developed student pre- and post-surveys to determine their knowledge and beliefs about climate change (Nation, Feldman, Smith, Wang, & Lou, 2016), and multiple-choice items that were incorporated into the district-wide marine science midterm and final exams.

The end of unit (EOU) surveys were in two parts. The first included Likert-type questions about the materials. Students were also asked the following open-ended questions: In what specific way do you feel that the combination of eBook story, computer game, and hands-on lab in this unit helped you to learn climate change science better and/or increase your interest in this unit? Or are there other things that could be added to the combination that would help make the unit more interesting?

Data were also collected from online surveys, midterm, and final exam to measure student responses to the materials and further understanding of climate change science. The CHANGE student survey, which was administered at the beginning and end of the 2014-2015 and 2015-16 academic years asked students about the causes and implications of climate change, and students' likelihood to take mitigative actions to help reduce the impacts of climate change on the built and natural world. The midterm and final exam were distributed to all marine science students in the district, including those how had not participated in the piloting of CHANGE. An item analysis was performed to determine whether or not experience of the CHANGE curriculum had a significant impact on student understanding of climate change science. In addition to these surveys and exams, data were collected for formative purposes from classroom observations and focus groups made up of the pilot teachers.

Data Analysis

Quantitative data gathered from the CHANGE student survey, midterm, and final exam results were analyzed using SAS (Statistical Analysis System). For the CHANGE student survey analysis, t-tests were used to compare the means from the initial data with the end of year data. Results from the midterm and final exam were also analyzed using a t-test to examine the difference of scores between students who experienced the CHANGE curriculum and the students who did not.

Open-ended qualitative data from the EOU surveys were analyzed using thematic coding to determine how the students feel about the combination of eBook story, computer game, and hands-on lab. After similar codes were grouped together, frequency distributions were generated for the codes (Patton, 2015). Classroom observations and focus groups data were analyzed in a similar fashion. CHANGE project leadership and graduate students engaged in continuous review of data to improve the quality of the findings and conclusions throughout the data analysis.

RESULTS

End of unit (EOU) survey data were collected for all three of the CHANGE topics described in this chapter. However, because of a technical problem with the online survey, we do not have responses to the open-ended or content questions for the topic on HABs. The EOU survey for the piloting of the sea level materials indicated that on the average students (n=117) agreed with the following statements:

- Playing the computer game helped me to engage with the content of this unit.
- I found the computer game relevant to learning the content of the unit.
- The combination of hands-on lab, eBook story and computer game made the unit more relevant to me.
- I felt more engaged in learning this unit because of the combination of hands-on lab, eBook story and computer game.

In addition, more than 75% of the students demonstrated that they gained an understanding of the different effects of the melting of sea and continental ice on sea level. For the coastal sediments topic the EOU data indicated that the students (n=109), on the average, agreed with the statements above. In addition, about three-fourths of them correctly estimated the ancient sea level and knew what types of data geologists use to determine past sea level.

The EOU data from the HAB unit focused on the eBook and the games. Overall the students (n=137) found the games to be fun and informative. In addition, we have formative data from extensive observations that we made of the implementation of the unit in three different schools. The observation data suggest that the unit was successful in engaging the students in authentic research practices, and that they came to understand the relationships between GCC and HABs.

For sea level and coastal sediments, we used a grounded theory approach (Corbin & Strauss, 2015) to analyze the open-ended question in which we asked the students how the combination of SEGs and hands-on activities helped them to learn the topics. For the sea level topic, the students were overwhelmingly positive about the combination with only 22 of 121 respondents indicating any negative aspect at all. For example, one student wrote, "I'd rather just read about things instead of all the labs and games." Another responded, "It was a little helpful but should be more hands-on and directly related to the topics."

Of the positive responses about the sea level topic, we coded 32 responses in which students stated that the combination was fun (n=6), interesting (n=13), or simply that they liked it (n=13).[3] More elaborate reasons included that it helped them to learn (n=26), there was a real world connection (n=4), the combination provided them with a variety of learning experiences (n=10), and there were multiple modes for learning and learning styles (n=15). Exemplars for these codes can be found in Table 2.

We found similar results from the open-ended responses to the same question in the end of unit survey for the coastal sediments topic. In this case, 10 of the 99 responses were negative, such as "It honestly needs to be more appealing" and

Table 2. Exemplars of codes used for analysis of sea level open-ended question from the EOU survey

Code	Exemplar	N
Multiple modes for learning, including learning styles	All the features put together gave me a diversity and variety of learning styles	25
Combination more engaging	I think it helped engage myself and other students because it offered different things to do throughout the lesson.	11
Real world connection	The combination of this trio helped me learn about climate change better because I could apply it to my everyday life	4
Helped me learn	All in all, I felt like I learned a lot about climate change in this unit	26
Positive (no reason given)	The combination was lovely.	13
Negative (various reasons)	I'd rather just read about things, instead of all the labs and games	22
Fun	It was fun and I could 'physically' engage in the unit.	6
Interesting	The mix of things made learning more interesting.	13

"More games that are for fun and not just lesson based". In addition, there were 15 responses that were either not relevant, unclear, or missing. The remainder of the coded responses were positive about the SEGs and hands-on activities (n=79). Of these, a total of 37 were that the combination was fun, interesting, or simply that they liked the combination. Students responded that the combination helped them to learn about the relationship between coastal sediments and climate change (n=8), it was engaging (n-9), or that it provided them with a variety of experiences that worked with multiple learning styles (n=13). Exemplars of these codes are in Table 3.

The remaining data that we analyzed relate more broadly to the implementation of the CHANGE curriculum materials. For example, in the beginning and end of the year surveys, students were asked, "How likely do you think it is that human actions have changed global climate?" We found that students were ranking humans as a likely contributor to climate change significantly higher ($p<0.01$) after experiencing the CHANGE curriculum. CHANGE items were added to the midterm and final exams of the marine science course for the 2014-15 and 2015-16 academic years. Therefore, all students enrolled in the course responded to those items. We analyzed the data for the midterm and final of 2014-15 and the midterm for 2015-16. The midterm for 2014-15 was used solely to test the items. For the final given that year we found that the students enrolled in the classes that used the CHANGE materials (n=176) scored significantly higher ($p<0.001$) than the students who did not experience the CHANGE materials (n=789). For the 2015-16 midterm CHANGE students (n=312) performed significantly higher ($p<0.001$) than the non-CHANGE students (n=1502).

Table 3. Exemplars of codes used for analysis of coastal sediments open-ended question from the EOU survey

Code	Exemplar	N
Multiple modes for learning, including learning styles	This combination overall worked really well to help learn the science of climate change as well as increase in the interest in the unit. This is because it reaches out to all types of learners. For example: visual, auditory, and kinesthetic learners. The combination together has a stronger impact.	23
Combination more engaging	I felt engaged in what I was doing. It made me feel that if someone needed help with something I would be able to assist them with my background knowledge on geology so it's successful.	7
Real world connection	The story along with the games gives an interesting real life situation.	
Helped me learn	I feel that the course was making a genuine interest in trying to help me learn, unlike most lessons where it's just a PowerPoint followed by a test.	14
Positive (no reason given)	I think the combination works just fine	11
Negative (various reasons)	It honestly needs to be more appealing.	10
Fun	We don't get bored half way through the lesson and we're always engaged in the unit.	1
Interesting	I became more interested because it helped me get a better understanding through a different method.	18

The 2014-15 final exam included questions about the HAB topic and the 2015-16 midterm included questions about the sea level and coastal sediments topics.

In addition to the results from the student data reported above, our observations of classrooms suggest that independent of which was done first by the students – the SEG or the hands-on activity – they were better able to complete the second activity in the ways intended by the CHANGE project. In general, the students were more familiar with the content and methods, and were better able to make arguments in whichever came second. Finally, data from the teacher focus groups indicate that they believe that the combination of SEGs and hands-on activities improve student learning by reinforcing the science content.

FUTURE RESEARCH DIRECTIONS

In our introduction to this chapter we argued that there has been little change in the ways in which science is taught in the US and elsewhere even though there has been a concerted effort to implement inquiry teaching and learning methods, as well as the use of ICT. Our study supports our belief that this is due to several factors, including the demands of accountability systems, and the resilience of traditional conceptions of what counts as school science. However, outside of schools, people engage in inquiry and ICT has become an integral part of how they work, play and communicate with one another. The curriculum materials we describe in this chapter were designed to bring everyday activities like reading, playing games, and inquiring into local space into school science. Our study so far supports that it is possible to change how teachers teach and students learn within the context of an existing course by using ICT and inquiry to augment what is already happening in schools. However, there is still much research to be done. For example, we would like to gain a better understanding of how teachers' beliefs about what they teach, like climate change, affects their practice and what their students learn, believe and our willing to do. We would also like to uncover the cognitive mechanisms that connect the reading of a fictional eBook with embedded serious educational games and their interaction with hands-on activities, with the learning of science content and practices. Finally, all our research so far has been relatively short term. Can the types of changes that we have seen so far be sustained in schools? Will teachers continue to use the new methods and materials without our support? Will the school district continue to support the teaching of climate change given the new political landscape in the US? It is our plan to investigate these questions in our future research.

CONCLUSION

In the above three examples of CHANGE curriculum materials, we demonstrated the ways that SEGs and in-class hands-on activities can work together to promote inquiry using digital and physical models. The SEG for the sea level topic, which was a simulation and Type 2 digital model (DM2), engaged students in Level 2 inquiry to learn how an actual tide gauge works. This provided them with the knowledge that they needed to see and understand what the jug model was modeling. For the coastal sediments topic, the SEG allowed students to virtually experience what it is like to go on a geology field trip and examine specimens in situ. This provided them with the background to connect what they did in the lab with the sediment samples to how research is done in the field by geoscientists. The formative data from the piloting of both of these topics, as well as our other data, support the efficacy of the approach.

Finally, for the HAB topic, the SEG provides students with the necessary background information about how and why an HAB could occur in Tampa Bay. However, it did not provide the type of virtual experience that students had with the Water Gauge Warrior and Beach Rock Decoder games. We do not necessarily see this as a problem because in the HAB activity students engaged in a full range of science practices that include developing research questions and hypotheses, determining the best way to use their bioreactors to test their hypotheses, collecting and analyzing data, and coming to defensible conclusions. What is missing from the activity itself is the context and causes of HABs, which they get from the Whack Algae game. Although we did not have the same type of formative data as for the other topics, the exam results and classroom observations support the complementary nature of the SEGs and hands-on activity for the HAB topic.

During this academic year the CHANGE materials are being incorporated into the Marine Science courses in twenty-seven high schools in the district. All indications from our three years of developing and piloting the CHANGE materials suggest that if teachers use them as the pilot teachers did, there will be major changes in what happens in their classrooms. Because the CHANGE materials are only for about 15% of the class meetings we expect that much of their teaching would likely remain similar to what they have already been doing. However, we think that he would be excited to see how the SEGs and hands-on activities are integrated in the CHANGE curriculum materials, resulting in a transformation of how teachers teach and students learn science in their schools.

REFERENCES

Anderson, R. (2002). Reforming science teaching: What research says about inquiry. *Journal of Science Teacher Education, 13*(1), 1–12. doi:10.1023/A:1015171124982

Annetta, L. A., Cook, M., & Schultz, M. (2007). Video Games: A Vehicle for Problem-Based Learning. *e-Journal of Instructional Science and Technology, 10*(1).

Annetta, L. A., Frazier, W. M., Folta, E., Holmes, S., Lamb, R., & Cheng, M.-T. (2013). Science teacher efficacy and extrinsic factors toward professional development using video games in a design-based research model: The next generation of STEM learning. *Journal of Science Education and Technology, 22*(1), 47–61. doi:10.1007/s10956-012-9375-y

Bang, E., & Luft, J. A. (2013). Secondary Science Teachers Use of Technology in the Classroom during Their First 5 Years. *Journal of Digital Learning in Teacher Education, 29*(4), 118–126. doi:10.1080/21532974.2013.10784715

Bank, T. W. (2012). Turn down the heat: why a 4 C warmer world must be avoided. The World Bank, 74455.

Bell, R. L., Smetana, L., & Binns, I. (2005). Simplifying inquiry instruction. *Science Teacher (Normal, Ill.)*, *72*(7), 30–33.

Boykoff, M. T., & Boykoff, J. M. (2004). Balance as bias: Global warming and the US prestige press. *Global Environmental Change*, *14*(2), 125–136. doi:10.1016/j.gloenvcha.2003.10.001

Brulle, R. J. (2014). Institutionalizing delay: Foundation funding and the creation of US climate change counter-movement organizations. *Climatic Change*, *122*(4), 681–694. doi:10.1007/s10584-013-1018-7

Bybee, R. W. (2006). Scientific Inquiry And Science Teaching. In L. B. Flick & N. G. Lederman (Eds.), *Scientific Inquiry and Nature of Science: Implications for Teaching, Learning, and Teacher Education* (pp. 1–14). Dordrecht: Springer Netherlands.

Cartier, J., Rudolph, J., & Stewart, J. (2001). *The nature and structure of scientific models*. National Center for Improving Student Learning and Achievement in Mathematics and Science.

Corbin, J., & Strauss, A. C. (2015). *Basics of Qualitative Research: Techniques and Procedures for Developing Grounded Theory* (4th ed.). Thousand Oaks, CA: Sage Publications.

Crawford, B. A. (2007). Learning to teach science as inquiry in the rough and tumble of practice. *Journal of Research in Science Teaching*, *44*(4), 613–642. doi:10.1002/tea.20157

Crawford, C. (1984). *The art of computer game design*. Berkeley, CA: Osborne/McGraw-Hill.

Donahue, D. M. (1993). Serving Students, Science, or Society? The Secondary School Physics Curriculum in the United States, 193065. *History of Education Quarterly*, *33*(3), 321–352. doi:10.2307/368196

Donnelly, D., OReilly, J., & McGarr, O. (2013). Enhancing the Student Experiment Experience: Visible Scientific Inquiry through a Virtual Chemistry Laboratory. *Research in Science Education*, *43*(4), 1571–1592. doi:10.1007/s11165-012-9322-1

Feldman, A., Cooke, M. L., & Ellsworth, M. s. (2010). The Classroom Sandbox: A physical model for scientific inquiry. *Science Teacher (Normal, Ill.)*, *77*(9), 58–62.

Gee, J. P. (2003). *What video games have to teach us about learning and literacy*. New York: Palgrave Macmillan.

Gehrke, N. J., Knapp, M. S., & Sirotnik, K. A. (1992). In Search of the School Curriculum. *Review of Research in Education*, *18*, 51–110. doi:10.2307/1167297

Halfhide, T., Wilkie, A., & Ergas, S. (2016). *Indigenous Algal Growth on Municipal Sludge Centrate and Measuring Lipid Productivity Using Fluorospectroscopy and Gravimetric Analyses*. Paper presented at the Algal Biomass, Biofuels and Bioproducts (ABBB), San Diego, CA.

Hallar, A. G., McCubbin, I. B., & Wright, J. M. (2011). CHANGE: A place-based curriculum for understanding climate change at Storm Peak Laboratory, Colorado. *Bulletin of the American Meteorological Society*, *92*(7), 909–918. doi:10.1175/2011BAMS3026.1

Harrison, A. G., & Treagust, D. F. (2000). A typology of school science models. *International Journal of Science Education*, *22*(9), 1011–1026. doi:10.1080/095006900416884

Herman, B. C., Feldman, A., & Vernaza-Hernandez, V. (2015). Article. *International Journal of Science and Mathematics Education*, 1–21.

Higgins, D., & Higgins, H. (2001). Intermedia. *Leonardo, 34*(1), 49-54. Retrieved from Project MUSE database website: https://muse.jhu.edu/article/19618

Hofstein, A., Navon, O., Kipnis, M., & Mamlok-Naaman, R. (2005). Developing students ability to ask more and better questions resulting from inquiry-type chemistry laboratories. *Journal of Research in Science Teaching*, *42*(7), 791–806. doi:10.1002/tea.20072

IPCC. (2013). The physical science basis. Contribution of working group I to the fifth assessment report of the Intergovernmental Panel on Climate Change.

Jollands, N., Ruth, M., Bernier, C., & Golubiewski, N. (2005) Climate's long-term impacts on New Zealand infrastructure—a Hamilton City case study. In Proceedings ecological economics in action. New Zealand Centre for Ecological Economics.

Kaiser, D., & Wasserman, L. (2016). The Rockefeller Family Fund vs. Exxon. *The New York Review of Books*, *63*(19). Retrieved from http://www.nybooks.com/articles/2016/12/08/the-rockefeller-family-fund-vs-exxon/

Kohut, A., Keeter, S., Doherty, C., & Dimock, M. (2009). Scientific achievements less prominent than a decade ago: Public praises science; scientists fault public, media. Washington, DC: The Pew Research Center for the People & the Press. Retrieved from http://www. people-press. org/files/legacy-pdf/528. pdf

Kuş, Z. (2015). Science and social studies teachers' beliefs and practices about teaching controversial issues: Certain comparisons. *JSSE-Journal of Social Science Education*, *14*(3), 84–97.

Leiserowitz, A. A. (2005). American risk perceptions: Is climate change dangerous? *Risk Analysis*, *25*(6), 1433–1442. doi:10.1111/j.1540-6261.2005.00690.x PMID:16506973

Linn, M. C., Clark, D., & Slotta, J. D. (2003). WISE design for knowledge integration. *Science Education*, *87*(4), 517–538. doi:10.1002/sce.10086

Maibach, E., Roser-Renouf, C., & Leiserowitz, A. (2009). *Global Warming's Six Americas 2009: An Audience Segmentation Analysis. Yale Project on Climate Change*. New Haven, CT: Yale University and George Mason University.

McCright, A. M., & Dunlap, R. E. (2011). Cool dudes: The denial of climate change among conservative white males in the United States. *Global Environmental Change*, *21*(4), 1163–1172. doi:10.1016/j.gloenvcha.2011.06.003

Michael, D. R., & Chen, S. (2006). *Serious games: games that educate, train, and inform*. Boston, MA: Thomson Course Technology PTR.

Nation, M., Feldman, A., Smith, G. G., Wang, P., & Lou, Y. (2016). *Uncovering High School Students' Knowledge and Beliefs about Climate Change to Inform the Development of Embedded Curriculum Materials*. Paper presented at the Annual Meeting of the Association for Science Teacher Education, Reno, NV.

National Research Council. (1996). *National science education standards*. Washington, DC: National Association Press.

National Research Council. (2000). *Inquiry and the National Science Education Standards*. Washington, DC: National Association Press.

National Research Council. (2012). *A Framework for K-12 Science Education: Practices, Crosscutting Concepts, and Core Ideas*. Washington, DC: National Academies Press.

NGSS Lead States. (2013a). *Next Generation Science Standards: For States, by States*. Washington, DC: The National Academies Press.

NGSS Lead States. (2013b). Next Generation Science Standards: For States, by States: Vol. 2. *Appendixes*. Washington, DC: The National Academies Press.

Nuffield Foundation. (2016). *A short history of curriculum development at the Nuffield Foundation*. Retrieved from http://www.nuffieldfoundation.org/short-history-curriculum-development-nuffield-foundation

OECD. (2013). *Draft PISA 2015 Science Framework*. Retrieved from http://www.oecd.org/pisa/pisaproducts/pisa2015draftframeworks.htm

Olson, J. (1981). Teacher influence in the classroom: A context for understanding curriculum translation. *Instructional Science*, *10*(3), 259–275. doi:10.1007/BF00139803

Patton, M. Q. (2015). *Qualitative Research and Evaluation Methods* (4th ed.). Thousand Oaks, CA: SAGE Publications, Inc.

Pruneau, D., Liboiron, L., Vrain, E., Gravel, H., Bourque, W., & Langis, J. (2001). People's idea about climate change: A source of inspiration for the creation of educational programs. *Canadian Journal of Environmental Education*, *6*, 121–138.

Schwarz, C., & Passmore, C. (2012, July 14). *Preparing for NGSS: Developing and Using Models. NSTA Web Seminars*. Retrieved from http://learningcenter.nsta.org/products/symposia_seminars/ngss/files/PreparingforNGSS--DevelopingandUsingModels_9-25-2012.pdf

Smith, G. A. (2007). Place-based education: Breaking through the constraining regularities of public school. *Environmental Education Research*, *13*(2), 189–207. doi:10.1080/13504620701285180

Smith, G. G. (2012). Computer-Game Play as Imaginary Stage for Reading: Implicit Spatial Effects of Computer Games Embedded in Hard Copy Books. *Journal of Research in Reading*, *35*(1), 1–19.

Smith, G. G., Li, M., Drobisz, J., Park, H.-R., Kim, D., & Smith, S. D. (2013). Play games or study? Computer games in eBooks to learn English vocabulary. *Computers & Education, 69*, 274–286. doi:10.1016/j.compedu.2013.07.015

Stake, R., & Easley, J. (1978). *Case studies in science education* (Vols. 1-2). Center for Instructional Research and Curriculum Evaluation, University of Illinois at Urbana-Champaign.

Subramaniam, K. (2012). How WebQuests Can Enhance Science Learning Principles in the Classroom. *The Clearing House: A Journal of Educational Strategies, Issues and Ideas, 85*(6), 237–242. doi:10.1080/00098655.2012.698323

Tamir, P., & Jungwirth, E. (1975). Students growth as a result of studying BSCS biology for several years. *Journal of Research in Science Teaching, 12*(3), 263–279. doi:10.1002/tea.3660120309

University Corporation for Atmospheric Research (UCAR). (2015). *How much has the global temperature risen in the last 100 years?* Boulder, CO: University Corporation for Atmospheric Research.

Zacharia, Z. C., Manoli, C., Xenofontos, N., de Jong, T., Pedaste, M., van Riesen, S. A., & Tsourlidaki, E. (2015). Identifying potential types of guidance for supporting student inquiry when using virtual and remote labs in science: A literature review. *Educational Technology Research and Development, 63*(2), 257–302. doi:10.1007/s11423-015-9370-0

KEY TERMS AND DEFINITIONS

Climate: The average weather over a region of the earth over a long period of years.

Climate Change: Changes to the earth's climate, usually referring to the effects of increasing amounts of heat trapping gases in the atmosphere due to human actions.

Harmful Algal Blooms (HABs): A sudden increase in the amount of algae in a body of water that typically has negative effects of the health of human and other organisms.

Information and Communications Technology (ICT): An umbrella term that refers to contemporary electronic devices or applications that includes the Internet, mobile telephones, computers, videoconferencing, and so on.

Inquiry: In the context of this chapter, inquiry is the method used by scientists as they seek new knowledge about the natural world.

Models: A representation of a phenomenon that can be physical, digital, or cognitive.

Sea Level Rise: The increase of the global mean sea level due to climate change.

Serious Educational Games (SEGs): Games in which education is the primary goal, rather than entertainment.

ENDNOTES

1. The acronym was derived from the title of the project, although it does not match the letters exactly.
2. The term "intermedia" was defined by Dick Higgins (Higgins & Higgins, 2001) as works of art that "fall conceptually between media that are already known" (p. 52).
3. The total n is greater than 121 because some responses received multiple codes.

Chapter 3
Using Virtual Labs in an Inquiry Context:
The Effect of a Hypothesis Formulation Tool and an Experiment Design Tool on Students' Learning

Tasos Hovardas
University of Cyprus, Cyprus

Nikoletta A. Xenofontos
University of Cyprus, Cyprus

Zacharias C. Zacharia
University of Cyprus, Cyprus

ABSTRACT

The present study employed two different Go-Lab scaffolds/tools. These tools were used by primary school students to carry out successive learning tasks during experimentation. The first tool assisted learners in formulating hypotheses, while the second tool guided students in designing experiments (EDT). Both tools were designed to take into account the trade-offs between structuring and problematizing student inquiry. Our aim was to investigate the effect of each tool separately, as well as the combined effect of the tools in supporting student work. Participants were 41 fifth graders from two classes of a public primary school in Cyprus. They were randomly assigned to four conditions: Condition 1 involved use of both tools,

DOI: 10.4018/978-1-5225-2528-8.ch003

Condition 2 included the hypothesis tool only, Condition 3 included the EDT only, and Condition 4 had no tools provided. Conditions including one of the two tools outperformed the condition with no tools in the corresponding skill scaffolded by the tool. The cumulative effect of both tools seems to have been greater than the effect of each tool separately.

INTRODUCTION

The aim of this chapter was to examine the effect of two scaffold tools, namely a formulating hypotheses tool and an experiment design tool, on students' learning. This work builds on the need for designing computer-supported learning platforms that promote students' knowledge and skills. Previous research has highlighted the benefits that accumulate from the use of such learning environments (e.g., de Jong, van Joolingen, Giemza, Girault, Hoppe, Kindermann, et al., 2010; de Jong, Weinberger, van Joolingen, Ludvigsen, Girault, Kluge et al., 2012). However, it has also stressed the need for providing guidance to the students when using these learning environments due to the complexity involved, not only in their architecture/ organization, but also in the pedagogy accompanying the enactment of the learning activities (e.g., Hovardas, 2016). In this case, we test the effectiveness of the aforementioned scaffolding tools in an inquiry science learning context.

BACKGROUND

The optimal degree of guidance for supporting student inquiry in science education has long been debated (Arnold, Kremer, & Mayer, 2014). Previous research has highlighted the possibility that guided inquiry could be beneficial for learners, for instance, in improving science process skills (e.g., Kirschner, Sweller, & Clark, 2006; Koksal & Berberoglou, 2014). However, there is always the need to engage students as active learners in inquiry-based science instruction, capable of taking over responsibility for a range of tasks (Minner, Jurist Levy, & Century, 2010). This unresolved controversy over emphasis on guidance, at the one extreme, and openness, at the other, has been also reflected in the design of computer-supported learning environments. In this case, guidance is taken over by software scaffolds, which aim to structure student tasks in order to decrease complexity and offload certain aspects of a variety of tasks (de Jong, 2006; Pea, 2004; Reiser, 2004; Reiser, Tabak, Sandoval, Smith, Steinmuller, & Leone, 2001; Simons & Klein, 2007; van Joolingen, 1999). If technology can narrow down the multiplicity of potential routes students might follow, then student effort can be devoted to following these more tractable trajectories. However, a rigidly structured learning activity sequence would reduce student ownership and responsibility of their inquiry (e.g., Chang, Chen, Lin, & Sung, 2008).

The challenge of configuring the optimal balance between guidance and openness in inquiry learning in computer-supported learning environments translates into tension between structuring student work, on the one hand, and "problematizing" student inquiry, on the other (Reiser, 2004). Eliminating task complexity, overall, might reduce student active engagement and lock them into unproductive pathways. After any learning gain has been accomplished, the tasks that follow should challenge students to move on at a higher level, beyond their current expertise (Kalyuga, 2007). In contrast to structuring, which reduces complexity, problematizing student inquiry introduces complexity (Reiser, 2004), at least up to a point, so that the difficulty students are confronted with always surpasses the knowledge and skills they have already acquired. By adding such challenge, learner focus is usually re-directed towards parts of the task that otherwise might not be addressed (Reiser, 2004). Despite the unsettled theoretical and methodological interplay between structuring and problematizing student work, to the best of our knowledge, the question of how to problematize inquiry has not yet received the attention it deserves in the relevant literature (see the work of Wieman on PhET simulations for a notable exception, e.g., Wieman, Adams, & Perkins, 2008).

The contrast between structuring and problematizing student inquiry is pronounced in procedures that involve a series of interrelated tasks to be completed (Reiser, 2004). Such a situation is encountered in scientific experimentation, which involves identifying variables, formulating hypotheses, designing and executing experiments, gathering, analysing, and interpreting data (e.g., Germann, Aram, & Burke, 1996; Hofstein, Navon, Kipnis, & Mamlok-Naaman, 2005; Kremer, Specht, Urhahne, & Mayer, 2014, van Joolingen & Zacharia, 2009). Students face quite a few obstacles in designing and executing valid experiments (de Jong, 2006; Reiser, 2004; Zacharia, Manoli, Xenofontos, de Jong, Pedaste, van Riesen, et al., 2015); such obstacles include, among others, classifying variables as dependent, independent and controlled, and planning experimental trials (e.g., Arnold et al., 2014; Chinn & Malhotra, 2002; De Boer, Quellmalz, Davenport, Timms, Herrmann-Abell, Buckley, Jordan, et al., 2014; Lin & Lehman, 1999; Roberts & Gott, 2003; van Joolingen & de Jong, 1991). Due to the modular and difficult nature of experimentation, obstacles have been identified even among older students (Arnold et al., 2014; Furtak, 2006; Germann et al., 1996; Kirschner et al., 2006). Therefore, it should not be surprising that experimentation is the part of the inquiry cycle that has been most often supported by software scaffolds (Zacharia et al., 2015).

Research on the impact of software scaffolds on experimentation has delivered mixed results (for a comprehensive review of support of student inquiry in computer-supported learning environments see Zacharia et al., 2015). More to the point, previous studies have been confined to examining separate tools and scaffolds, whereas experimentation involves a set of stages that need to be effectively executed.

PURPOSE OF THE STUDY

The present study employed two different tools, which were developed within the Go-Lab project (http://www.go-lab-project.eu/), and which were used by primary school students to carry out successive learning tasks during experimentation. The first tool assisted learners in formulating hypotheses ("Hypothesis Scratchpad"; http://www.golabz.eu/app/hypothesis-tool), while the second tool guided students in designing experiments ("Experiment Design Tool"; http://www.golabz.eu/apps/experiment-design-tool). Both tools were designed to take into account the trade-offs between structuring and problematizing student inquiry, as previously discussed [for analogous considerations see Etkina, Karelina, Ruibal-Villasenor, Rosengrant, Jordan, & Hmelo-Silver (2010) and Jordan, Ruibal-Villasenor, Hmelo-Silver, & Etkina (2011)]. The core objective of the study was to investigate the effect of each tool separately as well as the combined effect of the tools in supporting student work. In this regard, we aimed at answering the following research questions:

1. Does the use or no use of either of the aforementioned tools differentiate students' inquiry performance (for the inquiry performance dimensions see Table 2)?
2. For students, who used both tools, does this aggregate tool usage yield higher learning gains as compared to using each tool separately?
3. Can students apply the inquiry skills they had acquired during the study's treatment in a novel inquiry context?
4. Can a valid hypothesis correlate with a valid experimental design and, further, can a valid experimental design be accompanied by effective execution of the experiment in a virtual lab?
5. What is the relative weight of each variable studied (i.e., variables reflecting student performance and referring to software scaffolds and to learning products during the learning activity sequence) across conditions (i.e., participants given each tool separately; those given both tools; and those given no tools)?

Methods

Learning Environment

The Graasp authoring tool was used to create an online Inquiry Learning Space (ILS) (for details on ILSs, see: de Jong, Sotiriou, & Gillet, 2014; Govaerts, Cao, Vozniuk, Holzer, Zutin, Ruiz, et al. 2013; Rodríguez-Triana, Holzer, Vozniuk, & Gillet, 2015), following the inquiry cycle design framework (Pedaste et al., 2015). The Graasp authoring tool allows teachers to embed resources offered via the Go-Lab platform (e.g., software scaffolds and virtual or remote laboratories) in subsequent phases of student inquiry (see http://www.golabz.eu/video/create-ils-graasp for detailed instructions of how to create an Inquiring Learning Space from Graasp). An ILS is an online computer-supported learning environment, which is designed as a template within the Go-Lab project. The ILS is structured around a virtual or remote laboratory (http://www.golabz.eu/labs) and provides software scaffolds for students undertaking learning tasks (http://www.golabz.eu/apps). The content of the ILS referred to electrical circuits and included the Electrical Circuit Lab (Figure 1), which is available on the Go-Lab platform (http://www.golabz.eu/lab/

Figure 1. Two screenshots of the electrical circuit lab, depicting two bulbs connected in series (a) and three bulbs connected in parallel (b)
http://www.golabz.eu/lab/electrical-circuit-lab

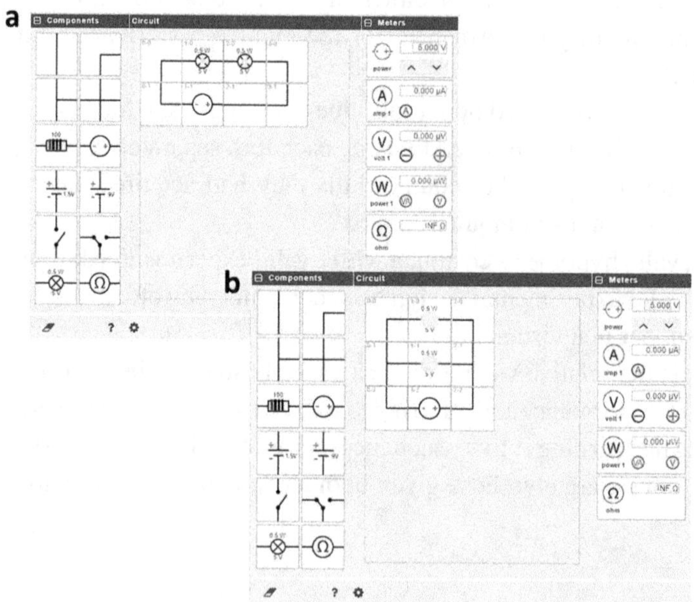

electrical-circuit-lab). The focus of the study was on the effect of two tools, namely, the Hypothesis Scratchpad (HS) and the Experiment Design Tool (EDT), both when these tools are used separately, and when they are present together in the learning environment. In the latter case, the combined effect of both tools was examined. Therefore, four different versions of the same ILS were developed. The first version included both the HS and the EDT, the second and third versions included only the HS or the EDT, respectively, while the fourth version included neither the HS nor the EDT. Whenever a software scaffold was absent, it was replaced by an Input Box (http://www.golabz.eu/apps/input-box), which is a simple note-taking application and does not provide the specific scaffolding functionalities of either the HS or the EDT.

The Hypothesis Scratchpad

Terms needed for formulating a hypothesis were given in the upper part of the HS (Figure 2). Students could drag and drop predefined conditionals and concepts in the space provided by the tool to create a hypothesis in the form of an "if…then" statement. Students could also create their own words or phrases in order to use them in their hypotheses by typing them in the gray box in the tool. Students who used the Input Box instead of the HS formulated their hypotheses without receiving any support in the form of keywords.

The Experiment Design Tool

The EDT (Figure 3) included, first, a classification task, where students had to distinguish which variable to vary (independent variable), which variables to keep constant (control variables), and which variable to measure (dependent variable). To do so, students dragged pre-set variables from the left side of the tool´s interface

Figure 2. A screenshot of the hypothesis scratchpad
http://www.golabz.eu/app/hypothesis-tool

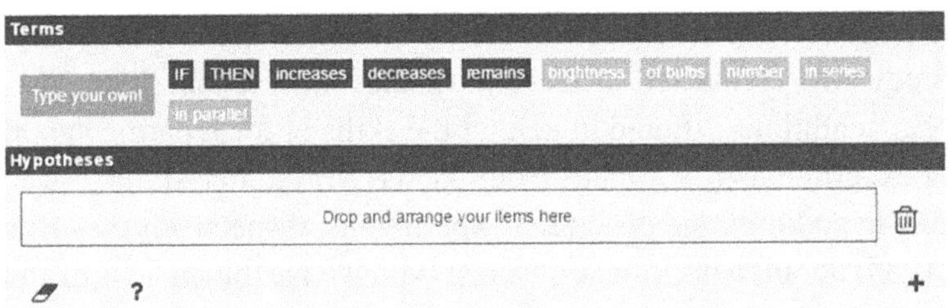

Figure 3. A screenshot of the experiment design tool
http://www.golabz.eu/apps/experiment-design-tool

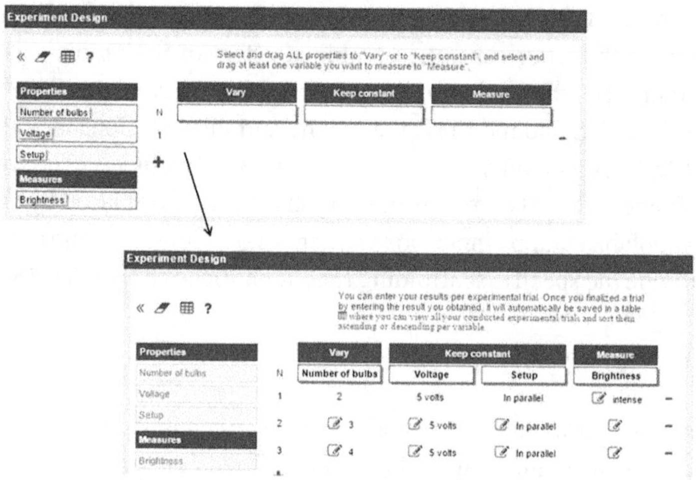

and dropped them in the proper column. A second task involved the arrangement of experimental trials to be undertaken. Students had to specify the values of each variable in each experimental trial they added to their experimental design. Students who used the Input Box instead of the EDT completed their experimental designs without receiving any support in terms of classifying variables or setting values for their experimental trials.

Participants

Participants were 41 fifth graders (10-11 years old) from two classes of a public primary school in Larnaca, Cyprus. They were randomly assigned to four conditions (Table 1): Condition 1 involved both tools (HS+EDT; 11 students; 6 boys, 5 girls), Condition 2 included the HS only (12 students; 6 boys, 6 girls), Condition 3 included the EDT only (9 students; 4 boys, 5 girls), and Condition 4 involved neither the HS nor the EDT (9 students; 5 boys, 4 girls). Students in all conditions were taught about the simple electrical circuit at the beginning of the school year, while the current study was conducted at the end of the same school year. Two pre-tests showed that conditions differed in neither the skills of formulating hypotheses and designing experiments (two scales from the TIPSII inquiry skills test, focused on formulating hypotheses and designing experiments, respectively; see Burns, Okey, & Wise, 1985) nor in their prior knowledge (knowledge dimensions examined were "remember", "understand" and "apply"; for a detailed description of pre-tests see the sub-section on Data collection and analyses). All students had basic computer and processing skills.

Table 1. Conditions, tools involved, and number of participants

Conditions	Tools Involved	Number of Participants
Condition 1	Hypothesis Scratchpad and Experiment Design Tool	11 (6 boys and 5 girls)
Condition 2	Hypothesis Scratchpad	12 (6 boys and 6 girls)
Condition 3	Experiment Design Tool	9 (4 boys and 5 girls)
Condition 4	No tool	9 (5 boys and 4 girls)

Learning Activity Sequence

The learning activity sequence started with an Orientation phase, where students were first reminded of the simple electrical circuit, and then, they were introduced to circuits connected in series and in parallel, through videos, diagrams and text (see Figure 4 for a complete flow of inquiry phases).

Figure 4. Flow of phases and sub-phases of student inquiry. Rectangles depict phases (in bold) or sub-phases, dark rhombuses learning products and white rhombuses any reference material offered to students by the teacher or the learning environment. Arrows show the sequence of phases (For a complete account of the inquiry-based learning framework and the succession of phases and sub-phases along the inquiry cycle see Pedaste et al. 2015.)

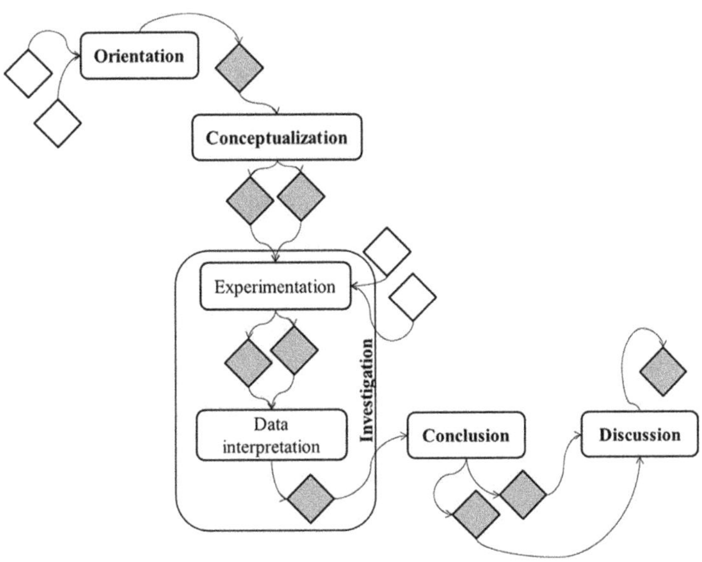

This preliminary set of learning activities ended with a problem presented to students, which referred to how light fixtures in a house are connected. After the Orientation phase, students proceeded to the Conceptualization phase, where they had to predict how the brightness of bulbs connected in series and in parallel would differ from the brightness of bulbs in a simple electrical circuit. Afterwards, students formulated hypotheses on how the brightness of bulbs would be impacted when adding more bulbs to circuits in series and in parallel. The next phase involved designing an experiment and executing the experiment in the Electrical Circuit Lab (Investigation phase). When the experimentation procedure had been completed, students responded to several questions in order to interpret their results (see Appendix 1). The last activity in the Investigation phase was an examination of what would happen when a bulb in a circuit, connected either in series or in parallel, burned out but was not removed from the circuit. In the Conclusion phase, students were asked to provide an answer for the initial problem stated in the ILS, which was about how the light fixtures in a house are connected. Students were prompted to provide enough evidence to justify their answer. The inquiry cycle ended with the Discussion phase, in which students responded to several reflection questions. They were asked to describe the steps that they went through in order to address the initial problem presented in the lesson. They were further asked if they had completed all the activities in the learning environment and if they could think of any activity which could have been done in a different way.

Procedure

The implementation lasted for three class meetings. In the first meeting (80 minutes), each condition completed pre-tests and undertook a familiarization activity with the tools and the virtual lab to be used, in a different context (weather) than the one encountered later on in the ILS (electrical circuits). Students were instructed explicitly about how to formulate a hypothesis in the form of an "if…then…" statement. Then they had the opportunity to create their own hypotheses regarding how weather would affect children's decision to play indoors or outdoors. Students were also explicitly taught the VOTAT strategy ("Vary One Thing At a Time"; also referred to as the "control-of-variables" strategy – CVS), in order to execute fair experiments, and they tried to set up a fair experiment for the hypotheses they had previously formulated in the weather context. VOTAT is a heuristic in designing experiments, where manipulating one independent variable at a time allows learners to attribute any change in the dependent variable to the independent variable which was manipulated (Glaser, Schauble, Raghavan, & Zeitz, 1992; Lin & Lehman, 1999; Tsirgi, 1980; Klahr, & Nigam, 2004; Veermans, van Joolingen & de Jong, 2006).

A demonstration of the Electrical Circuit Lab followed, where each student had the opportunity to create several circuits in the lab (see Figures 1a and 1b).

In the second meeting (80 minutes), students went through the ILS. Attention was paid to time-on-task effects so that participants in each condition spent about the same time to accomplish the entire learning activity sequence. The teacher mainly provided technical support to students when it was necessary; for instance, when students had accidentally exited the learning environment, the teacher would assist them with re-entering the ILS. For content-specific issues, the recommendation to students was to go through the instructions and hints included in the learning environment. In the third meeting (50 minutes), students in every condition used the HS and the EDT to formulate hypotheses and design experiments, respectively, in two new learning contexts. The first context was about rolling marbles in an inclined ramp and the second context addressed the solubility of sugar in water.

Data Collection and Analyses

Assessment of students' prior knowledge and inquiry skills involved two different instruments administered in a pre-test format. The knowledge test (see Appendix 2) consisted of four items, which corresponded to three cognitive processes termed "remember", "understand" and "apply", based on the revised Bloom's (1956) taxonomy, as presented by Anderson and Krathwohl (2001) and further elaborated in the reports edited by de Jong (2014) and Zervas (2013). For the inquiry skills test, items included in the TIPSII were selected and translated into Greek (Burns et al., 1985). The inquiry skills test consisted of 5 multiple-choice items, where 3 items referred to "identifying and stating hypothesis", and another 2 items referred to "designing investigations". The selection of TIPSII items was based on the appropriateness of wording and content in relation to student age. Both tests were scored blind to the students' assigned condition. A rubric was used for evaluation of the open-ended items on the knowledge test, and two independent coders scored 20% of the data. The inter-rater agreement between the coders was found to be high (Cohen's Kappa = 0.87).

Apart from the pre-tests, computer screen-captured data were collected for all students during the completion of the ILS (second meeting) and during the activities in the new learning contexts (third meeting) by means of RiverPast software (see http://river-past-screen-recorder-pro.soft112.com/ for a technical description of the software). Data analysis of this material focused on the learning products constructed by students as they progressed along the learning activity sequence (see in this regard Hovardas, 2016). Specifically, the activity sequence involved students' hypotheses, students' experimental designs, correspondence of students' hypotheses with their experimental designs, and correspondence of students' use of the virtual laboratory with their experimental designs. All the variables referring to software scaffolds and learning products constructed by students during the learning activity sequence are shown in Table 2. Coding schemes for scoring learning products were developed and two coders independently rated 10% of each category of learning products.

Table 2. Variables referring to software scaffolds and learning products during the learning activity sequence

Variable Code	Description of Variable and Values	Measure	Range (Min-Max)
ScoreHypo	Maximum score across hypotheses formulated by students in the ILS; "0" = no dependent variable included or invalid dependent variable (i.e., one that cannot be tested in the Electrical circuit lab); "1" = valid dependent variable but missing or invalid independent variable; "2" = valid dependent and independent variable	Scale	0-2
EDT_VOTAT	VOTAT strategy implemented in the ILS; "0" = no implementation; "1" = partial implementation; "2" = full implementation (e.g., across all hypotheses)	Scale	0-2
EDT_Trials	Experimental trials planned in the ILS; "0" = no planning; "1" = partial planning; "2" = full planning (at least two trials were planned for each hypothesis)	Scale	0-2
HS_EDT	Correspondence between hypotheses and experimental designs in the ILS; "0" = no correspondence; "1" = partial correspondence; "2" = full correspondence (e.g., across all hypotheses)	Scale	0-2
EDT_Trials_ Lab	Correspondence between experimental designs and circuits in the lab in the ILS; "0" = no correspondence; "1" = partial correspondence; "2" = full correspondence (e.g., across all trials)	Scale	0-2
postScoreHypo	Maximum score across hypotheses formulated by students in the new learning contexts; "0" = no dependent variable included or invalid dependent variable (i.e., one that cannot be tested in the provided labs); "1" = valid dependent variable but missing or invalid independent variable; "2" = valid dependent and independent variable	Scale	0-2
postEDT_ VOTAT	VOTAT strategy implemented in the new learning contexts; "0" = no implementation; "1" = partial implementation; "2" = full implementation (e.g., for both novel contexts)	Scale	0-2
postEDT_ Trials	Experimental trials planned in the new learning contexts; "0" = no planning; "1" = partial planning; "2" = full planning (e.g., for both novel contexts)	Scale	0-2
postHS_EDT	Correspondence between hypotheses and experimental designs in the new learning contexts; "0" = no correspondence; "1" = partial correspondence; "2" = full correspondence (e.g., for both novel contexts)	Scale	0-2

Note: HS = Hypothesis Scratchpad; EDT = Experiment Design Tool; ILS = Inquiry Learning Space; VOTAT = Vary-One-Thing-At-a-Time.

Inter-rater reliability (proportion of agreement) amounted to over 85% across all categories, while divergences between raters were settled through discussion.

Non-parametric tests and analyses were used to investigate trends in data, which involved *Kruskal-Wallis* tests, *Mann-Whitney* tests, *Wilcoxon* two-related samples tests, and *Spearman's* rank correlations. In addition, a correspondence analysis was performed to examine the relative weight of each variable studied across conditions.

RESULTS

Research Question 1

Table 3 presents mean values for all variables studied, by assigned condition and learning context. An overview of the table indicates that there were significant differences among conditions in most cases (i.e., in both the learning activity sequence and the new learning contexts). More specifically, conditions providing one of the two tools outperformed the condition with no tools in the corresponding skill pertaining to that tool. With regard to student performance when working in the ILS, students who used the HS only (Condition 2; HS only) scored higher for their hypotheses (Table 3; "ScoreHypo") than students who did not use this tool

Table 3. Mean values of variables studied, by condition and learning context

	Condition[1]				[2]Kruskal-Wallis χ^2
	HS and EDT	**HS Only**	**EDT Only**	**No Tool**	
Learning Activity Space					
ScoreHypo	1.73	1.82	1.22	1.00	11.69**
EDT_VOTAT	1.45	0.00	2.00	0.00	35.32***
EDT_Trials	1.73	0.75	1.44	0.67	10.95*
HS_EDT	1.36	0.83	0.56	0.67	5.09^{ns}
EDT_Trials_Lab	1.73	1.08	1.22	0.67	10.03*
New Learning Contexts					
postScoreHypo	2.00	1.42	0.67	0.78	12.90**
postEDT_VOTAT	1.55	0.00	1.33	0.00	30.57***
postEDT_Trials	1.82	0.67	0.89	0.44	13.90**
postHS_EDT	1.27	0.50	0.22	0.38	11.11*

1: HS = Hypothesis Scratchpad; EDT = Experiment Design Tool; all values range between 0 (min) and 2 (max)

2: ns = non-significant; * $p < 0.05$; ** $p < 0.01$; *** $p < 0.001$

(Condition 4; no tool) (*Mann-Whitney* $Z = -2.73$, $p < 0.01$). Additionally, students who used the EDT only (Condition 3; EDT only) showed increased implementation of the VOTAT heuristic (Table 3; "EDT_VOTAT"), compared to students who did not use this tool (Condition 4; no tool) (*Mann-Whitney* $Z = -4.12$, $p < 0.001$).

Research Questions 2 and 3

The combined effect of both tools seems to have outweighed the effect of each tool separately in the new contexts provided to students for experimentation after they concluded activities in the ILS. This was especially evident in the case of formulating hypotheses (Table 3; "postScoreHypo") (*Mann-Whitney* $Z = -2.05$, $p < 0.05$, for the difference between Conditions 1 and 2; *Mann-Whitney* $Z = -3.16$, $p < 0.01$, for the difference between Conditions 1 and 3), planning experimental trials (Table 3; "postEDT_Trials") (*Mann-Whitney* $Z = -3.18$, $p < 0.01$, for the difference between Conditions 1 and 2; *Mann-Whitney* $Z = -2.48$, $p < 0.05$, for the difference between Conditions 1 and 3) and the correspondence between student hypotheses and experimental designs (Table 3; "postHS_EDT") (*Mann-Whitney* $Z = -2.26$, $p < 0.05$, for the difference between Conditions 1 and 2; *Mann-Whitney* $Z = -2.81$, $p < 0.01$, for the difference between Conditions 1 and 3).

Overall, the condition that employed both tools (Condition 1; HS+EDT) showed increased scores for most of the studied variables in the new learning contexts, namely, the new contexts given to students after they had exited the ILS, in comparison to the values recorded in the ILS (Table 3). This finding provides an indication that the beneficial effect of software scaffolding on student inquiry performance was either maintained or even increased across the learning contexts for Condition 1. The opposite can be observed for the other three conditions, where students scored lower across all variables in the new learning contexts as compared to the ILS. Students who used both tools (Condition 1) outperformed students who used only one tool or no tool across all dimensions examined in the new learning contexts. However, *Wilcoxon* two-related-samples tests revealed that there were no significant time trends for any condition from the ILS to the new learning contexts.

Research Question 4

Scores for the condition with both tools (Condition 1) showed significant correlations among various parameters studied. Hypothesis scores correlated positively with correspondence of hypotheses and experimental designs (*Spearman's rho* = 0.74, $p < 0.05$), while planning of experimental trials correlated positively with correspondence of student use of the virtual laboratory with their experimental designs (*Spearman's rho* = 0.98, $p < 0.001$). Another significant correlation was between planning experimental trials and employing the VOTAT heuristic in the new learning contexts (*Spearman's rho* $= 0.71, p < 0.05$). No such correlations were found in the other conditions. All of these results offer a strong indication that for Condition 1 (HS+EDT), student competence on a learning task could have positively catalyzed performance on forthcoming learning tasks in either the ILS or in the new learning contexts. Namely, using both tools seemed to offer an added value within the learning process, as compared to using one tool only. In other words, the use of both tools not only reinforced separate dimensions of student performance but it also initiated a co-evolvement of skills.

Research Question 5

A correspondence analysis was performed in order to investigate which variables were most pronounced according to tool usage (i.e., Condition 1: HS+EDT, Condition 2: HS, Condition 3: EDT, Condition 4: no tool). This analysis aimed at revealing aspects of the student performance (i.e., variables plotted by the analysis) across varying tool usage and acrossconditions. The biplot of the analysis is shown in Figure 5.

Along the biplot, white cycles denote variables under study (i.e, variables depicting various dimensions of student performance) are positioned according to tool usage (i.e., conditions students had been assigned to with regard to tool usage). The closer a specific dimension of student performance (i.e., variable) is placed to a condition reflecting tool usage, the more expressed this dimension had been for that condition. On the positive part of the first axis we can observe that Condition 3 (EDT only) is related to the implementation of the VOTAT heuristic in the ILS as well as in the new learning contexts ("EDT_VOTAT" and "postEDT_VOTAT", respectively). On the negative part of the first axis, Condition 2 (HS only) is characterized by relatively increased scores for hypotheses in both the ILS and the new learning contexts ("ScoreHypo" and "postScoreHypo", respectively). Condition 1 (HS + EDT) features in the negative part of the second axis and it is distinguished by a marked correspondence between hypothesis formulation and experimental designs in the new learning context ("postHS_EDT") as well as by increased validity in planning experimental trials in the new learning contexts ("postEDT_Trials"). The relative position of conditions and variables on the biplot implies that each tool separately fostered performance on learning tasks related to its scaffolding properties. Namely, the HS reinforced student ability to adequately formulate hypotheses, while the EDT enhanced student competence in implementing the VOTAT heuristic. Indeed, this was the case in the learning context of the ILS as well as in the new learning contexts. The condition that combined both tools (Condition 1; HS + EDT) was marked by relatively increased inter-contextual transfer. The latter means that the learning gains in that condition were highly probable to be also detected in an upcoming learning

Figure 5. Bi-plot of correspondence analysis depicting variables studied (white circles: ScoreHypo; EDT_VOTAT; EDT_Trials; HS_EDT; EDT_Trials_Lab; postScoreHypo; postEDT_VOTAT; postEDT_Trials; postHS_EDT) and conditions (black boxes: HS + EDT; HS; EDT; no tool). Overall, the closer a variable is to a condition, the more this condition is characterized by increased values of this variable, as compared to the other conditions. The first axis accounted for 88% of total variance, while the second axis added another 11%.

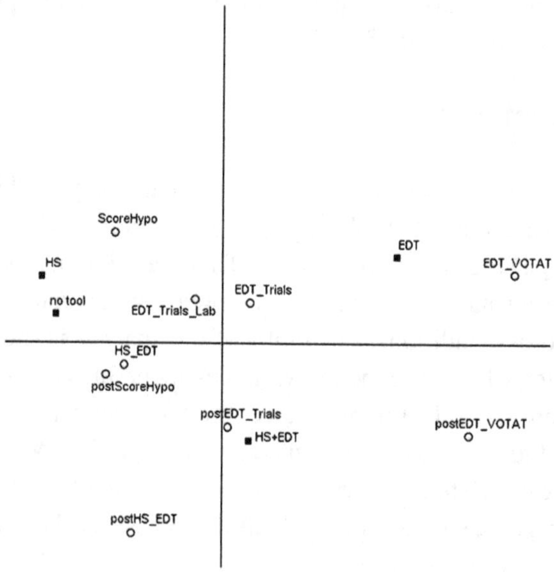

context. That condition was also characterized by inter-task transferability of learning gains. For instance, an increased validity in student hypotheses was highly possible to be accompanied later on by a valid experimental design.

FUTURE RESEARCH DIRECTIONS

Previous research has highlighted the benefits of storing, retrieving, and exchanging learning products in computer-supported inquiry learning environments (e.g., de Jong et al., 2010, 2012). It has also stressed the potential of using learning products for diagnosing student performance and enacting formative assessment (e.g., Hovardas, 2016). This study's findings also concur with these claims. However, this research domain is still in need of further research. There are quite a few unresolved issues. For example, there is a need to examine whether learning products, which have been constructed and stored during a learning activity sequence, might themselves serve as scaffolds for student inquiry either during upcoming activities or when students would encounter new learning contexts.

Future research could also use learning products to investigate additional functionalities, which software scaffolds could carry out. For example, the HS can include varying numbers of words (variables involved in the phenomenon under study plus conditionals necessary to interrelate variables) to support students in formulating hypotheses. Varying the number of words provided to students might be a way to vary the degree of scaffolding the HS offers to students, perhaps with fading of the scaffolding as learners gain experience. This would mean that the HS could include fewer words for more experienced learners (see in this regard Kalyuga & Sweller, 2004; Reisslein, Atkinson, Seeling, & Reisslein, 2006; Seufert & Brünken, 2006). In the case of the EDT, fading in and out of scaffolding could involve the number of variables offered to students to design their experiments. Since most analogous tools in computer-supported learning environments are delivered in a "one-for-all fashion" (Kalyuga, 2007), future research might screen optimal timing for introducing and removing scaffolding, as well as varying of scaffolding based on student experience and competence (see, for instance, de Jong, 2006). If learners proceed effectively in a learning activity sequence even after scaffolding has been removed, this would provide a substantial indication that the scaffolded skill has been acquired. Future research might further examine possible effects on cognitive load and demands on working memory imposed by different scaffolding configurations for tools.

CONCLUSION

A first major result of the study was that each software scaffold separately succeeded in supporting students in the corresponding learning tasks, which addressed the first research question. Such an encouraging finding must be attributed to the design and properties of each tool. Further, it implies that both tools may have adequately handled the two contrasting needs of structuring student inquiry on the one hand, and problematizing student work, on the other. An explanation for that delicate and balanced contribution might be that both tools were fairly effective in offering procedural information to students, namely, information on what students should do and how they should address the corresponding learning tasks (see, for instance, Arnold et al., 2014). The implication coming out of this finding, for instruction purposes, is that specific configurations of software scaffolds might prove adequate both for decreasing complexity in student inquiry, as well as for actively engaging students in subject matter. A related explanation might refer to the tools' allowance for serial processing of learning tasks (see, for instance, the VOTAT strategy that was related to the use of the EDT), where the complexity in a learning activity sequence might be addressed by partitioning tasks into smaller-scale, shorter assignments that need to be dealt with one after the other (Clarke, Ayres, & Sweller, 2005; Pollock, Chandler, & Sweller, 2002). Such a serial processing of learning tasks might have kept essential cognitive load within the limits that learners could manage based on their cognitive capacity and the processing limitations of working memory (Kalyuga, 2007; Sweller, van Merrienboer, & Paas, 1998). It might also have kept students alert and motivated to encounter upcoming tasks, which could have adequately touched upon the dimension of problematizing student inquiry.

With regard to the second research question, the results presented in this paper indicate that students using both tools benefited more as compared to students who used one tool only. This finding is a clear sign of a combined effect, where the impact of both software scaffolds might have outweighed the impact of each tool separately. Concerning the third research question, the combined effect of tool usage was also detected in new inquiry contexts, which is a strong indication of inter-contextual transfer of skill gains. An additional combined effect was implied by significant correlations among studied parameters in the condition that incorporated both tools (fourth research question). These correlations indicate that using both tools might not only increase the corresponding skills separately (i.e., formulating hypotheses and designing experiments) but that it might also result in skill gains that spread across learning tasks as the learning activity sequence unfolds. Namely, skill gains in formulating hypotheses might pass on to designing experiments and the latter might pass on to handling a virtual lab. Although previous research may have indicated such a linkage (e.g., Arnold et al., 2014; Veermans, van Joolingen, & de Jong, 2006), the results of this study offer the first empirical validation of this transfer. It is also highlighted that the results imply that transfer could involve two different scales. The first addresses transition between different contexts (inter-contextual transfer), while the second is anchored within a single learning activity sequence and refers to gains being transferred from a learning task to a forthcoming learning activity (inter-task transfer). Results concerning transfer were further validated by the correspondence analysis (fifth research question).

ACKNOWLEDGMENT

This study was conducted in the context of the research project Global Online Science Labs for Inquiry Learning at School (Go-Lab), which is funded by the European Community under the Information and Communication Technologies (ICT) theme of the 7th Framework Programme for R&D (Grant agreement no.: 317601).

REFERENCES

Anderson, L. W., & Krathwohl, D. R. (2001). *A taxonomy for learning, teaching, and assessing: A revision of Bloom's taxonomy of educational objectives*. New York: Addison Wesley Longman.

Arnold, J. C., Kremer, K., & Mayer, J. (2014). Understanding students experiments - What kind of support do they need in inquiry tasks? *International Journal of Science Education, 36*(16), 2719–2749. doi:10.1080/09500693.2014.930209

Bloom, B. S. (1956). *Taxonomy of educational objectives. Handbook I: The cognitive domain*. New York: David McKay.

Burns, J., Okey, J., & Wise, K. (1985). Development of an integrated process skill test: TIPS II. *Journal of Research in Science Teaching, 22*(2), 169–177. doi:10.1002/tea.3660220208

Chang, K.-E., Chen, Y.-L., Lin, H.-Y., & Sung, Y.-T. (2008). Effects of learning support in simulation-based physics learning. *Computers & Education, 51*(4), 1486–1498. doi:10.1016/j.compedu.2008.01.007

Chinn, C. A., & Malhotra, B. A. (2002). Epistemologically authentic inquiry in schools: A theoretical framework for evaluating inquiry tasks. *Science Education, 86*(2), 175–218. doi:10.1002/sce.10001

Clarke, T., Ayres, P., & Sweller, J. (2005). The impact of sequencing and prior knowledge on learning mathematics through spreadsheet applications. *Educational Technology Research and Development, 53*(3), 15–24. doi:10.1007/BF02504794

De Boer, G. E., Quellmalz, E. S., Davenport, J. L., Timms, M. J., Herrmann-Abell, C. F., Buckley, B. C., & Flanagan, J. C. et al. (2014). Comparing three online testing modalities: Using static, active, and interactive online testing modalities to access middle school students understanding of fundamental ideas and use of inquiry skills related to ecosystems. *Journal of Research in Science Teaching, 51*(4), 523–554. doi:10.1002/tea.21145

de Jong, T. (2006). Computer simulations – Technological advances in inquiry learning. *Science, 312*(5773), 532–533. doi:10.1126/science.1127750 PMID:16645080 de Jong, T. (Ed.). (2014). *Preliminary inquiry classroom scenarios and guidelines. D1.3. Go-Lab Project (Global Online Science Labs for Inquiry Learning at School).* Retrieved from http://www.go-lab-project.eu/sites/default/files/files/deliverable/file/Go-Lab%20D1.3.pdf

de Jong, T., Sotiriou, S., & Gillet, D. (2014). Innovations in STEM education: The Go-Lab federation of online labs. *Smart Learning Environments*, *1*(1), 1–16. doi:10.1186/s40561-014-0003-6

de Jong, T., & van Joolingen, W. R. (1998). Scientific discovery learning with computer simulations of conceptual domains. *Review of Educational Research*, *68*(2), 179–201. doi:10.3102/00346543068002179

de Jong, T., van Joolingen, W. R., Giemza, A., Girault, I., Hoppe, U., Kindermann, J., & van der Zanden, M. et al. (2010). Learning by creating and exchanging objects: The SCY experience. *British Journal of Educational Technology*, *41*(6), 909–921. doi:10.1111/j.1467-8535.2010.01121.x

de Jong, T., Weinberger, A., van Joolingen, W. R., Ludvigsen, S., Ney, M., Girault, I., & Zacharia, Z. et al. (2012). Designing complex and open learning environments based on scenarios. *Educational Technology Research and Development*, *60*, 883–901. doi:10.1007/s11423-012-9258-1

Etkina, E., Karelina, A., Ruibal-Villasenor, M., Rosengrant, D., Jordan, R., & Hmelo-Silver, C. E. (2010). Design and reflection help students develop scientific abilities: Learning in introductory physics laboratories. *Journal of the Learning Sciences*, *19*(1), 54–98. doi:10.1080/10508400903452876

Furtak, E. M. (2006). The problem with answers: An exploration of guided scientific inquiry teaching. *Science Education*, *90*(3), 453–467. doi:10.1002/sce.20130

Germann, P. J., Aram, A., & Burke, G. (1996). Identifying patterns and relationships among the responses of seventh-grade students to the science process skill of designing experiments. *Journal of Research in Science Teaching*, *33*(1), 79–99. doi:10.1002/(SICI)1098-2736(199601)33:1<79::AID-TEA5>3.0.CO;2-M

Glaser, R., Schauble, L., Raghavan, K., & Zeitz, C. (1992). Scientific reasoning across different domains. In E. De Corte, M. Linn, H. Mandl, & L. Verschaffel (Eds.), *Computer-based learning environments and problem solving* (pp. 345–373). Berlin: Springer-Verlag. doi:10.1007/978-3-642-77228-3_16

Govaerts, S., Cao, Y., Vozniuk, A., Holzer, A., Zutin, D. G., Ruiz, E. S. C., . . . Gillet, D. (2013). Towards an online lab portal for inquiry-based STEM learning at school. In J.-F. Wang & R. Lau (Eds.), *Advances in web-based learning – ICWL 2013, 12th International Conference, Kenting, Taiwan, October 6-9, 2013, Proceedings* (pp. 244-253). Berlin: Springer. doi:10.1007/978-3-642-41175-5_25

Hofstein, A., & Lunetta, V. N. (2003). The laboratory in science education: Foundations for the twenty-first century. *Science Education*, *88*(1), 28–54. doi:10.1002/sce.10106

Hofstein, A., Navon, O., Kipnis, M., & Mamlok-Naaman, R. (2005). Developing students ability to ask more and better questions resulting from inquiry-type chemistry laboratories. *Journal of Research in Science Teaching*, *42*(7), 791–806. doi:10.1002/tea.20072

Hovardas, T. (2016). A learning progression should address regression: Insights from developing non-linear reasoning in ecology. *Journal of Research in Science Teaching*, *53*(10), 1447–1470. doi:10.1002/tea.21330

Jordan, R. C., Ruibal-Villasenor, M., Hmelo-Silver, C. E., & Etkina, E. (2011). Laboratory materials: Affordances or constraints? *Journal of Research in Science Teaching*, *48*(9), 1010–1025. doi:10.1002/tea.20418

Kalyuga, S. (2007). Expertise reversal effect and its implications for learner-tailored instruction. *Educational Psychology Review*, *19*(4), 509–539. doi:10.1007/s10648-007-9054-3

Kalyuga, S., & Sweller, J. (2004). Measuring knowledge to optimize cognitive load factors during instruction. *Journal of Educational Psychology*, *96*(3), 558–568. doi:10.1037/0022-0663.96.3.558

Kirschner, P. A., Sweller, J., & Clark, R. E. (2006). Why minimal guidance during instruction does not work: An analysis of the failure of constructivist, discovery, problem-based, experiential, and inquiry-based teaching. *Educational Psychologist*, *41*(2), 75–86. doi:10.1207/s15326985ep4102_1

Klahr, D., & Nigam, M. (2004). The equivalence of learning paths in early science instruction: Effect of direct instruction and discovery learning. *Psychological Science*, *15*(10), 661–667. doi:10.1111/j.0956-7976.2004.00737.x PMID:15447636

Koksal, E. A., & Berberoglou, G. (2014). The effect of guided inquiry instruction on 6th grade Turkish students achievement, science process skills, and attitudes toward science. *International Journal of Science Education*, *36*(1), 66–78. doi:10.1080/09500693.2012.721942

Kremer, K., Specht, C., Urhahne, D., & Mayer, J. (2014). The relationship in biology between the nature of science and scientific inquiry. *Journal of Biological Education*, *48*(1), 1–8. doi:10.1080/00219266.2013.788541

Lin, X., & Lehman, J. D. (1999). Supporting learning of variable control in a computer-based biology environment: Effects of prompting college students to reflect on their own thinking. *Journal of Research in Science Teaching*, *36*(7), 837–858. doi:10.1002/(SICI)1098-2736(199909)36:7<837::AID-TEA6>3.0.CO;2-U

Minner, D. D., Jurist Levy, A., & Century, J. (2010). Inquiry-based science instruction – What is it and does it matter? Results from a research synthesis years 19842002. *Journal of Research in Science Teaching*, *47*(4), 474–496. doi:10.1002/tea.20347

Pea, R. D. (2004). The social and technological dimensions of scaffolding and related theoretical concepts for learning, education, and human activity. *Journal of the Learning Sciences*, *13*(3), 423–451. doi:10.1207/s15327809jls1303_6

Pedaste, M., Mäeots, M., Siiman, L. A., de Jong, T., van Riesen, S. A. N., Kamp, E. T., & Tsourlidaki, E. et al. (2015). Phases of inquiry-based learning: Definitions and the inquiry cycle. *Educational Research Review*, *14*, 47–61. doi:10.1016/j.edurev.2015.02.003

Pollock, E., Chandler, P., & Sweller, J. (2002). Assimilating complex information. *Learning and Instruction*, *12*(1), 61–86. doi:10.1016/S0959-4752(01)00016-0

Reiser, B. J. (2004). Scaffolding complex learning: The mechanisms of structuring and problematizing student work. *Journal of the Learning Sciences*, *13*(3), 273–304. doi:10.1207/s15327809jls1303_2

Reiser, B. J., Tabak, I., Sandoval, W. A., Smith, B. K., Steinmuller, F., & Leone, A. J. (2001). BGuILE: Strategic and conceptual scaffolds for scientific inquiry in biology classrooms. In S. M. Carver & D. Klahr (Eds.), *Cognition and instruction: Twenty-five years of progress* (pp. 263–305). Mahwah, NJ: Lawrence Erlbaum Associates, Inc.

Reisslein, J., Atkinson, R. K., Seeling, P., & Reisslein, M. (2006). Encountering the expertise reversal effect with a computer-based environment on electrical circuit analysis. *Learning and Instruction, 16*(2), 92–103. doi:10.1016/j.learninstruc.2006.02.008

Roberts, R., & Gott, R. (2003). Assessment of biology investigations. *Journal of Biological Education, 37*(3), 114–121. doi:10.1080/00219266.2003.9655865

Rodríguez-Triana, M. J., Holzer, A., Vozniuk, A., & Gillet, D. (2015). Orchestrating inquiry-based learning spaces: An analysis of teacher needs. In F. W. B. Li, R. Klamma, M. Laanpere, J. Zhang, B. F. Manjón, & R. W. H. Lau (Eds.), *Advances in web-based learning -- ICWL 2015, 14th International Conference, Guangzhou, China, November 5-8, 2015, Proceedings* (pp. 131-142). Cham, Switzerland: Springer. doi:10.1007/978-3-319-25515-6_12

Seufert, T., & Brünken, R. (2006). Cognitive load and the format of instructional aids for coherence formation. *Applied Cognitive Psychology, 20*(3), 321–331. doi:10.1002/acp.1248

Simons, K. D., & Klein, J. D. (2007). The impact of scaffolding and student achievement levels in a problem-based learning environment. *Instructional Science, 35*(1), 41–72. doi:10.1007/s11251-006-9002-5

Sweller, J., van Merrienboer, J., & Paas, F. (1998). Cognitive architecture and instructional design. *Educational Psychology Review, 10*(3), 251–296. doi:10.1023/A:1022193728205

Tsirgi, J. E. (1980). Sensible reasoning: A hypothesis about hypotheses. *Child Development, 51*(1), 1–10. doi:10.2307/1129583

van Joolingen, W. R. (1999). Cognitive tools for discovery learning. *International Journal of Artificial Intelligence in Education, 10*, 385–397.

van Joolingen, W. R., & de Jong, T. (1991). Supporting hypothesis generation by learners exploring an interactive computer simulation. *Instructional Science, 20*(5-6), 389–404. doi:10.1007/BF00116355

van Joolingen, W. R., & Zacharia, Z. C. (2009). Developments in inquiry learning. In N. Balacheff, S. Ludvigsen, T. de Jong, & S. Barnes (Eds.), *Technology-enhanced learning* (pp. 21–37). Dordrecht, The Netherlands: Springer. doi:10.1007/978-1-4020-9827-7_2

Veermans, K., van Joolingen, W. R., & de Jong, T. (2006). Use of heuristics to facilitate scientific discovery learning in a simulation learning environment in a physics domain. *International Journal of Science Education*, *28*(4), 341–361. doi:10.1080/09500690500277615

Wieman, C. E., Adams, W. K., & Perkins, K. K. (2008). PhET: Simulations that enhance learning. *Science*, *322*(5902), 682–683. doi:10.1126/science.1161948 PMID:18974334

Zacharia, Z. C., Manoli, C., Xenofontos, N., de Jong, T., Pedaste, M., van Riesen, S. A. N., & Tsourlidaki, E. et al. (2015). Identifying potential types of guidance for supporting student inquiry when using virtual and remote labs in science: A literature review. *Educational Technology Research and Development*, *63*(2), 257–302. doi:10.1007/s11423-015-9370-0

Zervas, P. (Ed.). (2013). *The Go-Lab inventory and integration of online labs – labs offered by large scientific organisations. D2.1. Go-Lab Project (Global Online Science Labs for Inquiry Learning at School).* Retrieved from http://www.go-lab-project.eu/sites/default/files/files/deliverable/file/Go-Lab-D2.1.pdf

KEY TERMS AND DEFINITIONS

Coherence in an Inquiry-Based Learning Activity Sequence: Coherence in an inquiry-based learning activity sequence refers to whether valid hypotheses lead to an acceptable and complete experimental design, and if the latter is followed by thorough execution of the experiment after having adequately planned experimental trials.

Combined Effect of Software Scaffolds: Scaffolds could be combined. However, a combined usage of scaffolds might outperform the sum of the effects on student performance when each scaffold is used separately.

Inter-Contextuality: Inter-contextuality concerns the application of acquired knowledge and skills in a novel learning context (not previously experienced). If students apply their knowledge and skills adequately in this new context, this is a quite clear and strong indication of acquired learning gains.

Introducing and Removing Scaffolding (Fading In / Fading Out): Guidance provided to students by software scaffolds can gradually fade in or fade out, depending on students' experience. Less experienced students receive more support, whereas more experienced students (i.e., with higher prior knowledge and skills) could succeed in effectively undertaking learning tasks with lesser or no support.

Problematizing Student Inquiry: Problematizing student inquiry involves challenging students and attracting their attention to parts of an inquiry-based learning task or activity sequence that might otherwise have been missed.

Structuring Student Inquiry in Computer-Supported Learning Environments: To structure student inquiry in computer-supported learning environments, computer-based guidance is suggested (e.g., scaffolds, heuristics, prompts). The idea is to remove the embedded complexity in an inquiry-based learning process.

Transfer of Knowledge or Skills: Students' ability to apply/transfer knowledge or skills in a new learning situation.

APPENDIX 1

Questions provided to students as prompts for data interpretation.

1. The brightness of the bulbs, which are connected in series, is (1) higher, (2) lower or (3) the same as the brightness of the bulb in a simple electrical circuit?
2. When bulbs are added in series, their brightness (1) increases, (2) decreases or (3) remains the same?
3. The brightness of the bulbs, which are connected in parallel, is (1) higher, (2) lower or (3) the same as the brightness of the bulb in a simple electrical circuit?
4. When bulbs are added in parallel, their brightness (1) increases, (2) decreases or (3) remains the same?

APPENDIX 2

Figure 6.

Knowledge test

Note: In order to complete this test, you will need approximately 20 minutes. You need to respond to all items (1-6). The results of this test will not count to your total score in the lesson. Your input will be used anonymously for research purposes.

1. In which of the following the bulb will light up? Please choose **one** answer.

A) B) C) D)

a) A
b) B
c) C
d) D
e) C και D
f) A, C και D
g) None of them

Explain your reasoning:

2. Look at the following circuits:

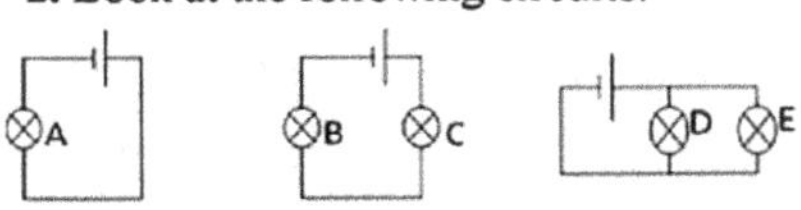

Note: The symbol ⊣⊢ represents the battery which is connected to the circuit. An identical battery is connected to all the circuits above.

3. What will happen if the middle bulb burns out (and if it will not be removed from the circuit)? Please choose **one** answer.

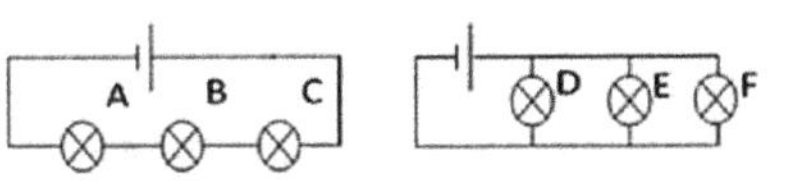

Note: The symbol ⊣⊢ represents the battery which is connected to the circuit. An identical battery is connected to all the circuits above.

a) A and C will not light up, D and F will have the same brightness
b) All the bulbs (A, C, D and F) will not light up
c) A and C will have the same brightness, but less than D and F, which will have the same brightness
d) All the bulbs (A, C, D and F) will have the same brightness
e) D and F will not light up, while A and C will have the same brightness

4. Multiple electrical sockets are used for the operation of multiple electrical appliances. How are the electrical appliances connected in multiple sockets? Explain your reasoning.

Chapter 4
Collaborative Learning and Assessment of Science:
Structuring Effective Groups

Yigal Rosen
Harvard University, USA

ABSTRACT

In order to understand potential applications of collaborative problem solving (CPS) assessment tasks, it is necessary to examine empirically the multi-faceted student performance that may be distributed across collaboration methods and purposes of the assessment. Ideally, each student should be matched with various types of group members and must apply the skills in varied contexts and tasks. One solution to these assessment demands is to use computer-based (virtual) agents to serve as the collaborators in the interactions with students. This paper proposes human-to-agent (H-A) approach for formative CPS assessment and describes an international pilot study aimed to provide preliminary empirical findings on the use of H-A CPS assessment to inform collaborative learning. Overall, the findings showed promise in terms of using H-A CPS assessment task as a formative tool for structuring effective groups in the context of CPS online learning.

INTRODUCTION

This chapter is capitalizing on the affordances of digital tools to deepen and extend the kinds of science learning and assessment highlighted in the Framework for K–12 Science Education and the Next Generation Science Standards (National Research Council, 2012), Programme for International Student Assessment (PISA) Framework

DOI: 10.4018/978-1-5225-2528-8.ch004

for Science and Collaborative Problem Solving (OECD, 2013a, 2013b). On one hand, a scientifically literate person is willing to engage in reasoned discourse about science and technology which requires the competencies to explain phenomena scientifically, evaluate and design scientific enquiry, interpret data and evidence scientifically, and analyze and evaluate data, claims and arguments in a variety of representations and draw appropriate scientific conclusions (OECD, 2013a). On the other hand, collaborative problem solving (CPS) competency is characterized by the capacity of an individual to effectively engage in a process whereby two or more agents attempt to solve a problem by sharing the understanding and effort required to come to a solution and pooling their knowledge, skills, and efforts to reach that solution (OECD, 2013b). That is, the OECD frameworks for assessment build upon an individual assessment of problem-solving, which was already defined and well understood in earlier PISA assessments (Chauncey, & Azevedo, 2010), and conjoins that definition with a new domain framework of collaboration made operational in a simulated collaborative context. An agent could be considered either a human agent or a computer agent that interacts with the student. The competency is typically assessed by evaluating how well the individual collaborates with agents during the problem-solving process. This includes establishing and maintaining shared understanding, taking appropriate actions to solve the problem, and establishing and maintaining group organization. CPS refers to problem-solving activities that involve collaboration among a group of individuals (O'Neil, Chuang, & Baker, 2010; Zhang, 1998). It is a conjoint construct consisting of collaboration, or: "coordinated, synchronous activity that is the result of a continued attempt to construct and maintain a shared conception of a problem" (Roschelle, & Teasley, 1995, p. 70), and problem solving, or: "cognitive processing directed at achieving a goal when no solution method is obvious to the problem solver" (Mayer, & Wittrock, 1996). CPS has both cognitive and social aspects, and the outcomes from a CPS tasks are generally the results of the interaction of both. Liu, et al. (2015) proposed a conceptual model that documents a matrix of individual and social skills involved in CPS, which provides a basis for designing rich performance assessments tasks in the context of science. The individual cognitive skills are used to complete tasks independently of other team members. In this individual dimension of the CPS skills contextualized to inquiring knowledge, the following cognitive skills: conceptual understanding and inquiry skills in science (e.g., data collection, data analysis, prediction making, evidence-based reasoning). The second dimension in the CPS skills matrix is social skills, which are often acquired through social interactions with peers and can affect both group and individual performance. There are four major categories of social skills, namely sharing information, assimilating and accommodating knowledge, regulating problem-solving activities, and maintaining a positive communication atmosphere.

In this research, an operational definition of Science CPS is based on PISA framework and refers to "the capacity of an individual to effectively engage in a group process whereby two or more agents attempt to solve a science problem by sharing knowledge and understanding, organizing the group work and monitoring the progress, taking actions to solve the problem, and providing constructive feedback to group members." This definition treats the competency as conjoint dimension collaboration skills and the skills needed to solve a science problem. For the assessment design purposes, the focus is on individual capacities within collaborative situations. Thus, the effectiveness of collaborative science problem solving depends on the ability of group members to collaborate and to prioritize the success of the group over individual successes. At the same time, this ability is still a trait in each of the individual members of the group.

Paralleling the use of group settings in the classroom for communication and collaboration, measuring students' CPS skills requires group-work on performance assessments. Assessments, then, can extend beyond the measurement of purely individual performance and involve individuals working in groups.

The complex problem space of CPS enables consideration of the importance of the context of assessment, the role of assessment in promoting higher levels of knowledge and performance, and the role of assessment in determining what someone knows and can do. The proposed proficiency descriptors in Table 1 aimed to guide efforts in designing assessment tasks for CPS skills (Rosen, & Mosharraf, 2015). The CPS proficiencies in Table 1 can be measured by observing whether the student responds to questions and requests for information, acknowledges other team members' actions, takes appropriate actions to solve the problem, works in harmony with the team members, and provides constructive feedback to the team members in order to optimize the CPS process. If such acts are not observed, the student exhibits performance below a proficient level.

If the student responds but does not initiate questions or requests for information, does not always take appropriate actions or work in a balanced manner with the team members, and provides feedback that represents only basic dimensions of collaboration, then the student meets minimal standards. If the student initiates questions and requests, proactively helps the team in organizing the work, acknowledges contributions made by others, and responds positively to feedback, then the student is above the minimal standard. These interpretations of pro-active versus responsive student's behaviors during collaboration are in line with PISA 2015 CPS assessment (OECD, 2013b), and were used as one of the key guiding principles in assessment tasks design.

Table 1. Collaborative problem solving proficiency descriptors

Proficiency	Establishing and Maintaining Shared Understanding	Organizing the Group Work and Monitoring the Progress	Taking Actions to Solve the Problem	Providing Constructive Feedback to Group Members
Low	The student responds to or generates information that has little relevance to the task, yet the student contributes when explicitly or repeatedly prompted. Student takes actions that create additional misunderstandings of shared knowledge.	The student operates individually, often not in concert with the appropriate role for the task. Student's actions and communications suggest that the student does not understand the roles of the other team members.	The student's actions contribute minimally to achieving group goals. Student performs actions that are inappropriate for the distribution of tasks.	The student's feedback includes minimal to no acknowledgment of other team members' contribution to the task. Student proposes no pathways for teamwork improvement when prompted and pays little to no attention to feedback from other team members.
Medium	The student responds to most requests for information. The student does not always proactively take the initiative to overcome difficult barriers in collaboration. Student generates and responds to inquiries to clarify problem goals, problem constraints, and task requirements.	The student participates in assigned roles and contributes to the overall strategies for solving the problem. Student acknowledges or confirms roles taken by other group members. Student responds appropriately when asked to complete the student's role assignment.	The student selects actions that contribute to achieving group goals and occasionally initiates actions. Student acknowledges completion of actions when prompted. Student participates in modification of actions without initiating the modifications.	The student's feedback acknowledges contributions made by team members and draw attention to feedback from other team members but without implementing the improvements proposed by others as appropriate. Student responds appropriately when asked to discuss possible improvements to the teamwork.
High	The student proactively takes the initiative in requesting information from others and responds to requests for information. Student detects deficits (gaps or errors) in shared understanding when needed and takes the initiative to perform actions and communication to solve the deficits.	The student acts as a responsible team member and proactively takes the initiative to solve difficult barriers in collaboration. Student monitors the actions of others on the team and actively inquiries about the tasks and plans to be completed by team members, and effectively responds to conflicts, changes in the problem situation, and new obstacles to goals.	The student initiates unprompted actions and identifies efficient pathways to goal resolution. Actions fully comply with the planned distribution of roles and tasks. Student inquiries about the actions, tasks, and plans to be completed by members of the group to solve the problem when contextually appropriate.	The student actively acknowledges contributions made by team members. Student detects gaps in teamwork and discusses possible pathways to improvement. The student responds positively to feedback from other team members and takes the initiative to implement the improvements proposed by others as appropriate.

BACKGROUND

Development of a standardized computer-based assessment of collaborative problem solving skills, specifically for large-scale assessment programs, remains challenging. Unlike some other skills, collaborative problem solving typically requires using complex performance tasks, grounded in varied educational domains, with interaction among students. These factors can affect the level of control that can be applied to ensure accurate assessment of students.

Student performance in CPS can be assessed through a number of different methods. These include measures of the quality of the solutions and the objects generated during the collaboration (Avouris, Dimitracopoulou, & Komis, 2003); analyses of log files, intermediate results, paths to the solutions (Adejumo et al., 2008), team processes and structure of interactions (O'Neil, Chung, & Brown, 1997); and quality and type of collaborative communication (Cooke et al., 2003; Foltz, & Martin, 2008; Graesser, Jeon, & Dufty, 2008). To ensure valid measurement on the individual level, each student should be paired with partners displaying various ranges of CPS characteristics (Rosen, & Foltz, 2014). This way each individual student will be situated fairly similarly to be able to show his or her proficiency in CPS. Educators are urged to carefully consider group composition when creating collaborative groups or teams (Fall, Webb, & Chudowsky, 1997; Rosen, & Rimor, 2009; Webb, 1995; Wildman et al., 2012). Additionally, students should act in different roles (e.g., team leader) and be able to work collaboratively in various types of environments. Another challenge in CPS assessment refers to the need for synthesizing information from individuals and teams along with actions and communication dimensions (Laurillard, 2009; O'Neil, Chen, Wainess, & Shen, 2008; Rimor, Rosen, & Naser, 2010). Communication among the group members is central in the CPS assessment and it is considered a major factor that contributes to the success of CPS (Fiore & Schooler, 2004; Dillenbourg & Traum, 2006; Fiore et al., 2010).

Another key consideration in CPS assessment is the development of situations where collaboration is critical to performing successfully on the task. Such situations require interdependency between the students where information must be shared and team organization (Griffin, Care, & McGaw, 2012). For example, dynamic problem situations can be developed where each team member has a piece of information and only together can they solve the problem (called hidden-profile or jigsaw problems, Aronson, & Patnoe, 2011). Similarly, ill-defined tasks can be developed using such tasks as group bargaining where there are limited resources but a group must converge on a solution that satisfies the needs of different stakeholders. Finally, information between participants may also be conflicting, requiring sharing of the information and then resolution in order to determine what information best solves the problem

(Rosen, 2014). CPS tasks may include one or more of these situations, while the common factors in all these collaborative tasks are handling discord, disequilibrium, and group think. Usually, a group member cannot complete the task without taking actions to ensure that a shared understanding is established. Thus, a key element in CPS tasks is interdependency between the group members so that grounding is both required and observable.

HUMAN-TO-AGENT APPROACH IN CPS ASSESSMENT

Simulated team members for collaboration with a preprogrammed profile, actions and communication would potentially provide the coverage of the full range of collaboration skills with sufficient control. In the Human-to-Agent (H-A) approach, CPS skills are measured by pairing each individual student with a computer agent or agents that can be programmed to act as team members with varying characteristics relevant to different CPS situations. Group processes are often different depending on the task and could even be competitive. Use of computer agents provides an essential component of non-competitiveness to the CPS situation, as a student experiences it. Additionally, if the time-on-task is limited, taking the time to explain to each other may lower group productivity. As a result of these perceived constraints, a student collaborating in Human-to-Human (H-H) mode may limit significantly the extent to which CPS dimensions, such as shared understanding, are externalized through communication with the partner. The agents in H-A communication can be developed with a full range of capabilities, such as text-to-speech, facial actions, and optionally rudimentary gestures. In its minimal level, a conventional communication media, such as text via emails, chat, or graphic organizer with lists of named agents can be used for H-A CPS purposes. However, CPS in H-A setting deviates from natural human communication delivery.

Current intelligent agents have limited conditional branching and typically focus on only a single topic or subject matter (Graesser, 2016).

The dynamics of H-H interaction (timing, conditional branching) cannot be perfectly captured with agents, and agents cannot adjust to idiosyncratic characteristics of humans. For example, human collaborators can propose unusual, exceptional solutions; the characteristic of such a process is that it cannot be included in a system following an algorithm, such as H-A interaction. If educators rely on CPS teaching of students in H-A interactions exclusively, there may be the risk that these students will build up expectations that do exactly follow such an algorithm.

It should be noted that in both settings, CPS skills should be measured through a number of tasks, where each task represents a phase in the problem solving and collaborative process and can contain several steps. CPS tasks may include: consensus building, writing a joint document, making a presentation, or Jigsaw problems. In each task, the student is expected to work with one or more team members to solve a problem, while the team members should represent different roles, attitudes and levels of competence in order to vary the collaborative problem solving situation the student is confronted with.

EXPLORING FORMATIVE USES OF CPS ASSESSMENT

The study builds on previous research efforts to explore the feasibility of H-A approach for CPS assessment (Rosen, & Foltz, 2014; Rosen, & Mosharraf, 2015) and extends the effort by deploying H-A assessment in online CPS learning context. Collaborative learning is a powerful educational intervention, especially when dealing with a complex problem that multiple perspectives or specific expertise (Fawcett, & Garton, 2005; Hattie, 2009; Roseth et. al., 2006). Team members can assist in creating a rich learning environment by providing social comparisons, cognitive restructuring, giving feedback, emotional support and deliberate practice. However, research has also indicated that there are challenges in ensuring that the collaborative experience is successful for all participants, as individuals can dominate and marginalize others (Arvaja, et al., 2002), students can spend too much time off-task (Mueller, & Fleming, 2001), and struggling team members can limit overall success (Chiu, & Khoo, 2003). Team composition is one of the key factors in in collaborative problem solving (Dillenbourg, 1999; Rosen, & Rimor, 2009). Team performance is compromised to the extent that the division of labor is unintelligent, subgoals are not achieved, the group goals are blocked, and there are communication breakdowns. Collaborative problem solving tasks with high interdependency are very sensitive to group composition. One team member who has low competency can dramatically decrease the performance of the entire team and force other team members to compensate in order to achieve team goals. An overly strong leader can prevent other team members from manifesting their talents. A meaningful collaborative interaction rarely emerges spontaneously, but requires careful structuring of the collaboration to promote constructive interactions.

In order to better understand possible implications of CPS assessment for formative purposes, H-A assessment task was used to inform group composition in Animalia science CPS learning project. More specifically, the aim of the study was to explore the differences in student CPS performance and group engagement between low-CPS and mixed-CPS groups conditions assigned based on H-A standardized CPS pre-test. The hypothesis was that given H-A pre-test measure of CPS proficiency, students participating in mixed-CPS groups would benefit from initial higher collaborative interaction that would in turn result in a meaningful advantage over the students participating in low-CPS groups. It should be noted that this study is among the first systematic attempts to investigate different facets of the CPS assessments and possible effects on teaching and learning.

METHOD

The study participants included 220 students, all 14 years old, from Israel, Italy, Bulgaria, Czech Republic, and Spain. The study and student engagement with Animalia activities were conducted during February to April, 2015. Recruitment of participating schools was achieved through collaboration with World ORT and the Ministry of Education based on the following criteria: (a) population of 14-year-old students proficient in English, (b) interest and commitment of teachers in engaging their students in an international online CPS learning and assessment project, and (c) sufficient technology infrastructure (e.g., access to computers, high-speed Internet).

Participating students were divided into two conditions based on their CPS proficiency, as measured by H-A standardized CPS pre-test Zoo Quest:

1. Low-CPS condition (108 students): 4-person teams, all low-CPS proficiency, two each from two different schools.
2. Mixed-CPS condition (112 students): 4-person teams, 2 mid.-high and 2 low CPS proficiency members in each team, two each from two different schools.

Of the total students who participated in low-CPS condition, 52 were boys (48.1%) and 56 were girls (51.9%). In mixed-CPS condition, 55 were boys (49.1%), and 57 were girls (50.9%).

Collaborative Problem Solving Pre-Test

Standardized H-A CPS assessment task Zoo Quest was used to inform grouping in the Animalia project (Rosen, & Foltz, 2014; Rosen, & Mosharraf, 2015). In this H-A CPS computer-based assessment task, the student was asked to collaborate with a partner (computer-driven agent) to find the optimal conditions for an animal at the zoo (26 years of life expectancy). The task was designed to measure the CPS skills of the student to establish and maintain shared understanding, take appropriate actions to solve the problem, monitor progress, and provide feedback to the partner (OECD, 2013b). The task demands were developed in line with PISA (OECD, 2013b) design considerations for CPS assessment tasks. The assessment is focused on how well the individual interacts with agents during the course of problem solving; this includes achieving a shared understanding of the goals and activities as well as efforts to solve the problem and pooling resources. An agent could be considered either a human or a computer agent that interacts with the student. In both cases, an agent has the capability of generating goals, performing actions, communicating messages, sensing its environment, and adapting to changing environments. Some of the CPS skills are reflected in actions that the student performs, such as making a decision by choosing an item on the screen, or selecting values of parameters in a simulation. Other skills require acts of communication, such as asking the partner questions, answering questions, making claims, issuing requests, giving feedback on other agents' actions, and so on.

In this task, the student was able to select different types of food, life environments, and extra features, while both partners were able to see the selections made and communicate through a phrase-chat (selections from predefined 4-5 options). An animal's life expectancy under the given conditions was presented after each trial of the conditions. The student and the partner were prompted to discuss how to reach better conditions for an animal at the beginning of the task. By the end of the task, the student was asked to rate the partner (1-3 stars) and provide written feedback on the partner's performance. The students were limited to complete the task within 5 minutes. The task was programmed in such a way that at least two attempts for problem solving and at least one communication act with a partner were required to be able to complete the assessment task. The task and scoring technique were

Figure 1. Solving a problem in Zoo Quest human-to-agent collaborative assessment task

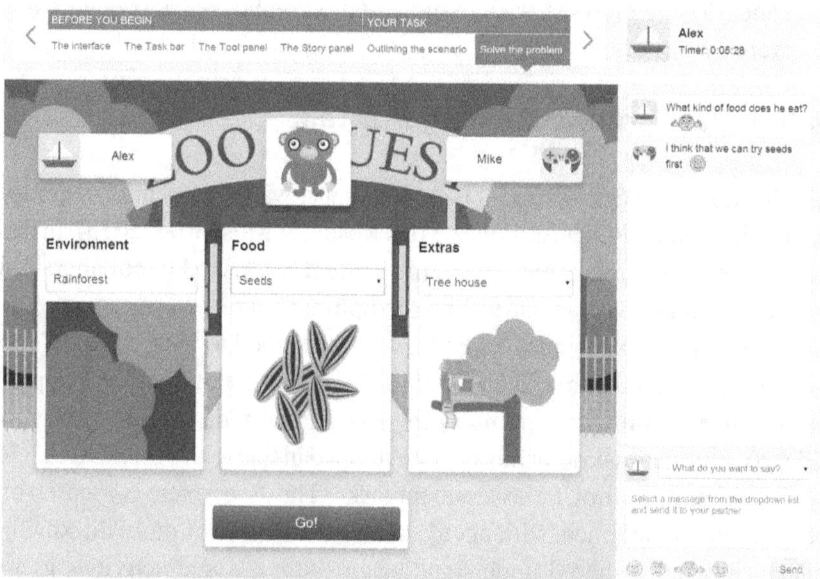

checked through a series of previous studies and proficiency levels were established (Rosen, & Foltz, 2014; Rosen, & Mosharraf, 2015).

The phrases presented at the chat were based on a pre-programmed decision-tree and situated in order to allow the student to authentically communicate with a partner and to be able to cover the CPS measures defined for the task. The computer agent was programmed to act with varying characteristics relevant to different CPS situations (e.g., agree or disagree with the student, contributing to solving the problem or proposing misleading strategies, etc.). Each student (regardless of student's actions) was provided with the same amount of help and misleading strategies proposed by the agent. This approach provided each individual student with similar optimal chances to show his or her CPS skills.

Certain questions or responses, however can lead to different sets of available responses during conversation (e.g., asking a question), and so the person initially asking questions may not do so for the entire conversation. Only the student can submit guesses for the conditions and complete the task though. As the task progresses, some of the possible replies may be marked as "hidden", which will mean they do not show again until the leader has submitted the attempt at solving the problem. If no replies are available then a secondary list of replies are available – this will be the case when all conditions have been agreed upon, and the focus will change to submitting. If the students have submitted their attempt at solving the problem, then some additional statements may become available to be said. There are various types of sentences that can be used through the phrase-chat.

The computer agent was programmed to act with varying characteristics relevant to CPS dimensions: problem solving, shared understanding, and group organization. The problem-solving dimension was represented by the extent to which the agent helped the student to reach the optimal conditions for the animal (i.e., environment, food, and extra features). Shared understanding interaction was focused on agent's responses to grounding questions from the student. Group organization dimension was represented by agent's responses to student's questions related to progress monitoring. The agent was programmed to agree with student's selections in the first stage. Later in the process the agent occasionally disagreed with the student (every even selection per variable made by student excluding optimal selection), confirmed selections made by the student (every odd selection per variable made by student or in case of optimal selection), or proposed alternative solution (each two tryouts of all selected variables, excluding optimal selection). This standardized medium-level collaborative behavior of computer agent provided each individual student with similar optimal chances to show his or her CPS skills.

Scoring Considerations

The target CPS skills for this assessment consisted of shared understanding, taking appropriate actions to solve the problem, establishing and maintaining group organization, and providing feedback (e.g., OECD, 2013b). In line with CPS dimensions, scores for student performance in the assessment task broken down as being: shared understanding (40 points), problem solving (26 points), and group organization represented by monitoring progress (26 points) and providing feedback (8 points). The different scales for scores across CPS measures result from different range of scorable situations or data points in each category. First, students in this task must be able to establish, monitor, and maintain the shared understanding throughout the problem solving task by responding to requests for information, sending important information to the partner about tasks completed, establishing or negotiating shared meanings, verifying what each other knows, and taking actions to repair deficits in shared knowledge. Second, students must be able to identify the type of activities that are needed to solve the problem and to follow the appropriate steps to achieve a solution. Moreover, in this task students must be able to help organize the group to solve the problem, consider the resources of group members, understand their own role and the roles of the other agents, follow the rules of engagement for their role, monitor the group organization, and help handle communication breakdowns, conflicts, and obstacles. Finally, the students must be able to reflect on the success of the CPS process and the outcome.

Student scores in the first three CPS dimensions were generated automatically based on a predefined programmed sequence of possible optimal actions and communication that was embedded into the assessment task. The problem-solving dimension was scored as one point per each year of the animal's life expectancy that was achieved by selecting the variables. Students that were not fully engaged in an effort to reach the optimal life expectancy were scored lower in problem solving score, than the students that reached the solution. Shared understanding score consisted of a number of grounding questions that were initiated by a student in appropriate situations (e.g., explaining the reason for a variable selection, questioning "What can we do to reach better conditions for the animal?") and appropriate responses to the grounding questions made by the partner. For example, students that made their own selections with low or no communication with the partner were scored low in shared understanding measure. Monitoring progress score was created based on communication initiated by the student prior to the submission of the selected variables (e.g., questioning "Are you ready to go ahead with our plan?" before clicking on "Go") and the statements made by the student based on the life expectancy results that were achieved (e.g., "Should we keep this selection or try again?"). For example, students that clicked on "Go" without being engaged in confirmation of the selection made before submission were scored low in progress monitoring score.

Scoring of student feedback was provided independently by two teachers from participating schools in the United States. The teachers were trained through a one-day workshop to consistently evaluate whether student's feedback indicated both successful and challenging aspects of working with the partner on the task, and acknowledged the contributions the partner made toward reaching a solution. Spelling and grammar issues did not affect student score. Overall, the scoring strategy was discussed with a group of ten teachers from participating countries in order to achieve consensus on CPS scoring strategy and reduce cultural biases as much as possible. Inter-coded agreement of feedback scoring was 92%.

It should be noted that for more comprehensive assessment purposes (e.g., PISA), multiple assessment tasks that vary across the four dimensions presented in Table 1 are needed in order to provide a comprehensive inference on student CPS proficiency.

Project Animalia: Online Collaborative Science Problem Solving

An investigation to look further at the premises of human and computer agent technology in the context of learning and assessing collaborative science problem solving skills was undertaken through collaboration with World ORT and the Ministry of Education in Israel. Animalia is an online mini-course designed to promote students' collaborative problem solving skills in the context of complex ecosystems (Bakken, Bielinski, Johnson, & Rosen, 2015). Animalia is a simulated fictitious village with a serious environmental problem. The fish in a local river are dying and the cause(s) of this problem is unknown. A team of scientists from an international organization (i.e. students) must determine the root cause of the problem and recommend sound, scientifically based solutions. The foundation of the problem-solving activity relies on understanding specific science concepts such as pH, oxygen saturation, nitrogen levels, the biological effects of pollutants, ecology, frog life cycles, algae physiology, as well as cause and effects in ecosystems. To create a more realistic scenario, team members will be given multiple and sometimes misleading, irrelevant, or contradictory pieces of information about what was happening in Animalia. It was up to the students to determine the authenticity of the data and to judge the credibility of the source. Each team member have had access to only some information about the overall situation so that determining the root cause of the problems in Animalia require each team member to share his/her information with the others. During the first task, each student was assigned a specific scientist role and given a written description of that scientist's area of expertise. Of the four science roles, two were chemistry-related (Water and Soil Scientist, and Farm and Industry Scientist) and two were biology-related (Plant Scientist and Animal Scientist). Each role determined what information a student would receive at various points throughout the project. This was an essential part of the task's design: to promote ongoing communication by giving each student unique information and making the need to share this information with the team apparent. If each student did not share the information that he or she acquired in an assigned role, the problem itself would be much more difficult to solve successfully.

To foster collaboration in this problem-solving task, each episode consisted of either a shared document that required contributions from each team member or a set of questions designed to promote discussion. Team members were encouraged to distill the information they receive into its essential components and to hypothesize what they think are the causes of the problem(s). Team members were also given feedback on each other's ideas and then form a consensus opinion. Finally, each team summarized findings and recommendations in a written report and an oral presentation. The teams described the cause(s) and provided recommendations to solve the environmental problems in Animalia. Figures 2-3 display examples of the screens in Animalia.

Learning was assessed through collaborative tasks (written report and oral presentation), and science concepts test (20-item multiple choice test) that were administered at the beginning and at the end of the mini-course (see Bakken, Bielinski, Johnson, & Rosen, 2015, for details). Collaborative problem solving proficiency was assessed at the beginning of Animalia through H-A CPS task Zoo

Figure 2. Outlining tasks in Animalia structuring effective groups episode 2

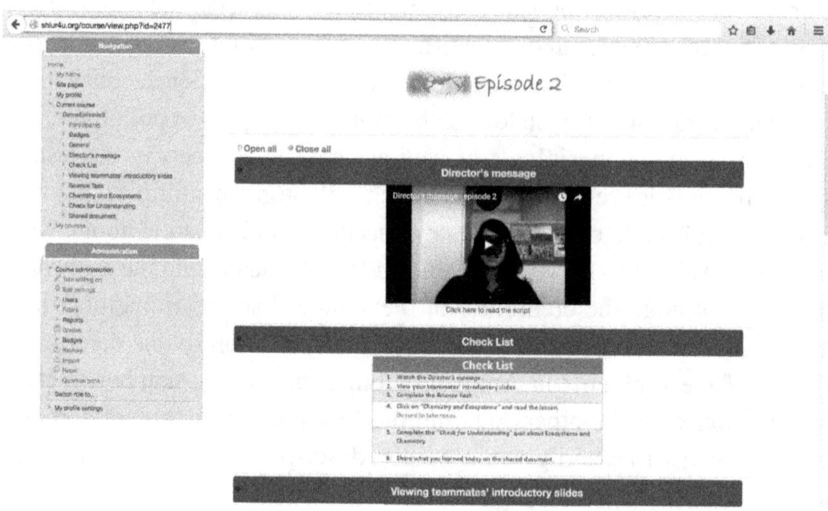

Figure 3. Outlining tasks in Animalia episode 6

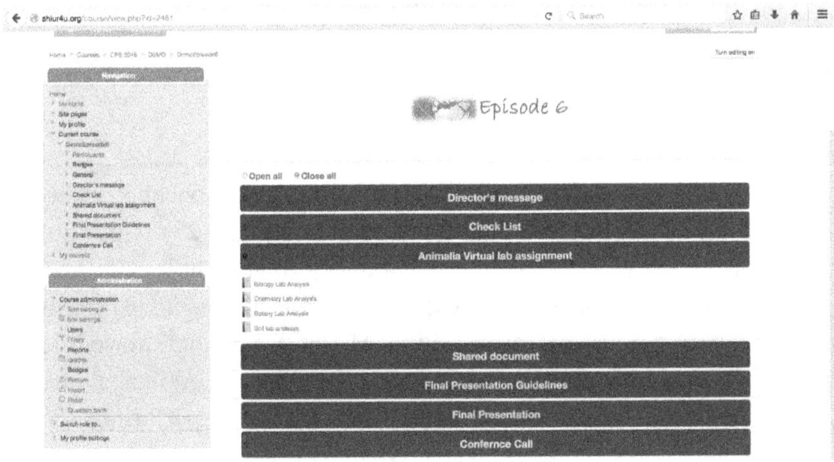

Quest (Rosen, & Foltz, 2014; Rosen, & Mosharraf, 2015). Additionally, students completed two surveys in which they were asked to rate their team members' overall engagement in CPS. Students rated their group engagement with respect to effort, communication, and willingness to work with the group in the future (see Bakken, Bielinski, Johnson, & Rosen, 2015, for details).

The development team designed this mini-course to be completed in approximately 14 hours. Even though this is a collaborative task, due to the constraints of physical distance and time zone differences, students typically worked independently and communicate asynchronously (at different times) with their team members in other locations. The only two times students were online at the same time was when the teams discussed townspeople interview videos and gave their final presentations using Google Hangouts. Animalia content was provided in English, and English was spoken between partner schools. This gave students the real-world experience of collaborating with people from different cultures, with different levels of English proficiency, different levels of science knowledge, and across time zones. The content and related activities were provided online via a designated Moodle website. Each team have had its own "course" in the Moodle system which enabled team members to share information with each other, but prevented them from seeing information from other teams. Each session was available to students sequentially during the project. When they logged in, students saw only those sessions that have been released up to that point. Across sites, teams were asked to stick to a common schedule so that all team members are progressing through the project work at the same pace. Animalia includes several professional development sessions to guide teachers on implementation of the course.

RESULTS

No significant differences were found in Animalia science concepts pre-test scores between participants in the two conditions (t(218)=1.1, p=.27). This similarity in student science knowledge allowed comparability of student results in Animalia between the two conditions. In order to explore possible differences in two conditions, an analysis of variance was performed. First, MANOVA results showed significant difference between Mixed-CPS and Low-CPS conditions (Wilks' Lambda=.87, F(3,217)=4.5, p<.01). Hence, individual post-hoc t-tests were performed. The results indicated that students who collaborated in Mixed-CPS groups showed significantly higher level of performance in science concepts post-test (ES=.6, t(218)=4.5, p<.01), writing Animalia CPS cause-effect report (ES=.5, t(218)=4.3, p<.01), and presenting their findings to their classmates (ES=.5, t(218)=3.9, p<.01).

Students completed two surveys in which they were asked to rate their team members' overall ability to collaborate on a Likert scale 1-5 (from Strongly Disagree to Strongly Agree). Percentage of students that indicated high levels of group engagement (Strongly Agree and Agree) in mid.- and end-course survey for both conditions are provided in Figures 4-5.

Figure 4. Comparison of student feedback on group engagement at Animalia mid.-course survey

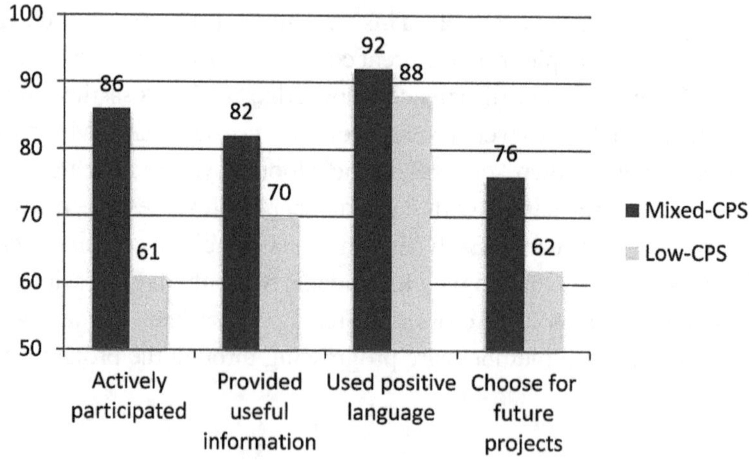

Figure 5. Comparison of student feedback on group engagement at Animalia end-course survey

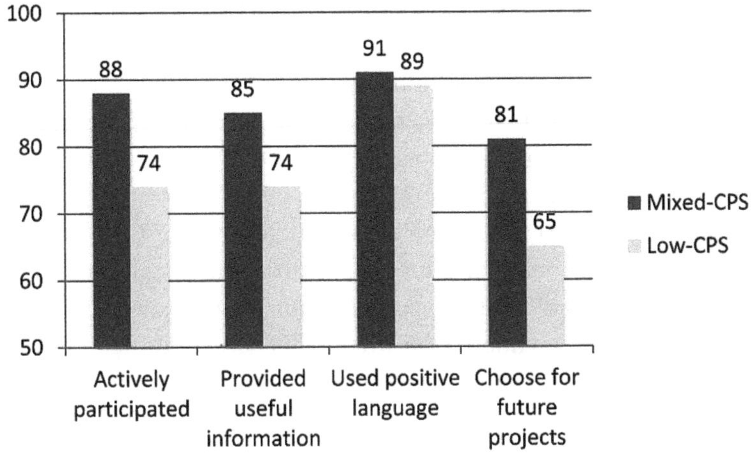

The results suggest that overall student engagement in mixed-CPS groups was higher than in low-CPS groups consistently early in the course and at the end.

DISCUSSION

Policymakers, researchers, and educators are engaged in vigorous debate about assessing collaborative science problem solving skills on an individual level in valid, reliable and scalable ways. Key assessment design and psychometric challenges include influence on individual and group scores of the roles that individuals take on or are assigned; status, interaction style, personality, and knowledge and skill differences in group composition; and consistency of individual behavior in different groups, for different tasks, on different occasions. Analyses of the list of challenges facing CPS in formative assessment programs suggests that H-A approach should be further explored. The goal of this study was to explore ways of using CPS assessment for online CPS learning and assessment science CPS program. More specifically, H-A CPS assessment was used for student grouping. Careful consideration of group composition is one of the key principles when creating collaborative groups (Rosen, & Rimor, 2009; Webb, et al., 1998; Wildman et al., 2012). The composition of the group can vary along a great many dimensions including group-member knowledge and skill level, gender, personality, motivation, status, popularity, ethnic background, race, and other demographic characteristics. Emerging evidence shows that cognitive abilities, organizing skills, cooperativeness, and team orientation greatly influences team functioning and success (National Research Council, 2013, Mathieu, et al, 2014). These influences appear in assessments studies as well, such as high-ability students showing higher scores in homogeneous groups and in high-functioning heterogeneous groups than in heterogeneous groups. For example, group members failing to answer each other's questions, failing to explain their answers, insulting others (Webb, Nemer, & Zuniga, 2002).

In this study two conditions were created based on the CPS pre-test results. Students in each of these conditions were exposed to identical learning and assessment tasks and were able to collaborate and communicate by using identical methods and resources. However, while in the Mixed-CPS condition teams were able to benefit from mid. to high proficiency team members, students participated in the Low-CPS condition were constantly facing deficits in CPS skills. Although these skills were further developed through the participation in the Animalia project, the findings indicated that students participated in Low-CPS condition were placed in a disadvantage, compared to their peers in the Mixed-CPS condition. In other words, H-A standardized CPS pre-test allowed to create more balanced groups that in turn worked more effectively to achieve their goals.

FUTURE RESEARCH DIRECTIONS

Among the questions that remain to be answered are included, the extent to which interacting with computer agents generalizes to interacting with human partners, whether using computer agents as partners produces CPS experiences that are comparable from test taker to test taker, how many standardized group compositions are needed for generalizable scores, and how to select group compositions to represent the target domain of group compositions.

One of the major limitations of this preliminary study is relatively small and non-representative sample of 14-years-old students. However, due to lack of empirical research in the field of computer-based assessment of CPS skills, it is necessary to conduct small-scale pilot studies in order to inform a more comprehensive approach of CPS assessment. Further studies could consider including a representative sample of randomly assigned students in a wider range of ages and backgrounds. Additionally, future studies could consider exploring differences in student performance in a wide range of CPS assessment tasks and formative applications.

CONCLUSION

Qualitatively different assessment approaches could work differently for different purposes. Formative assessment program which have adopted rich training on the CPS construct for its teachers may consider the H-H approach for CPS assessment as a more powerful and authentic tool to inform teaching and learning, while H-A may be implemented as a formative scalable tool across a large district or in international settings. Non-availability of students with a certain CPS level in a class may limit the fulfillment of assessment needs, but technology with computer agents can fill the gaps. In many cases, using simulated computer agents instead of relying on peers is not merely a replacement with limitations, but an enhancement of the capabilities that makes independent assessment possible. CPS assessment methods described in this article offer one of the few examples today of a direct, standardized assessment targeting problem solving and collaboration skills. CPS brings new challenges and considerations for the design of effective assessment approaches because it moves the field beyond standard item design tasks. The assessment must incorporate concepts of how humans solve problems in situations where information must be shared and considerations of how to control the collaborative environment in ways sufficient for valid measurement of individual and team skills. The quality and practical feasibility of these measures are not yet fully documented. However, these measures rely on the abilities of technology to engage students in interaction, to simulate others with whom students can interact, to track students' ongoing responses, and to draw inferences from those responses. Group composition is one of the important issues in large-scale assessments of collaborative skills (Webb, 1995; Wildman et al., 2012). Overcoming possible bias of differences across groups by using computer agents or other methods becomes even more important within international large-scale assessments (i.e., PISA) and online learning (e.g., massive open-access online courses) where cultural boundaries are crossed. The results of this study suggest that computer agent technique is one of the promising directions for CPS assessment tasks that should be further explored. However, as discussed in this article, each mode of collaboration involves limitations and challenges. Further research is needed in order to establish comprehensive validity evidence and generalization of findings in the context of H-A CPS assessment and its uses to inform collaborative learning.

REFERENCES

Adejumo, G., Duimering, R. P., & Zhong, Z. (2008). A balance theory approach to group problem solving. *Social Networks*, *30*(1), 83–99. doi:10.1016/j.socnet.2007.09.001

Aronson, E., & Patnoe, S. (2011). Cooperation in the classroom: The jigsaw method. London: Pinter & Martin, Ltd.

Arvaja, M., Hakkinen, P., Rasku-Puttonen, H., & Etelapelto, A. (2002). Social processes and knowledge building during small group interaction in a school science project. *Scandinavian Journal of Educational Research*, *46*(2), 161–179. doi:10.1080/00313830220142182

Avouris, N., Dimitracopoulou, A., & Komis, V. (2003). On analysis of collaborative problem solving: An object-oriented approach. *Computers in Human Behavior*, *19*(2), 147–167. doi:10.1016/S0747-5632(02)00056-0

Chauncey, A., & Azevedo, R. (2010). Emotions and motivation on performance during multimedia learning: How do I feel and why do I care? In V. Aleven, J. Kay, & J. Mostow (Eds.), *Intelligent Tutoring Systems* (Vol. 6094, pp. 369–378). Springer Berlin Heidelberg. doi:10.1007/978-3-642-13388-6_41

Chiu, M. M., & Khoo, L. (2003). Rudeness and status effects during group problem solving: Do they bias evaluations and reduce the likelihood of correct solutions? *Journal of Educational Psychology*, *95*(3), 506–523. doi:10.1037/0022-0663.95.3.506

Cooke, N. J., Kiekel, P. A., Salas, E., Stout, R., Bowers, C., & Cannon-Bowers, J. (2003). Measuring team knowledge: A window to the cognitive underpinnings of team performance. *Group Dynamics*, *7*(3), 179–219. doi:10.1037/1089-2699.7.3.179

Dillenbourg, P. (Ed.). (1999). *Collaborative learning: Cognitive and computational approaches*. Amsterdam: Pergamon, Elsevier Science.

Dillenbourg, P., & Traum, D. (2006). Sharing solutions: Persistence and grounding in multi-modal collaborative problem solving. *Journal of the Learning Sciences*, *15*(1), 121–151. doi:10.1207/s15327809jls1501_9

Fall, R., Webb, N., & Chudowsky, N. (1997). *Group discussion and large-scale language arts assessment: Effects on students' comprehension*. CSE Technical Report 445. Los Angeles, CA: CRESST.

Fawcett, L. M., & Garton, A. F. (2005). The effects of peer collaboration on childrens problem solving ability. *The British Journal of Educational Psychology*, *75*(2), 157–169. doi:10.1348/000709904X23411 PMID:16033660

Fiore, S., Rosen, M., Smith-Jentsch, K., Salas, E., Letsky, M., & Warner, N. (2010). Toward an understanding of macrocognition in teams: Predicting process in complex collaborative contexts. *The Journal of the Human Factors and Ergonomics Society*, *53*(2), 203–224. doi:10.1177/0018720810369807

Fiore, S., & Schooler, J. W. (2004). Process mapping and shared cognition: Teamwork and the development of shared problem models. In E. Salas & S. M. Fiore (Eds.), *Team cognition: Understanding the factors that drive process and performance* (pp. 133–152). Washington, DC: American Psychological Association. doi:10.1037/10690-007

Foltz, P. W., & Martin, M. J. (2008). Automated communication analysis of teams. In E. Salas, G. F. Goodwin, & S. Burke (Eds.), *Team effectiveness in complex organizations and systems: Cross-disciplinary perspectives and approaches* (pp. 411–431). New York: Routledge.

Graesser, A. C. (2016). Conversations with AutoTutor help students learn. *International Journal of Artificial Intelligence in Education*, *26*(1), 124–132. doi:10.1007/s40593-015-0086-4

Graesser, A. C., Jeon, M., & Dufty, D. (2008). Agent technologies designed to facilitate interactive knowledge construction. *Discourse Processes*, *45*(4), 298–322. doi:10.1080/01638530802145395

Griffin, P., Care, E., & McGaw, B. (2012). The changing role of education and schools. In P. Griffin, B. McGaw, & E. Care (Eds.), *Assessment and teaching 21st century skills* (pp. 1–15). Heidelberg, Germany: Springer. doi:10.1007/978-94-007-2324-5_1

Hattie, J. A. C. (2009). *Visible learning: A synthesis of 800+ meta-analyses on achievement*. London: Routledge.

Laurillard, D. (2009). The pedagogical challenges to collaborative technologies. *International Journal of Computer-Supported Collaborative Learning*, *4*(1), 5–20. doi:10.1007/s11412-008-9056-2

Liu, L., Hao, J., von Davier, A., Kyllonen, P., & Zapata-Rivera, D. (2015). A tough nut to crack: Measuring collaborative problem solving. In Y. Rosen, S. Ferrara, & M. Mosharraf (Eds.), Handbook of Research on Technology Tools for Real-Life Skill Development (pp. 344-359). Hershey, PA: Information Science, IGI Global.

Mathieu, J. E., Tannenbaum, S. I., Donsbach, J. S., & Alliger, G. M. (2014). A review and integration of team composition models: Moving toward a dynamic and temporal framework. *Journal of Management*, *40*(1), 130–160. doi:10.1177/0149206313503014

Mayer, R. E., & Wittrock, M. C. (1996). Problem-solving transfer. In D. C. Berliner & R. C. Calfee (Eds.), *Handbook of educational psychology* (pp. 47–62). New York: Macmillan Library Reference USA, Simon & Schuster Macmillan.

Mueller, A., & Fleming, T. (2001). Cooperative learning: Listening to how children work at school. *The Journal of Educational Research*, *94*(5), 259–265. doi:10.1080/00220670109598761

National Research Council. (2012). *A Framework for K-12 Science Education: Practices, Crosscutting Concepts, and Core Ideas*. Washington, DC: National Academies Press.

National Research Council. (2013). *New directions in assessing performance of individuals and groups: Workshop summary*. Washington, DC: National Academies Press.

O'Neil, H. F. Jr, Chen, H. H., Wainess, R., & Shen, C. Y. (2008). Assessing problem solving in simulation games. In E. L. Baker, J. Dickieson, W. Wulfeck, & H. F. O'Neil (Eds.), *Assessment of problem solving using simulations* (pp. 157–176). Mahwah, NJ: Lawrence Erlbaum Associates.

O'Neil, H. F. Jr, & Chuang, S. H. (2008). Measuring collaborative problem solving in low-stakes tests. In E. L. Baker, J. Dickieson, W. Wulfeck, & H. F. O'Neil (Eds.), *Assessment of problem solving using simulations* (pp. 177–199). Mahwah, NJ: Lawrence Erlbaum Associates.

O'Neil, H. F. Jr, Chuang, S. H., & Baker, E. L. (2010). Computer-based feedback for computer-based collaborative problem solving. In D. Ifenthaler, P. Pirnay-Dummer, & N. M. Seel (Eds.), *Computer-based Diagnostics and Systematic Analysis of Knowledge* (pp. 261–279). New York: Springer-Verlag. doi:10.1007/978-1-4419-5662-0_14

O'Neil, H. F., Chung, G., & Brown, R. (1997). Use of networked simulations as a context to measure team competencies. In H. F. O'Neil Jr., (Ed.), *Workforce readiness: Competencies and assessment* (pp. 411–452). Mahwah, NJ: Lawrence Erlbaum Associates.

OECD. (2013a). *PISA 2015 Science Framework*. OECD Publishing.

OECD. (2013b). *PISA 2015 Collaborative Problem Solving Framework*. OECD Publishing.

Rimor, R., Rosen, Y., & Naser, K. (2010). Complexity of social interactions in collaborative learning: The case of online database environment. *Interdisciplinary Journal of E-Learning and Learning Objects*, *6*, 355–365.

Roschelle, J., & Teasley, S. D. (1995). The construction of shared knowledge in collaborative problem-solving. In C. E. O'Malley (Ed.), *Computer-supported collaborative learning* (pp. 69–97). Berlin: Springer-Verlag. doi:10.1007/978-3-642-85098-1_5

Rosen, Y. (2014). Comparability of conflict opportunities in human-to-human and human-to-agent online collaborative problem solving. *Technology, Knowledge and Learning*, *19*(1-2), 147–174. doi:10.1007/s10758-014-9229-1

Rosen, Y., & Foltz, P. (2014). Assessing collaborative problem solving through automated technologies. *Research and Practice in Technology Enhanced Learning*, *9*(3), 389–410.

Rosen, Y., & Mosharraf, M. (2015). Computer agent technologies in collaborative assessments. In Y. Rosen, S. Ferrara, & M. Mosharraf (Eds.), Handbook of Research on Technology Tools for Real-Life Skill Development (pp. 319-343). Hershey, PA: Information Science, IGI Global.

Rosen, Y., & Rimor, R. (2009). Using collaborative database to enhance students' knowledge construction. *Interdisciplinary Journal of E-Learning and Learning Objects*, *5*, 187–195.

Rosen, Y., & Rimor, R. (2012). Teaching and assessing problem solving in online collaborative environment. In R. Hartshorne, T. Heafner, & T. Petty (Eds.), Teacher education programs and online learning tools: Innovations in teacher preparation (pp. 82-97). Hershey, PA: Information Science Reference, IGI Global.

Roseth, C. J., Fang, F., Johnson, D. W., & Johnson, R. T. (2006, April). *Effects of cooperative learning on middle school students: A meta-analysis*. Paper presented at the Annual Meeting of the American Educational Research Association, San Francisco, CA.

Webb, N. M. (1995). Group collaboration in assessment: Multiple objectives, processes, and outcomes. *Educational Evaluation and Policy Analysis*, *17*(2), 239–261. doi:10.3102/01623737017002239

Webb, N. M., Nemer, K. M., Chizhik, A. W., & Sugrue, B. (1998). Equity issues in collaborative group assessment: Group composition and performance. *American Educational Research Journal*, *35*(4), 607–651. doi:10.3102/00028312035004607

Webb, N. M., Nemer, K. M., & Zuniga, S. (2002). Short circuits or superconductors? Effects of group composition on high-achieving students science performance. *American Educational Research Journal*, *39*(4), 943–989. doi:10.3102/00028312039004943

Wildman, J. L., Shuffler, M. L., Lazzara, E. H., Fiore, S. M., Burke, C. S., Salas, E., & Garven, S. (2012). Trust development in swift starting action teams: A multilevel framework. *Group & Organization Management*, *37*(2), 138–170. doi:10.1177/1059601111434202

Zhang, J. (1998). A distributed representation approach to group problem solving. *Journal of the American Society for Information Science*, *49*(9), 801–809. doi:10.1002/(SICI)1097-4571(199807)49:9<801::AID-ASI5>3.0.CO;2-Q

KEY TERMS AND DEFINITIONS

Agent: Either a human or a computer-simulated participant in a collaborative problem solving group.

Collaboration: Coordinated, synchronous activity that is the result of a continued attempt to construct and maintain a shared conception of a problem.

Collaborative Problem Solving: The capacity of an individual to effectively engage in a group process whereby two or more agents attempt to solve a problem by sharing knowledge and understanding, organizing the group work and monitoring the progress, taking actions to solve the problem, and providing constructive feedback to group members.

Computer Agent: An avatar with a preprogrammed profile, actions and communication. Computer agents can be capable of generating goals, performing actions, communicating messages, sensing environment, adapting to changing environments, and learning.

Problem Solving: Cognitive processing directed at achieving a goal when no solution method is obvious to the problem solver.

Chapter 5
Learning About Genetics in an Elementary Classroom Using a Web-Based Inquiry Science Environment (WISE) Unit

Amal Ibourk
Florida State University, USA

Michelle Williams
Williams Learning Solutions, Inc., USA

Merle Heidemann
Michigan State University, USA

ABSTRACT

Genetics is an increasingly important topic in today's society, yet continues to be an ongoing challenge to the science education community and an important aspect of school student learning. This chapter examines how STEMGenetics, an inquiry-based technology-enhanced online unit, engages upper elementary students in building a better understanding of the concepts of inheritance of traits, variation of traits, and the life cycle of plants. Using pre/post measures, supplemented with qualitative analyses of students' responses, the authors show how upper elementary students' understanding of inheritance and variation of traits, as well as the life cycle of plants, resulted in greater learning gains. In addition, engaging in learning technologies, such as dynamic visualizations, provided the students opportunities to interact with the scientific phenomena and enhanced their reasoning about inheritance of traits and its relationship to the life cycle of plants.

DOI: 10.4018/978-1-5225-2528-8.ch005

INTRODUCTION

In issues ranging from the consumption of genetically modified foods to getting tested for one's personal risk of certain illnesses, understanding genetics plays a vital role in everyday health-related decision-making (Costa-Font, Gil, & Traill, 2008; Kesselheim, Cook-Deegan, Winickoff, & Mello, 2013; Stark & Pompeii, 2010). Despite the demonstrated importance of understanding genetics and other heredity-related topics, numerous studies have shown that these ideas are difficult for students at all ages to grasp (e.g., Ayuso & Banet, 2002; Haga, 2006; Duncan & Reiser, 2007; Duncan, Rogat, & Yarden, 2009; Lewis & Wood-Robinson, 2000; Tsui & Treagust, 2007). Developing a full understanding of how genetics, the environment, and the interaction between these two factors can affect organisms' traits is a key expectation for students in third grade and beyond. In the U.S., these expectations are reflected in the Framework for K-12 Science Education (2012) developed by the National Research Council (NRC) of the National Academy of Sciences and the Next Generation Science Standards (NGSS, 2013). The key ideas for the upper elementary students found in the grade band (3 -5) under the NRC Framework are that offspring inherit genetic information, that variations in this information result in variation in traits, and that the environment also plays a role in influencing traits and thus enhancing observed variation (i.e., even identical twins are usually not entirely identical).

Given the complex nature of genetics and the extent of persistent non-normative ideas, technology-enhanced instruction has "tremendous" potential for promoting student learning around complex and abstract science topics such as genetics (Banet & Ayuso, 2000; Roseman, Linn, & Koppal, 2008; Shear, Bell, & Linn, 2004; Songer, 2007; Tsui & Treagurst, 2007). Technology-enhanced inquiry curricula and instruction can lead to improvement in the K-12 students' learning outcomes (Edelson, Gordin, & Pea, 1999; Kali, Linn, & Roseman, 2008; Krajcik, Blumenfeld, Marx, Bass, Fredericks, & Soloway, 1998; Linn, Davis, & Bell, 2004; Reiser, Tabak, Sandoval, Smith, Steinmuller, & Leone, 2001; Songer, Lee, & Kam, 2002). These inquiry-based materials help students develop deep understandings of complex science ideas that are standards-based (Lee, Linn, Varma, & Liu, 2010; Linn & Hsi, 2000; Roschelle, Pea, Hoadley, Gordin, & Means, 2000; Slotta & Linn, 2009;) For instance, computer-based environments such as Geniverse (https://concord.org/teaching-genetics/dragons) engage students in complex reasoning about genetics while experimenting with breeding different virtual dragons. This in turn provides students with a mode of instruction that is web-based and technology enhanced, allowing them to manipulate different dynamic representations while potentially deepening their understanding of concepts of genetics.

This study investigated how upper elementary students' understanding of the core ideas of *inheritance of traits* and *variation of traits* changed as they participated in STEMGenetics, an inquiry-based technology-enhanced online unit that uses the Web-based Inquiry Science Environment (WISE, Linn, Davis, & Bell, 2004). Using a technology-based learning environment can support inquiry teaching (e.g. Marx, Blumenfed, Krajcik, & Soloway, 1998). The animation and simulation-based contexts in WISE provide multiple rich occasions to press for building and developing reasoning and explanations. Although technology-based learning environments can create additional demands for teachers and students (e.g. Fishman, Best, Marx, & Tal, 2001; Slotta, 2004; Williams, Linn, Ammon, & Gearhart, 2004), such as requiring new logistical practices, they can also offer new representations of science content and ways to explore complex questions using technology.

The rationale for the design of the Grade 5 STEMGenetics curriculum and assessment materials was based on a few research studies on young students. For example, many 16-year-old students do not believe that plants are capable of sexual reproduction (Lewis & Wood-Robinson, 2000; Okeke & Wood-Robinson, 1980; Engle Clough & Wood-Robinson, 1985). Therefore, the STEMGenetics curriculum has grade five students, upper elementary students, learn about sexual reproduction in plants.

The purpose of the study was to address the scientifically inaccurate ideas that students have around the concepts of *inheritance of traits* and *variation of traits,* based on the few research studies on young kids. By the end of the unit, students should understand that 1) plants of the same species vary in how they look because they have different genetic information; 2) plants get half of their genetic information from each parent; 3) and the environment can also affect how plants look because it can change how the traits are expressed. Specifically, the key research question we sought to answer for this chapter was:

- How do upper elementary school students' understanding of *inheritance of traits* and *variation of traits* changes during the enactment of a technology-enhanced inquiry-based unit?

BACKGROUND

Overview of the STEMGenetics Project

The major goals of the STEMGenetics research project were three-fold: 1) to articulate a detailed standards- and research-based conceptual framework for describing students' conceptions of genetics and how students develop a full understanding of genetics across grade spans (upper-elementary, middle and high school); 2) to develop innovative instructional materials with embedded assessments that provide rich information about students' conceptual understanding of genetics, along with teacher professional development to help practitioners make decisions about what to do next in instruction; and 3) to examine the implementation of these instructional materials and assessments to investigate students' understanding of genetics concepts.

Most research on students' understandings of genetics is at the secondary level (Duncan et al., 2009; Venville, Gribble, & Donovan, 2005; Hickey, Kindfield, Horwitz, & Christie, 2003). Yet U.S. national and U.S. state standards indicate that students should be introduced to genetics-related ideas as early as elementary school (NGSS, 2013; Michigan Department of Education, 2007; NRC, 2012; Texas Education Agency, 2010). This study sought to add further to the research on upper-elementary school students' understanding of genetics concepts (Venville, Gribble, & Donovan, 2005).

Students at the upper elementary level have difficulty understanding the contributions of both parents to the genetic makeup of their offspring (Wood-Robinson, 1994). For example, students tend to believe that a particular trait is inherited from one parent; boys believe they will inherit more traits from the father than from the mother (Clough & Wood-Robinson, 1985, Kargbo, Hobbs, & Erickson, 1980). Other studies have found that some students struggle to explain what happens when the traits of the parents and the offspring differ, and to connect the relationship between genetic information and expression of traits (Banet & Ayuso, 2000; Lewis & Kattmann, 2004; Terwogt, Stegge, & Rieffe, 2003). For example, students have problems understanding that the traits of parents are not always expressed in the offspring. Students aged 15-19 may be able to use terms, such as gene and genetic information, but are not able to make a clear distinction between the two (Lewis & Kattman, 2004). Lastly, students also demonstrate inaccurate scientific ideas about the environment and inherited traits (Kargbo et al. 1980). Some students believe that a change caused by the environment may be passed to the offspring. They expect, for example, that a dog with the tail bobbed may transmit this trait to the puppies. These non-normative ideas are persistent, not only for young students. If these ideas are addressed at earlier ages, it would help better to support students' understanding of *inheritance of traits* and *variation of traits* at middle school, high school, and college.

STEMGenetics Curriculum Design

Knowledge Integration

In this study, the Knowledge Integration (KI) framework was used to inform both the curriculum and the assessment design of the unit. The KI framework views science learning as a process of eliciting, adding, distinguishing, and organizing ideas to explain scientific phenomena (Linn & Hsi, 2000; Linn, Eylon, & Davis, 2004; Linn & Eylon, 2011; Kali, Linn, & Roseman, 2008; Gerard, Spitulnik, & Linn, 2010). This framework was also instrumental in helping design a unit that encourages learners to be actively involved in making sense of knowledge about genetics, in reflecting on their learning, and in connecting their classroom experiences with real life situations.

Design Knowledge for Promoting an Integrated Understanding of Trait Inheritance

The design of the STEMGenetics materials and assessments was also informed by the KI framework's four meta-principles, which are: (a) making thinking visible, (b) providing social supports, (c) making science accessible, and (d) promoting lifelong science learning. Drawing from these meta-principles, the unit featured a series of online and hands-on inquiry-based activities that were enhanced by dynamic visualizations and simulations that scaffolded students' learning. For example, using dynamic visualizations in the online WISE unit helped in making scientific phenomena around the topic of genetics accessible to students, and helped make students' thinking more visible. This framework also supported inquiry pedagogy, where students generate predictions before conducting experiments, record and interpret data, evaluate and critique scientific evidence, use evidence to support conclusions, and reconcile differences between new and prior science ideas.

Knowledge Integration Design Patterns

In this study, the Grade 5 STEMGenetics unit made use of two knowledge integration design patterns: *Predict, Observe, Explain* (POE) and *Explore a Simulation.* The POE design pattern involves introducing a demonstration of a scientific phenomenon, eliciting predictions, running the demonstrations, and asking the student to reconcile contradictions (Lynn & Eylon, 2006). The *Predict* stage draws on students' prior experiences and provides the opportunity to make students' thinking visible, and it elicits their initial ideas about a phenomenon. The *Observe* stage is an opportunity for students to learn through experiences and to add new ideas to their repertoire of existing ideas, which can be done through classroom activities or observations of the interactive visualizations in WISE. The *Explain* stage elicits students' understanding about the unit content as it stands after their experiences in the *Observe* stage, which may be changed from their prior understanding. In the *Explain* stage, students distinguish and sort between their old ideas and their new ideas, and they organize them to form a deeper understanding of the phenomena. The *Explore a Simulation* stage engages students with animation- and simulation-based contexts that provide multiple rich occasions to press for building and developing reasoning and explanations.

Because students build on their existing knowledge when learning new concepts, they are provided multiple opportunities to engage with new content so that they might revise their understandings about the concepts of *inheritance of traits* and *variation of traits* (Bransford, Brown, Cocking, & NRC, 1999). In practice, this takes the form of a succession of iterative cycles through which students are asked to respond to a question that is tied to one component of the design patterns. Subsequently, students engage in further learning that may inform additional opportunities to engage with the design pattern. Therefore, in the STEMGenetics unit, POE is not contained within one activity; rather, it drives the unit design as a whole. In the context of this unit, students observe the results of crossing Wisconsin Fast Plants® (WFP) of different stem colors, make a prediction about the stem color of offspring of the next generation, grow their own crossed (WFP), and then revise their original predictions.

As called for in the U.S. Framework for K-12 Science Education (National Research Council [NRC], 2012), the STEMGenetics Grade 5 unit explicitly integrates core ideas, practices, and crosscutting concepts within the genetics curricula. Table 1 provides a break-down of the primary and secondary goals adapted from the U.S. NRC Framework (2012) that the unit addresses. The unit also incorporates helping students learn the scientific practices of constructing scientific explanations and analyzing and interpreting data. It emphasizes the importance of students constructing explanations to describe patterns of trait inheritance in offspring as well as the use of simulations for exploring and testing predictions about the relationships between parental and offspring traits. The cross-cutting concept the unit addresses is the patterns through which students are investigating inheritance patterns.

Table 1. Unit learning goals

Primary Learning Goals	LS3.A: Characteristics of organisms are inherited from their parents. Characteristics can be modified by environmental influences. LS3.B: Offspring acquire information for their traits from their biological parents. Different organisms vary in how they look and function because they have different inherited information. LS3.B: The environment also affects the traits that an organism develops—differences in where they grow or in the food they consume may cause organisms that are related to end up looking or behaving differently.
Secondary Learning Goals	LS1.A: Plants and animals have both internal and external structures that serve various functions in growth, survival, behavior, and reproduction. (Boundary: Stress at this grade level is on understanding the macroscale systems and their function, not microscopic processes.) LS1.B: Reproduction is essential to the continued existence of every kind of organism. Plants and animals have unique and diverse life cycles that include being born (sprouting in plants), growing, developing into adults, reproducing, and eventually dying.

Dynamic Visualizations

Interactive dynamic visualizations were used to exemplify the concepts that students were learning in their classroom lessons, and to help make scientific phenomena more accessible to them. Dynamic visualizations are an important feature in the STEMGenetics WISE unit, because they fall under the *Explore a Simulation* stage in the unit design. They can help advance fifth graders' understanding of *inheritance of traits* and *variation of traits,* and they also have an explanatory value and function relative to the learning goals outlined in Table 1.

One example of a dynamic visualization was the pollination animation, which shows an example of a pollinator, a bee, going from flower to flower (see Figure 1). At first, the teacher facilitates the animation for the students by walking through the different steps of the animation. This interactive animation allows the students to add their ideas about pollination and plant reproduction after they engage with the different parts of the animation. As the bee continues its travels, it deposits a bit of pollen, gathered from a previous flower, onto another. This helps students see pollination in action and stimulates discussion of other means of pollination—flower-to-flower contact, wind, animals, etc.

In Figure 1(a), the students read the text that tells them that the pollen contains genetic information that is passed down to the next generation. The zoom button next to the image of the flower allows them to zoom further into the flower and to see where the pollen grains are located on the flower.

Figure 1. Pollination animation in WISE

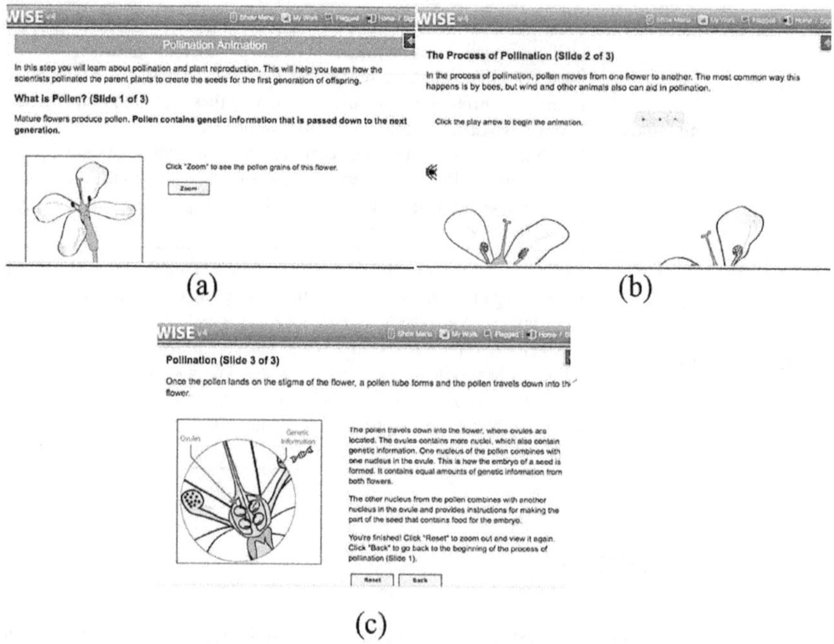

In Figure 1(b), students are presented with the initial process of the pollination animation. The play button of the animation shows the students how the bee, a pollinator, carries genetic information from one flower to another. The animation is placed strategically after the text that explains to them about the bee being a pollinator. This way, students are more likely to read the text before playing the button, and to have added some knowledge about the process of pollination to their existing knowledge about plants' life cycle and inheritance of traits before playing the animation.

In Figure 1(c), there are two buttons that the students are interacting with: the reset button and the back button. The reset button takes them back to slide 3, by which they are introduced to the process of fertilization. In slide 3, students get to see part of the process of fertilization in which the genetic information from the pollen travels down into the flower to where the other genetic information in the ovules is located. The back button, on the other hand, is a button that gives students the option to go back to the pollination animation and to revisit the beginning of the pollination process of the bee pollinating the flower.

In sum, the interactive pollination animation walks the students through the processes of pollination and fertilization and helps them make sense of the different steps that are involved with these processes, particularly the process of pollination. The benefits of this level of interactivity is that students are in control of their learning as they press the different buttons to view how the process of pollination starts, from the bee as a pollinator carrying the pollen from flower to flower. The students also interact with buttons that allow them to zoom in and see where the pollen is located, how the pollen travels down the pollen tube into the flower, and how the other parent's genetic information is located in the ovule.

Lesson Story: Predicting a New Generation of Fast Plants®

This unit is a 4-week unit that engages students in understanding inheritance by answering the driving question, *Why do plants of the same species vary in how they look*? The lesson story within the Grade 5 STEMGenetics unit revolves around a "family" of purple- and green-stemmed WFP. Students are shown a Parent Generation of two plants, one purple-stemmed and one green-stemmed, and they meet two scientists who have discovered that the offspring plants (the First Generation) resulting from the cross of these parents all had purple stems. Students then predict the stem color of the Second Generation of offspring, which they will grow in class. At the end of the unit, students identify patterns of variation based on data from their WFP experiments within and across classes. Figure 2 shows the Scientists' Fast Plants poster that the teacher used in class as a visual to refer back constantly to the driving question of the unit of *Why do plants of the same species vary in how they look*?

The unit also offers ways for students to build connections between what they were witnessing in their WFP and in their classroom activities. For example, students conduct a trait survey among their peers to see which traits (e.g. brown hair color or blue eye color or having a straight hairline or a widow's peak) are more common than others in their class. They also construct charts and graphs to make presentations to their classmates about their findings.

METHODS

Participants

This study was implemented in a suburban district in the Midwestern U.S., with one Grade 5 teacher and 59 students. The district agreed to use the STEMGenetics curriculum as a means of teaching the core science topic of inheritance, and of integrating technology into its instructional practice. The teacher had experience

Figure 2. Scientists' fast plants experiment

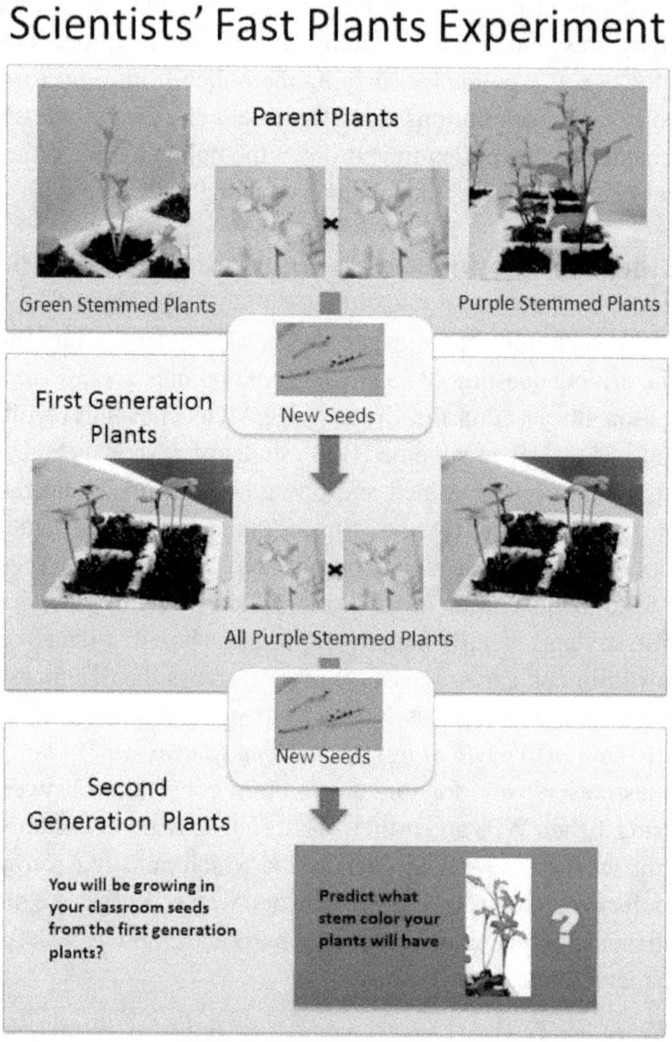

using WISE, and this was his second year teaching the unit. He taught the unit three times a week over a period of four weeks. When teaching the unit, he paired students into groups and made sure to model the different steps, such as the pollination animation step, before students interacted with the different online activities. The school district student population included: 60% Caucasian, 18% African American, 9% Asian, 7% Hispanic, 5% Multi-racial, and 1% American Indian and Pacific Islander. Twenty-five percent of the students were on free or reduced price lunch. There were approximately equal numbers of male and female students in the school.

Data Sources and Scoring

A test was administered as a pre-post assessment, to measure student learning on heredity prior to and after the 4-week STEMGenetics unit. The assessment was a 45-minute test of both multiple-choice (MC) and constructed-response (CR) questions, covering all the key learning goals of the unit. Table 2 lists the key learning goals by item type for the fifth-grade assessment. The total possible score for the 17 items was 33 points.

We used an adapted Knowledge Integration (KI) rubric from Linn, Lee, Tinker, Husic and Chiu (2006) to score pre- and post-test assessment items and to look at how students linked relevant or irrelevant ideas about *inheritance of traits* and *variation of traits* and used them to construct explanations. By drawing from the KI lens, we categorized the ideas that students articulated and the connections they drew between those ideas as constituting either an irrelevant, isolated, partial, or complete explanation. If students had irrelevant ideas in a given context, then their explanation would fall under an irrelevant explanation and would be scored as a 0. If students had a relevant idea but failed to recognize links with other relevant ideas, then their explanation would fall under an isolated explanation and would be scored as a 1. If students had more than one relevant idea but did not fully elaborate links between these ideas in a given context, then their explanation would fall under a partial explanation and would be scored as a 2. Lastly, if students elaborated a scientifically valid link between two ideas relevant to a given context, then their explanation would fall under a complete explanation, and would be scored as a 3. Table 3 shows Question 4 as an example, and the adapted KI rubric used to score Question 4. An example of a typical student response is also included. For question 4, the upper elementary students' progress level at this developmental stage based on the (NGSS, 2013) is a 2, or a Partial Understanding. By the end of this grade band (3-5), students understand that characteristics of organisms are inherited from their parents and that "reproduction is essential to the continued existence of every kind of organism" (NRC, 2012, p.146).

Table 2. Test specification for fifth-grade assessment

Learning Goals (Construct)	MC	MC+CR	CR	Total
Inheritance of Traits	5	2	2	9
Variation of Traits	1	-	5	6
Life Cycle	1	-	1	2
Total	7	2	8	17

Note. MC = Multiple Choice item and CR = Constructed Response item. Total refers to the number of items.

Table 3. Question 4 and the rubric

4. Some wild flowers need bees for pollination. One summer, there were no bees. What does this mean in terms of wildflowers' life cycle?			
Score	**KI Explanation**	**Description**	**Example of Student Work**
M	Blank/Invalid	Blank – no writing at all Invalid – symbol or IDK	I: ?, IDK
0	Irrelevant Students have irrelevant ideas in a given context.	Off task – write something that is not relevant to the item OR student repeats the question OR Student response indicates that the already existing wildflowers will die/stop growing or that pollination is needed for the wildflower to live/grow OR Student states that there will be no pollination but does not connect this to any larger outcomes [This is essentially reiterating the idea expressed in the first two sentences without connecting to the larger life cycle] OR Student expresses the non-normative idea that bees need the pollen to survive	Example: The flowers will not get pollination because if we do not have pollination we will not get food
1	Isolated Have relevant ideas but fail to recognize links between them. Make links between relevant and irrelevant ideas.	Student correctly responds that the flowers will not create seeds/offspring/new flowers, but does not explain why OR Student correctly identifies that the species will die out and/or that the life cycle will stop (e.g. eventually this particular wildflower species will die) OR Student explains that other pollinators can fill the role left by bees, so pollination will still happen, BUT does not connect this to the larger life cycle.	Example: That means no baby flowers were made
2	Partial Have relevant ideas but do not fully elaborate links between them in a given context.	Explains that the wild flower will not produce new seeds/reproduce/have another generation because insects/bees play a critical role in the pollination of plants or because pollination is necessary to create new seeds (but does not necessarily link this to the idea of the species dying out in the long term) OR Explains that the plant species will die out and/or that the life cycle will stop because insects/bees play a critical role in the pollination of plants (but does not necessarily link this to the short-term idea that new seeds can't be created)	Example: All the flowers would die out because flowers need pollen to fertilize the stigma which goes down to the ovary which fertilizes the eggs to make them seeds
3	Complete Students understand how two scientific concepts interact in a given context. Elaborate a scientifically valid link between two ideas relevant to a given context.	Student explains that the plant species will die out and/or that the life cycle will stop because insects/bees play a critical role in the pollination of plants and links it to the idea that new seeds can't be created. OR Explains that other pollinators can fill the role left by the bees AND connects this idea to the larger life cycle by identifying that seeds/offspring/new plants will still be made.	Example: All the flowers would die out because flowers need pollen to fertilize the stigma which goes down to the ovary which fertilizes the eggs to make them seeds.

There were 59 students selected for coding who had valid responses in both their pre- and post-test responses. Two raters with expertise in science content and teaching, scored each response to open-ended questions. Each rater was randomly assigned to the responses, and blind coding was achieved so that raters were not aware of whether the responses were from a pre- or post-assessment. The raters were trained on using the scoring guide to score, and agreement on the scores for the students' responses were ensured. Disagreements were resolved by discussions.

RESULTS

Paired sample t-tests were conducted to analyze students' performances on the pre- and post-assessment for this class. Students made significant improvement from the pre-test to the post-test. The results show that students achieved higher scores on the post-assessment overall, with a mean score of 13.93 on the pretest and a mean score of 18.17 on the posttest. Although the results were significant and the improvement is a start, the students were nowhere near where they should be. The results are presented in Table 4. Students achieved significantly higher overall scores on the post-assessment compared to the pre-assessment.

During observations made from the class and at the beginning of the Grade 5 STEMGenetics unit, the researchers noted that students' definitions of pollen revolved around "how it is something that makes you sneeze," "gives you allergies," or "is food for the bees." Before the unit, the fifth-grade teacher noticed that his students started with little knowledge about genetic inheritance, and they lacked understanding of genetic contribution. He remarked, "They actually know more about the allergy medicines advertised to alleviate pollen-caused symptoms!" (Fifth grade teacher, 2014). According to the fifth-grade teacher, his students lacked an understanding of how many organisms' traits, including those of plants, are determined by genetic information from two parents.

Pollination Animation and the Driving Question of the Unit

The pollination animation helped students make connections back to the driving question of the unit: *How do plants of the same species vary in how they look?* Students learned through this animation (see Figure 1) that through the processes of pollination and fertilization, seeds are formed and contain equal amounts of genetic information from two parent plants. The different inherited information from each parent explains the variation in traits. This step of the pollination animation in WISE

Table 4. T-test of G5 assessment

Item	Max Score	Mean Pre-Test	Mean Post-Test	Mean Difference	SD	df	t-Value	p-Value
1	1	0.90	0.93	.034	.263	57	1.000	.322
2	1	0.74	0.79	.052	.510	57	.772	.443
3	2	1.09	1.42	.321	.850	52	2.748	.008
4	2	0.24	0.70	.463	.818	53	4.161	.000*
5	3	1.30	1.59	.286	.825	55	2.592	.012
6	2	1.54	1.40	-.140	.549	56	-1.930	.059
7	3	0.48	1.27	.795	.851	43	6.198	.000*
8	1	0.69	0.84	.155	.586	57	2.015	.049
9	3	0.23	0.64	.404	.538	46	5.150	.000*
10	3	0.75	0.90	.157	.703	50	1.592	.118
11	2	1.02	0.98	-.035	.421	56	-.629	.532
12	2	1.60	1.74	.132	.833	52	1.154	.254
13	2	0.80	0.96	.157	.505	50	2.219	.031
14	2	1.30	1.35	.056	.763	53	.535	.595
15	2	0.76	1.20	.431	.781	50	3.944	.000*
16	1	0.63	0.77	.143	.520	55	2.057	.044
17	1	0.74	0.77	.035	.421	56	.629	.532
Total Score	33	13.93	18.17	4.241	3.321	57	9.727	.000*

Note. The Pre-test was administered before the unit. The Post-test was administered after the unit. SD= Standard Deviation. df= degrees of freedom. The number of test items was 17. The total possible score was 33. The maximum score for an item was a 3.

$*p < .001$

also brought the students back to the WFP Parents' Generation. The scientists had initially cross-pollinated two parents: a Fast Plant parent with a purple stem color and a Fast Plant parent with a green stem color, resulting in the First Generation being all purple-stem colored.

With their teacher, the students then discussed how they had initially made a prediction about the stem color of their Fast Plant, the Second Generation. According to the fifth-grade teacher, most of his students' initial ideas entailed "purple begets purple." A few of his students chose other alternatives such as: "all white because it's a trick," for example, or "half-and-half because it's only fair".

Pollination Animation and Growing a Second Generation of Fast Plants

By the time the students got to the pollination animation in the unit, their Second-Generation WFP had sprouted, and they were able to see the stem color of their plants. This is where the grand realization occurred, since these were the seeds from the First Generation (purple/purple) cross-pollination. When both purple and green stems reappeared in their Second-Generation plants, the researchers noted how the students were excited to share theories about how this happened. This was the real "seedbed" for discussion in the STEMGenetics unit. The fifth-grade teacher referred his students back to the pollination animation and how the pollen is the carrier of genetic information from one Parent Plant and that it combines with the genetic information of another Parent Plant. Connecting the reappearance of the green stem color in the second generation of offspring that they planted to the pollination animation helped the students make sense of the ways that two parents contribute genetically to an offspring.

In one of the steps in WISE, students are able to display an aggregate view of the distribution of their class's WFP stem color through a pie chart. What the pie chart showed was a greater distribution for the purple stem color trait, indicating that the purple stem color trait was more common than the green stem color trait. What the students came to understand by the end of the STEMGenetics unit was that their WFP, which are the Second-Generation Plants, expressed both traits, but that the purple stem color trait was more common than the green stem color. So, the seeds from the purple-stemmed plants, or First Generation plants, that the students as a class planted had inherited genetic information for the green stem trait from the Parent Plants the scientists had cross-pollinated. What the pollination animation further helped the students see then is that genetic information is passed down during reproduction and that the offspring have half of their genetic information for each trait from each parent.

Question 4 Results

For this chapter, the focus is on the findings from Question 4, an Open Response (OR) question that asked students to explain what would happen to the wildflowers' cycle if there were no bees one summer (see Table 3). This question was chosen because it aligns with the pollination animation and looks at how a technological tool such as the dynamic visualizations of the pollination animation can scaffold students in understanding the core ideas of *inheritance of traits* and *variation of traits* around the importance of the processes of pollination and fertilization, specifically pollination. The question looks at whether students understand the role of pollinators, in this case the bees, in the pollination process of the life cycle of the plants. Within this grade band for the NGSS (2013), students are expected to demonstrate understanding in relation to these core ideas that "plants and animals have traits inherited from parents and that variation of these traits exist in a group of similar organisms" (3-LS3-1). In addition, students are expected to understand the importance of reproduction in continuing the existence of the plant's next generation. What question 4 further looks at is whether students' reasoning includes that the bee is a pollinator that carries pollen, which contains the genetic information from one of the parent plants, from one flower to another, and the bee's essential role in reproduction, specifically in the process of pollination.

In essence, what Question 4 delved into was the concept of contribution of genetic information coming from each parent, but more importantly, the importance of the process of reproduction in carrying out this passage of genetic information. In the pollination animation, students learned about pollination and fertilization in plants as important processes in the life cycle of a plant. They were also the ones advancing the different segments and frames in the animation (see Figure 1). The animation not only explained these two processes but also emphasized that the pollen and the ovule contain genetic information about traits, thus addressing a scientifically wrong idea that students tend to have about traits being inherited from only one parent (Clough & Wood-Robinson, 1985; Kargbo et al., 1980). The animation also demonstrated that a new seed begins with the combination of genetic information from both flowers, and how the genetic information in the pollen was carried by the bee. During the lesson on the pollination animation, the fifth-grade teacher had the students watch the interactive animation in their respective groups that he assigned. After observing the pollination animation and reading through the different segments, the teacher reconvened the class into a whole classroom discussion, where they actively participated in making sense of the information presented to them about the processes of pollination and fertilization by answering the teacher's prompts of "where did the pollen come from" or "why is pollen important?" This active learning and sharing of ideas helped the students reflect on their own learning as they shared their ideas and co-constructed their knowledge.

Table 5 provides the T-test results for Question 4, while Table 6 and Table 7 provide qualitative examples of student responses before the unit and after the unit. Students' improvement on Question 4 from the pre-test to the post-test is statistically significant. The results show that students achieved higher scores on the post-assessment overall, where the mean test scores went from 0.24 to 0.70.

Table 5. T-test results for question 4

Item	Possible Max Score	Mean_ Post-Test	Mean_ Pre-Test	Mean_ Difference	SD	df	t-Value	p-Value
4	2	0.70	0.24	.463	.818	53	4.161	.000*

Note. The Pre-test was administered before the unit. The Post-test was administered after the unit. SD = Standard Deviation. df= degrees of freedom.

* $p < .001$

Table 6. Student examples of question 4: "Some wildflowers need bees for pollination. One summer, there were no bees. What does this mean in terms of wildflowers' life cycle?" KI Scores 0 -1

Student	Pre-Test Response	Post-Test Response
Maria	They all die. (Irrelevant Explanation) - KI Score 0	There would not be another generation of them. (Isolated Explanation) - KI Score 1
Kate	The life cycle will be thrown out of balance and the flower might not thrive as much as it could. (Irrelevant Explanation)- KI Score 0	There probably will not be as much reproduction if there were no bees. (Isolated Explanation) - KI Score 1

Note. The Pre-test responses are before the unit was administered. The Post-test responses are after the unit was administered.

KI Score of 0 = Irrelevant ideas in the student's response; KI Score of 1= One relevant idea in the student's response.

KI Scores 0-1

Most of the students achieved a KI score of a level 1 on the post-test, which shows an improvement in their understanding around the ideas of pollination and its role in the plant life cycle. When we examined the students' paired responses in detail for Question 4, many students demonstrated a deeper understanding of the genetics concepts of *inheritance of traits* on the post-instruction performance, compared to their prior-instruction performance. Students' responses in Table 6 shown above from the pre-assessment are irrelevant explanations. Many of their responses centered on relating the absence of bees to the growth and survival of the wildflowers. For example, Maria, referring to the wildflowers, stated that "[t]hey all die." Her explanation had no relevant ideas to what this means in terms of the life cycle of the wildflowers or how transfer of genetic information to the next generation may be impaired by the absence of the bees. However, on the posttest, Maria developed

Table 7. Student examples of question 4: "Some wildflowers need bees for pollination. One summer, there were no bees. What does this mean in terms of wildflowers' life cycle?" KI Scores 1-2

Student	Pre-Test Response	Post-Test Response
Adam	Because there were no bees one year the wild flower life cycle stopped that year and all the wildflowers died. (Isolated Explanation) KI Score 1	With no bees the pollen and the ovules can't mix so there would be no healthy seeds to start the next generation. The wildflowers would start to die with no seeds to start the next generation. There would be no wild flowers the next year or the year after that. (Partial Explanation) KI Score 2
Betty	Less pollen will be spread so since flowers need pollen to survive and there are no bees to carry it around the flowers will become sick. (Isolated Explanation) KI Score 1	There will be less flowers, because the bees are the cross pollinators they carry pollen from one flower to another. After that the pollen would go down the pollen tubes then the seeds would get the genetics in the pollen, but if the bees were gone almost none of this would happen. (Partial Explanation) KI Score 2

Note. The Pre-test responses are before the unit was administered. The Post-test responses are after the unit was administered.

KI Score of 1= One relevant idea in the student's response; KI Score of 2= Two relevant ideas in the student's response.

a much deeper understanding; her response showed progress in her understanding that the bees' absence may affect the next generation of wildflowers, and she stated that "that there would not be another generation of them." Similarly, Kate started out with irrelevant ideas in her pretest response. Her response included ideas around the life cycle being "thrown out of balance" and the flower not being able "to thrive." After the post-test, Kate rightfully mentioned that the "reproduction" of the wildflowers may be hindered by the absence of bees, and she showed how her reasoning deepened as she built an integrated knowledge and added a relevant idea about the impact of the absence of bees in relation to the wildflowers' reproduction.

KI Scores 1-2

Table 7 shows two examples of the few students who did reach a KI score of a 2. For example, Adam explained that the "life cycle will stop," which is one relevant idea, but he did not elaborate or connect it to what it means in terms of the passage of generational information. On the other hand, in the posttest, Adam correctly identified that “with no bees the pollen and ovules can’t mix,” and he connected it to fertilization and seed formation. Adam also explained how with no seed formation there would not be a next generation. In his post-test answers, he was able to make scientific connections and references to the “ovule,” which is part of what students see in the pollination animation. Betty, in the pretest, correctly identified one relevant idea that with no bees, there is a likelihood of “less pollen being spread.” Conversely, in the post-test, Betty identified the bees as “cross-pollinators” and their role of carrying pollen from one flower to another. Not only did Betty identify the bees as cross-pollinators, she also showed progress in her level of detail and conceptual understanding by elaborating on how the genetic information would travel down the pollen tubes and how “the seeds would get the genetics in the pollen.” Betty clearly stated in her answer that the genetic information from one flower is contained in the pollen, and that the seeds receive genetic information from the pollen.

As the KI scores indicate in Tables 6 and 7, the quality of students’ reasoning improved. By the end of the unit, students’ explanations had more relevant ideas around the scientific phenomena of the wildflower’s life cycle and its relationship to the plant’s inheritance of traits. The students also showed an improvement in their understanding of the importance of the role of the bee in the reproduction process, as well as its role, as a pollinator, in carrying the genetic information contained in the pollen from one plant to another.

DISCUSSION

These findings support prior studies that students have difficulties in understanding abstract concepts, including genetic inheritance (Lewis & Wood-Robinson, 2000; Slack & Stewart, 1990), and that using educational-technology-learning tools can help students build a better understanding of these abstract concepts around genetics (Kara &Yesilyurt, 2008). What this study further suggests is that interacting with technology and online visualizations provide an opportunity for students to make a deeper connection between the underlying genetic processes of pollination and fertilization and their connection to the inheritance of traits in plants. In addition, the visualizations, specifically the pollination animation, may have scaffolded students in seeing how genetic information is being transmitted from each parent, and that the pollen contains the genetic information of one of the parent plants.

Interacting with a technology-enhanced inquiry-based unit provided students with a more nuanced understanding of the concepts of *inheritance of traits* and *variation of traits*. The unit helped students view plants as sexually reproducing (i.e. having two parents that contribute to traits) as opposed to inheriting a trait from only one parent (Wood-Robinson, 1994). The increased knowledge integration scores that most students earned by the end of the unit in Question 4 showed a better understanding of the relationship between the life cycle of plants and the inheritance of traits. In addition, students developed a much deeper understanding of the role of pollen in relation to the inheritance of traits. Their understanding went beyond their initial ideas of pollen being "food for bees" or "a cause for allergies." Instead, students' reasoning about the inheritance of traits included the important role of pollen in carrying the genetic information from one of the parent plants.

The pollination animation can help students understand the importance of the pollinators such as bees in carrying the genetic information contained in the pollen (see Figure 1). After interacting with the pollination animation, the students were able to construct a deeper understanding of the importance of the role of pollinators, such as bees, in transporting one of the parent's genetic information from flower to flower. Most student post-unit responses showed relevant ideas about pollinators being important in the life cycle of plants. Furthermore, most of these revolved around how pollinators were essential for the next generation, for reproduction, or for carrying pollen from flower to flower. In addition, the students' reasoning about plant genetics by the end of the unit included an understanding that all organisms, including plants, have genetic information (Duncan et al., 2009) and that genetic information determines which traits are passed down from parents to offspring, and thus explains the reappearance of both green and purple stem colored plants in their Second-Generation Fast Plants.

Teachers who use technology-based learning tools, such as dynamic visualizations and animations, can better scaffold their students in developing an integrated knowledge of STEM content around the core ideas of *inheritance of traits* and *variation of traits*. Interacting with the pollination animation helped students add normative ideas about the processes of pollination and fertilization while building an integrated understanding of genetics in plants. Therefore, interacting with these learning technologies and dynamic visualizations allowed students to visualize and further understand these complex concepts.

FUTURE RESEARCH DIRECTIONS

Having students interact with dynamic visualizations and animations, while in groups, can support them in developing a deeper understanding of scientific phenomena by making abstract and difficult concepts that are hard to grasp more visible. In this study, the teacher provided guidance to the dynamic visualizations and pollination animation, and modeled the different steps to the students before having the student groups interact with the pollination animation. This teacher guidance helped students to construct meaning from these dynamic visualization interactions.

Although these technology-enhanced units are growing in popularity, the various ways they may be implemented may play an important role in determining whether students achieve learning gains. The authors realize that how teachers guide and facilitate conversations with their students is crucial in helping them build an integrated knowledge of scientific ideas while engaging with technology.

While this study illustrates that a technology-enhanced unit can help deepen the understanding of upper-elementary students' understanding around the concepts of *inheritance of traits* and *variation of traits*, two questions drive the next round of inquiry:

- How can teachers further support their students as they engage with a technology-enhanced unit?
- How can teachers help their students revisit their initial ideas as they engage with the dynamic visualizations?

The authors recognize that these technology-enhanced units help students build normative ideas about the scientific phenomena they are learning. However, it is important to consider how teachers may view them as also possibly increasing their instructional load (Peters & Slotta, 2010) and challenging them to implement at times (Slotta & Linn, 2009). Providing teachers with critical prompts and information that help them guide their students through the different dynamic visualizations and make sense of these technology-related interactions are ways of supporting teachers in orchestrating through these units. Lastly, providing professional development to teachers for these technology-enhanced units can guide them with ways where they see themselves as the facilitators in these technology-enhanced classroom activities and may ensure that students are in charge of their learning.

CONCLUSION

Overall, students benefitted from the Grade 5 STEMGenetics unit and from engaging with the WISE animations in a number of ways. They learned about the transmission of genetic material in plants, and that the pollen contains the genetic information of one of the parent plants. Moreover, they came to understand that offspring were the sum of the genetic information transmitted from both parents and that the combination of that genetic material and its manifestation were unique and different in every instance. Providing upper elementary school students with opportunities to learn about *inheritance of traits* and *variation of traits*, with the help of visualizations and animations, can serve as a foundation for more complex genetics concepts such as multiple allele inheritance, topics included in high school biology courses.

ACKNOWLEDGMENT

This material is based upon work supported by the National Science Foundation. Any opinions, findings, and conclusions or recommendations expressed in this material are those of the authors and do not necessarily reflect the views of the National Science Foundation.

REFERENCES

Ayuso, E., & Banet, E. (2002). Alternativas a la enseñanza de la Genética en educación secundaria. *Ensenañza de las Ciencias*, *20*(1), 133–157.

Banet, E., & Ayuso, E. (2000). Teaching genetics at secondary school: A strategy for teaching about the location of inheritance information. *Science Education*, *84*(3), 313–351. doi:10.1002/(SICI)1098-237X(200005)84:3<313::AID-SCE2>3.0.CO;2-N

Bransford, J., Brown, A. L., Cocking, R. R., & National Research Council (U.S.). (1999). *How people learn: Brain, mind, experience, and school*. Washington, DC: National Academy Press.

Clough, E. E., & Wood-Robinson, C. (1985). Children's understanding of inheritance. *Journal of Biological Education*, *19*(4), 305–310.

Cook-Deegan, R., & Heaney, C. (2010). Patents in genomics and human genetics. *Annual Review of Genomics and Human Genetics*, *22*(1), 383–425. doi:10.1146/annurev-genom-082509-141811 PMID:20590431

Costa-Font, M., Gil, J. M., & Traill, W. B. (2008). Consumer acceptance, valuation of and attitudes towards genetically modified food: Review and implications for food policy. *Food Policy*, *33*(2), 99–111. doi:10.1016/j.foodpol.2007.07.002

Duncan, R. G., & Reiser, B. J. (2007). Reasoning across ontologically distinct levels: Students understandings of molecular genetics. *Journal of Research in Science Teaching*, *44*(7), 938–959. doi:10.1002/tea.20186

Duncan, R. G., Rogat, A. D., & Yarden, A. (2009). A learning progression for deepening students understandings of modern genetics across the 5th–10th grades. *Journal of Research in Science Teaching*, *46*(6), 655–674. doi:10.1002/tea.20312

Edelson, D. C., Gordon, D. N., & Pea, R. D. (1999). Addressing the challenge of inquiry-based learning. *Journal of the Learning Sciences*, *8*, 392–450. doi:10.1080/10508406.1999.9672075

Engel Clough, E., & Wood-Robinson, C. (1985). Childrens understanding of inheritance. *Journal of Biological Education*, *19*(4), 304–310. doi:10.1080/00219266.1985.9654757

Fishman, B., Best, S., Marx, R., & Tal, R. (2001). *Design research on professional development in a systemic reform context.* Paper presented at AERA 2001, Seattle, WA.

Gerard, L. F., Spitulnik, M., & Linn, M. C. (2010). Teacher use of evidence to customize inquiry science instruction. *Journal of Research in Science Teaching*, *47*(9), 1037–1063. doi:10.1002/tea.20367

Haga, S. B. (2006). Teaching resources for genetics. *Nature Reviews. Genetics*, *7*(3), 223–229. doi:10.1038/nrg1803 PMID:16462854

Hickey, D. T., Kindfield, A. C. H., Horwitz, P., & Christie, M. A. T. (2003). Integrating curriculum, instruction, assessment, and evaluation in a technology-supported genetics learning environment. *American Educational Research Journal*, *40*(2), 495–538. doi:10.3102/00028312040002495

Kali, Y., Linn, M. C., & Roseman, J. E. (2008). *Designing coherent science education: Implications for curriculum, instruction, and policy*. New York: Teachers College Press.

Kara, Y., & Yesilyurt, S. (2008). Comparing the impacts of tutorial and edutainment software programs on students achievements, misconceptions, and attitudes towards biology. *Journal of Science Education and Technology*, *17*(1), 32–41. doi:10.1007/s10956-007-9077-z

Kargbo, D. B., Hobbs, D., & Erickson, G. L. (1980). Childrens beliefs about inherited characteristics. *Journal of Biological Education*, *14*(2), 137–146. doi:10.1080/00219266.1980.10668980

Kesselheim, A. S., Cook-Deegan, R. M., Winickoff, D. E., & Mello, M. M. (2013). Gene patenting – The Supreme Court finally speaks. *The New England Journal of Medicine*, *369*(9), 869–875. doi:10.1056/NEJMhle1308199 PMID:23841703

Krajcik, J. S., Blumenfeld, P. C., Marx, R. W., Bass, K. M., Fredricks, J., & Soloway, E. (1998). Inquiry in project- based science classrooms: Initial attempts by middle school students. *Journal of the Learning Sciences*, *7*(3-4), 313–350. doi:10.1080/10508406.1998.9672057

Lee, H. S., Linn, M. C., Varma, K., & Liu, O. L. (2010). How do technology-enhanced inquiry science units impact classroom learning? *Journal of Research in Science Teaching*, *47*(1), 71–90. doi:10.1002/tea.20304

Lewis, J., & Kattmann, U. (2004). Traits, genes, particles and information: Re-visiting students understandings of genetics. *International Journal of Science Education*, *26*(2), 195–206. doi:10.1080/0950069032000072782

Lewis, J., & Wood-Robinson, C. (2000). Genes, chromosomes, cell division, and inheritance do students see any relationship? *International Journal of Science Education*, *22*(2), 177–195. doi:10.1080/095006900289949

Lewis, J., & Wood-Robinson, C. (2000). Genes, chromosomes, cell division, and inheritance - do students see any relationship? *International Journal of Science Education*, *22*(2), 177–195. doi:10.1080/095006900289949

Linn, M. c., Davis, E. A., & Bell, P. (Eds.). (2004). *Internet environments for science education*. Mahwah, NJ: Lawrence Erlbaum Associates.

Linn, M. C., & Eylon, B.-S. (2006). Science education: Integrating views of learning and instruction. In P. A. Alexander & P. H. Winne (Eds.), *Handbook of Educational Psychology* (2nd ed.; pp. 511–544). Mahwah, NJ: Lawrence Erlbaum Associates.

Linn, M. C., & Eylon, B. S. (2011). *Science learning and instruction: Taking advantage of technology to promote knowledge integration*. New York: Routledge.

Linn, M. C., Eylon, B. S., & Davis, E. A. (2004). The knowledge integration perspective on learning. In M. C. Linn, E. A. Davis, & P. Bell (Eds.), *Internet environments for science education* (pp. 73–83). Mahwah, NJ: Erlbaum.

Linn, M. C., & Hsi, S. (2000). *Computers, teachers, peers: Science learning partners*. Mahwah, NJ: Lawrence Erlbaum Associates, Inc.

Linn, M. C., Lee, H. S., Tinker, R., Husic, F., & Chiu, J. L. (2006). Teaching and assessing knowledge integration in science. *Science*, *313*(5790), 1049–1050. doi:10.1126/science.1131408 PMID:16931740

Marx, R., Blumenfeld, P., Krajcik, J., & Soloway, E. (1997). Enacting project-based science. *The Elementary School Journal*, *97*(4), 341–358. doi:10.1086/461870

Michigan Department of Education. (2007). *Science grade level content expectations*. Retrieved July 9, 2014, from http://www.michigan.gov/mde

National Research Council. (2012). *A framework for K-12 science education: Practices, crosscutting concepts, and core ideas*. Washington, DC: The National Academies Press.

NGSS Lead States. (2013). *Next generation science standards: ForsStates, by states*. Washington, DC: The National Academies Press.

Okeke, E. A. C., & Wood-Robinson, C. (1980). A study of Nigerian pupils understanding of selected biological concepts. *Journal of Biological Education, 14*(4), 329–338. doi:10.1080/00219266.1980.9654318

Peters, V. L., & Slotta, J. D. (2010). Scaffolding knowledge communities in the classroom: New opportunities in the Web 2.0 era. In M. J. Jacobson & P. Reimann (Eds.), *Designs for learning environments of the future: International perspectives from the learning sciences* (pp. 205–232). Secaucus, NJ: Springer. doi:10.1007/978-0-387-88279-6_8

Reiser, B. J., Tabak, I., Sandoval, W. A., Smith, B. K., Steinmuller, F., & Leone, A. J. (2001). BGuILE: Strategic and conceptual scaffolds for scientific inquiry in biology classrooms. In S. M. Carver & D. Klahr (Eds.), *Cognition and instruction: Twenty-five years of progress* (pp. 263–305). Mahwah, NJ: Lawrence Erlbaum.

Roschelle, J., Pea, R., Hoadley, C., Gordin, D., & Means, B. (2000). Changing how and what children learn in school with computer-based technologies. *The Future of Children, 10*(2), 76–101. doi:10.2307/1602690 PMID:11255710

Roseman, J. E., Linn, M. C., & Koppal, M. (2008). Characterizing curriculum coherence. In Y. Kali, M. C. Linn, & J. E. Roseman (Eds.), *Designing coherent science education: Implications for curriculum, instruction, and policy* (pp. 13–36). New York: Teachers College Press.

Shear, L., Bell, P., & Linn, M. C. (2004). Partnership models: The case of the deformed frogs. In M. C. Linn, E. A. Davis, & P. Bell (Eds.), *Internet environments for science education* (pp. 289–311). Mahwah, NJ: Erlbaum.

Slack, S. J., & Stewart, J. (1990). High school students problem-solving performance on realistic genetics problems. *Journal of Research in Science Teaching, 27*(1), 55–67. doi:10.1002/tea.3660270106

Slotta, J. D. (2004). The Web-based inquiry Science environment (WISE): Scaffolding knowledge integration in the science classroom. In M. C. Linn, P. Bell, & E. Davis (Eds.), *Internet environments for science education* (pp. 203–232). Mahwah, NJ: Lawrence Erlbaum & Associates.

Slotta, J. D., & Linn, M. C. (2009). *WISE science: Web-based inquiry in the classroom*. New York: Teachers College Press.

Songer, N. B. (2007). Digital resources versus cognitive tools: A discussion of learning science with technology. In S. Abell & N. Lederman (Eds.), *Handbook of research on science education* (pp. 471–491). Mahwah, NJ: Erlbaum.

Songer, N. B., Lee, H., & Kam, R. (2002). Technology-Rich Inquiry Science in Urban Classrooms: What are the barriers to inquiry pedagogy? *Journal of Research in Science Teaching*, *39*(2), 128–150. doi:10.1002/tea.10013

Stark, L. A., & Pompei, K. (2010). Making genetics easy to understand. *Science*, *327*(5965), 538–539. doi:10.1126/science.1183029 PMID:20110495

Terwogt, M. M., Stegge, H., & Rieffe, C. (2003). Childrens understanding of inherited resemblance: The case of two parents. *International Journal of Behavioral Development*, *27*(4), 366–374. doi:10.1080/01650250344000037

Texas Education Agency. (2010). *Texas essential knowledge and skills*. Retrieved July 9, 2014, from http://tea.state.tx.us

Tsui, C., & Treagust, D. F. (2007). Understanding genetics: Analysis of secondary students conceptual Status. *Journal of Research in Science Teaching*, *44*(2), 205–235. doi:10.1002/tea.20116

Venville, G., Gribble, S. J., & Donovan, J. (2005). An exploration of young childrens understandings of genetics concepts from ontological and epistemological perspectives. *Science Education*, *89*(4), 614–633. doi:10.1002/sce.20061

Williams, M., Linn, M. C., Ammon, P., & Gearhart, M. (2004). Learning to teach inquiry science in a technology-based environment: A case study. *Journal of Science Education and Technology*, *13*(2), 189–206. doi:10.1023/B:JOST.0000031258.17257.48

Wood-Robinson, C. (1994). Young peoples ideas about inheritance and evolution. *Studies in Science Education*, *24*(1), 29–47. doi:10.1080/03057269408560038

KEY TERMS AND DEFINITIONS

Dynamic Visualizations: Online interactive visualizations that serve as learning technology tools representing and modeling a scientific phenomenon.

Knowledge Integration Framework: A framework that views science learning as a process of eliciting, adding, distinguishing, and evaluating ideas to explain scientific phenomena.

Web-Based Inquiry Science Environment (WISE): An online science learning platform that aligns with the Knowledge Integration design framework and features learning technologies and tools such as dynamic visualizations.

Chapter 6
STEM Inquiry Through INPLACE Mobile Games

Denise M. Bressler
Rutgers University, USA

ABSTRACT

To make STEM learning meaningful, students need to feel the relevance and authenticity of the learning activity. Games—particularly mobile games—offer a unique opportunity for students to be immersed in collaborative STEM inquiry. INPLACE mobile games combine the best practices of what the games and learning field knows about using the affordances of mobile devices to engage students, to support collaboration, and to promote authentic practice in a discipline. INPLACE is an acronym that stands for Interdependent, Networked, Participatory Learning, Augmented, Collaborative Experience. School Scene Investigators is a game series designed according to the INPLACE framework; students playing it demonstrated higher engagement and scientific inquiry than students in a control activity. Ultimately, INPLACE provides a design framework that teachers and researchers can use for building mobile games that heightens engagement and increases inquiry-based learning.

INTRODUCTION

In the past century, we have seen a societal paradigm shift that has rendered the traditional educational system obsolete. In the Industrial Age, leadership was autocratic and factory workers were compliant (Reigeluth, 1994). The factory model of education was sufficient at providing students with enough knowledge and skills to live out full lives as factory workers. However, with the onset of the Information

DOI: 10.4018/978-1-5225-2528-8.ch006

Age, the workplace is evolving (Kanter, 2001). Mass collaboration is changing how business is conducted, leadership is shared, competitive principles include openness and sharing, and knowledge workers expect a participatory democracy (Tapscott & Williams, 2006). As discussed by Thomas and Brown (2011), the world is changing—and in accordance, so is knowledge. We need a system that prepares students for working in the Information Age, not the Industrial Age (Reigeluth, 1994).

Cultivating human resources is particularly urgent in the fields of science and technology (National Research Council [NRC], 2007). Low-level skill sets will not solve critical scientific issues such as climate change and the energy crisis; the country's most pressing problems need innovative problem solvers (Gee, 2010). According to the report published by the NRC (2012), "Science is fundamentally a social enterprise, and scientific knowledge advances through collaboration and in the context of a social system with well-developed norms" (p. 27). The ability to collaboratively participate in inquiry-based problem solving is of the utmost importance in scientific careers. Unfortunately, students are not prepared for such collaborative work because school environments place too much emphasis on individual work (Bransford, Brown, & Cocking, 2000) and standardized testing (Ravitch, 2011).

The new digital medium that warrants the most educational potential at this moment is mobile technology, because we are at a tipping point. For years, mobile phones have been a vital part of teenagers' social lives (Katz, 2006); young people live in a mobile world, and their devices connect them to it. Mobile technology has had time to incubate outside the K-12 system; now, it is ready to be a "disruptive innovation" (Christensen, Horn, & Johnson, 2008). More and more schools are inviting students to *bring your own device* (Norris & Soloway, 2011). We need to create and enact curriculum activities that ensure students will garner important skills and benefit from the unique affordances of mobile technology—no more business as usual.

The body of research is growing that demonstrates how mobile technology can promote the types of skills desperately needed by those entering the scientific workforce. Research supports the effectiveness of mobile activities to scaffold collaborative problem solving (Bressler & Bodzin, 2013; Dunleavy, Dede, & Mitchell, 2009; Squire & Jan, 2007). Mobile devices also afford portability, making them effective tools for situated learning (Dede, 2009; Rosenbaum, Klopfer, & Perry, 2007) and science inquiry (Chang, Chen, & Hsu, 2011; Looi et al., 2011). Integrating mobile technologies into the classroom holds promise, but to be successful, it requires rethinking "business as usual" modes of instruction because mobile technologies can support collaborative learning and active, inquiry-based learning.

Game-based learning is one way to capitalize on the affordances of mobile technology while also resolving some of the issues that have led to the educational system's obsolescence. First, game-based learning is engaging, and "schools need to create intrinsically engaging methods for learning" (Christensen et al., 2008, p. 7). Everyone has a drive to learn; it starts in infancy (Gopnick, Meltzoff, & Kuhl, 1999), yet today's schools are unexciting to many students (Gatto, 2009; Willingham, 2010). The K-12 educational system needs to figure out how to help students discover that drive. According to Annetta and Cheng (2008), "educational games seem to be effective in enhancing motivation and increasing student interest in subject matter… [they] generate a much higher level of students' positive emotional engagement" (p. 2). Second, well-designed collaborative mobile games offer a unique way for students to socially construct knowledge. According to Klopfer (2008), students playing collaborative mobile learning games "help each other, observe each other, and act together to create communities as they learn to solve problems" (p. 223). Third, educational games show great potential to facilitate science learning (NRC, 2011; Steinkuehler & Squire, 2014) and researchers are building a body of research toward best practices.

INPLACE mobile games combine the best practices of what the games and learning field knows about using the affordances of mobile devices to engage students, to support collaboration, and to promote authentic practice in a discipline. INPLACE is an acronym that stands for **I**nterdependent, **N**etworked, **P**articipatory **L**earning, **A**ugmented, **C**ollaborative **E**xperience. Throughout the coming pages, this chapter will discuss the origins of the INPLACE design framework for mobile augmented reality (AR) games and then specify each aspect of the framework along with its theoretical foundation. Additionally, this chapter will share results of INPLACE games tested to date and build the case for using INPLACE mobile games as a means for collaborative science, technology, engineering and mathematics (STEM) inquiry in the future.

BACKGROUND

Educational video games are a nascent field of research (Young et al. 2012). Yet, according to the 2011 report *Learning Science through Computer Games and Simulations* published by the NRC, there is potential for games to advance many science learning goals.

Games for STEM Inquiry

Games can be designed to support STEM inquiry, but most games people play are the ones designed for commercial success. Can commercial games support STEM inquiry? *World of Warcraft (WoW)* is a commercial game-based virtual world that has been in existence for over ten years with millions of subscribers from around the world. Studies with *WoW* have actually shown that players in multiplayer games have a natural tendency to participate in scientific argumentation. Steinkuehler and Duncan (2008) examined online discussions in *WoW* and determined that 86% of the discussions displayed "social knowledge construction." Specifically, the online discussions included evidence-based argumentation and other scientific discursive practices. Because WoW was not designed to promote scientific practice explicitly, this discovery speaks to the natural tendency of players to engage in argumentation.

Of course, games can also be designed to support STEM inquiry, and they yield similar results. In order to support scientific practice, Sasha Barab and his research team designed *Taiga Park*, part of the *Quest Atlantis* game-based multiuser virtual world. In *Taiga Park*, players uncover and try to solve a water-quality problem that is causing the fish population to decline (Barab, Sadler, Heiselt, Hickey, & Zuiker, 2007). Substantial evidence supported the fact that players were engaged with the narrative of the game world; but more importantly, the built-in ambiguity of the fish scenario meant that students argued with evidence to substantiate their perspective in the game. Through their inquiry process, students in the game also learned the content; in other words, researchers found evidence that student learning within *Taiga Park* transferred to standardized test items (Barab et al., 2007). Further study with *Taiga Park* revealed that the experience of scientific practice during game play can also translate to performance-based tasks. With a performance-based transfer task that required causal understanding, dyads of students who experienced scientific practice in *Taiga Park* performed significantly better than control groups (Barab et al., 2009). In a later study, researchers from TERC designed *Martian Boneyards* in order to study scientific inquiry (Asbell-Clarke et al., 2012). They created a massively multiplayer online world where adventurers discovered bones near an abandoned science center; players were recruited to help solve the mystery by gathering data and analyzing it. Researchers took on the roles of virtual characters in the game in order to study the interactions and facilitate the evolution of the storyline. With *Martian Boneyards*, the research team found evidence that the game world nurtured sustained scientific inquiry (Asbell-Clarke et al., 2012). Specifically, the researchers found that players used language suggestive of evidence-based argumentation and that the inquiry process was highly collaborative.

Overall, research with game-based learning shows that learners are not simply acquiring facts; instead, they are participating in scientific practice. Games afford students this type of active participation in their learning—players *learn by doing*. The game world becomes the community of practice, and the player is the apprentice. In discussing an epistemic game called *Urban Science*, Shaffer et al. (2009) explained that "professional work and reflection on that work with peers and mentors in game form can be a powerful educational tool" (p. 4). Overall, educational games work well for promoting STEM inquiry because players engage in authentic practice by playing professional roles and working with scientific tools.

Mobile Games for STEM Inquiry

So what can mobile games add to the discussion of educational games? Basically, when students learn effectively with mobile devices, their learning becomes contextualized, connected, and dispersed through time; all of these themes afford a more personalized learning experience for the student (Crompton, 2013). Mobile learners can socially construct knowledge during the lived practice of their daily lives; they can be in the real world experiencing STEM inquiry. Additionally, when the mobile device is used with AR, it becomes a lens to another world; in other words, a virtual world can be layered into the player's real-world environment. It's no wonder then that collaborative mobile AR games, still in their infancy, show potential for promoting STEM inquiry such as scientific practice (Squire & Klopfer, 2007) and scientific literacy (Squire & Jan, 2007).

Although Pokémon Go has brought mobile AR games quickly into the public eye, it was over a decade ago that the Massachusetts Institute of Technology (MIT) started conducting research on mobile AR games. Their initial game, *Environment Detectives*, was one of the first games to enable students to conduct a scientific investigation using a mobile device in the real world. The game scenario was that a dangerous chemical had been discovered in the groundwater near MIT. Using handhelds with a global positioning system (GPS), students investigated the local area. Within the game, students were required to investigate the chemical spill and report back with the cause, a remediation plan, and potential health risks to the public. The project's purpose was to understand the learning affordances of mobile AR games and demonstrate the project's potential for success in formal learning (Klopfer, Squire, & Jenkins, 2002). In terms of STEM inquiry, researchers found that the form factors of the game tended to promote negotiation and discussion among the student groups.

To move the research agenda forward, several universities collaborated on a large-scale implementation of a mobile AR game called *Alien Contact* funded by the U.S. Department of Education. In this game, students discover that aliens have landed on Earth and they need to determine why the aliens are there. Working as teams, groups encounter a series of science and math puzzles in the AR world along with characters to interview. Every team member played a unique role and received different information to encourage everyone to communicate and share their knowledge. The research results culled from three different case study sites showed high engagement, especially among previously disengaged students (Dunleavy et al., 2009).

In early qualitative studies, evidence of scientific inquiry during mobile AR game play was found (Squire & Jan, 2007). Squire and Jan (2007) explored scientific argumentation using a case study comparison of three different groups playing *Mad City Mystery*. In the game, students worked in teams to explore a mysterious death that occurred near Lake Mendota. To solve the case, players collected data samples, interviewed virtual characters, and examined government documents. The researchers found that as students pieced together their explanations, all teams engaged in argumentation cycles. Specifically, the researchers reported that during the game, players weighed evidence, developed hypotheses, tested them against evidence, and generated evidence-based theories. Squire and Jan (2007) concluded that mobile AR "games could have promise for tools that develop scientific literacy" (p. 23).

In another qualitative case study of *Sick at South Beach*, Squire (2010) found similar results. In the game, players took on scientific roles in order to investigate why students got sick after spending time at the local beach. In order to have multiple data points in which to triangulate findings, researchers collected field notes, videotapes, and interviews; they also conducted a document analysis. Researchers noticed that student discussions were not directed by the teacher, but rather revolved around the problem at hand. In fact, in these problem-oriented discussions, Squire (2010) found that groups argued quite intensely for their role's scientific point of view. The players were no longer compliant consumers of information, but instead they were performing the actions of real scientists.

More recently, mobile AR games are showing significant learning gains when investigated quantitatively. As part of a microbiology course, students used their own mobile devices to investigate contagious outbreaks in the game called *Project Outbreak* (Bennani & Novak, 2016). Students adopted professional roles and worked in groups as they put their knowledge of microbiology to work in real-world scenarios. Anecdotally, faculty noticed that students who played *Project Outbreak* had better comprehension of course content than those who were traditionally taught. This finding was also supported quantitatively by significantly higher exam scores and final grades (Bennani & Novak, 2016). Overall project findings demonstrated that student retention increased along with students' success rates when *Project Outbreak* was included as part of the course.

Now, enough evidence surrounding mobile AR games has been collected that researchers are reviewing and reporting on the literature. The reviews are quite positive in terms of learning outcomes and scientific inquiry. Cheng and Tsai (2013) reviewed five studies from the field of location-based AR games and found that collaborative role-playing design encourages players to observe scientific phenomena, ask questions, investigate data, create hypotheses, and construct explanations—all important scientific practices. In a more recent review of the literature, Koutromanos, Sofos, and Avraamidou, (2015) documented that all seven of the studies they reviewed had an impact on student learning. They stated that most of the games they reviewed revolve around science topics because the inquiry-based learning of science lends itself well to this medium. Koutromanos et al. (2015) concluded that even though the findings are positive, there is still a lot of work to be done—there are many unanswered questions. The time is right to synthesize the literature base and develop a replicable design framework.

INPLACE MOBILE GAMES

As discussed earlier, game-based learning, especially with mobile AR games, holds great promise to support STEM inquiry. The next section details a design framework that brings together the design elements of mobile AR games that have been most successful; the theoretical underpinnings of the framework are embodied in the acronym INPLACE. The second section describes how the INPLACE framework was tested in several stages to confirm its ability to engage students and support STEM inquiry.

INPLACE: The Framework

The emerging field of mobile AR games has been exploring practical issues pertaining to using such games in the K-12 environment, as well as what design elements are most successful (Dunleavy & Dede, 2013). Research derived from these mobile AR games of the past has laid the foundation for the INPLACE design framework. INPLACE mobile games combine the best practices of what the games and learning field knows about using the affordances of mobile devices to engage students, to support collaboration, and to promote authentic practice in a discipline.

INPLACE mobile games provide location-specific information to players through the use of location-aware AR activated by GPS triggers or visual markers such as quick-response (QR) codes. This means that players experience content in context. As mentioned earlier, INPLACE is an acronym created by this chapter's author that stands for **I**nterdependent, **N**etworked, **P**articipatory **L**earning, **A**ugmented, **C**ollaborative **E**xperience. The following sections explain the theoretical underpinnings of the design framework.

Interdependent

Mobile AR games are most successful when collaboration is scaffolded through positive interdependence and jigsaw pedagogy. According to Johnson, Johnson, and Holubec (1993), "positive interdependence is successfully structured when group members perceive that they are linked with each other in a way that one cannot succeed unless everyone succeeds" (p. 9). The notion of jigsaw pedagogy is that each student in a group becomes an expert on one aspect of the activity and teaches it to the other group members (Aronson & Patnoe, 2011).

Researchers have found that interdependent roles are an effective way to scaffold collaborative problem solving within mobile AR games (Dunleavy et al. 2009; Squire & Jan, 2007); therefore, students not only work together in INPLACE games, but there is a positive interdependence between individuals. Essentially, the group success depends on the success of all the individuals. For example, in *Time Lab 2100*, students played in pairs to travel back in time to make changes in an attempt to decrease the impact of climate change; each player only received half of the dialogue from characters who offered expertise (Klopfer & Sheldon, 2010). Another way to have groups work together is to differentiate the roles. In *Sick at South Beach*, students were water chemists, public health doctors, or wildlife ecologists; each student developed a different expertise over the course of the curriculum and argued quite intensely for their point of view when it came time to develop a consensus (Squire, 2010).

Networked

INPLACE games are *networked* in both the technical and social sense of the word. First, the games are supported by a network of computers. Although software might reside on the mobile device, it is the fact that the mobile device is connected to the Internet that allows the game to distribute information to different players. Second, the games afford a rich network of interactions (Gee, 2003). Players are interacting with the storyline, characters, and tools, and players are interacting with each other in the real world. This fosters feelings of connection to the content and between the teammates. At the Boston Museum of Science, the research team of Klopfer, Perry, Squire, Jan, and Steinkuehler (2005) found that *Mystery at the Museum* fostered feelings of connection between the visitors and the museum. Feedback from players confirmed that they were engaged, not only with individual exhibits, but also with the larger context of the museum and with their fellow players.

Participatory Learning

Fundamentally, INPLACE mobile games shift learning from an acquisition model to a participatory model; to facilitate learning, students assume roles within the game. Lave and Wenger (1991) described how students begin learning within a community of practice: "it crucially involves participation as a way of learning—of both absorbing and being absorbed in—the culture of practice" (p. 95). In the INPLACE mobile games that have been developed and researched to date, the method of participatory learning that works is to have students assume roles within the game.

In *Alien Contact*, students became chemists, cryptologists, computer hackers, and FBI agents in order to determine why aliens landed in the school yard (Dunleavy et al. 2009). Researchers reported that the students were extremely engaged and found role playing to be very motivating. In fact, role playing is so powerful that when Facer et al. (2004) had students break out of their roles to reflect on game strategy, they deemed it a failure of the game because it did not capitalize on students' self-motivation or conceptual understanding from the viewpoint of their roles. The results from these two studies show that students are experiencing *projective identity* as discussed by Gee (2003). When players identify so intimately with the roles, the games can leverage projective identity in order to create compelling, participatory learning experiences.

Students playing professional roles within a game get to play, practice, and perform the skills necessary to do the job—free to learn through failure without the threat of a deadly mistake. Research has shown that when students take on professional roles in mobile AR games, they engage in scientific inquiry (Dunleavy et al., 2009; Rosenbaum et al., 2007) and argumentation (Squire & Jan, 2007) as they play. Researchers have even determined that when students embody authentic roles, it helps them understand complex scientific issues. Rosenbaum et al. (2007) found that the students understood complex causality after playing *Outbreak @ The Institute* where they took on roles as doctors, medical technicians, and public health experts in order to determine to cause of a disease outbreak. Rosenbaum et al. (2007) determined that "students saw their abilities within the game as tied to their roles in an authentic way" (p. 40).

When Squire and Jan (2007) studied the discourse between players in *Mad City Mystery*, which was designed with interdependent roles, they found that players engaged in scientific argumentation because of the task, roles, embedded resources, and context of the game. Specifically, Squire and Jan (2007) reported that players acted like scientists by weighing evidence, developing hypotheses, testing hypotheses against evidence, and generating evidence-based theories. Overall, the use of interdependent roles in INPLACE games scaffolds the social interaction and discourse necessary to build scientific knowledge, offering an unmatched way for students to socially construct knowledge.

Augmented (Reality)

Cheng and Tsai (2013) reviewed studies on AR in science education and concluded that most studies showed that learners had positive attitudes toward AR. Beyond just being "the new thing" or "something cool," AR provides an important component of the immersive feel of the game. According to Squire and Jan (2007), these games augment reality because they are "games played in the real world with the support of digital devices (PDAs, cellphones) that create a fictional layer on top of the real world context" (p. 6). INPLACE mobile games use location-aware AR activated by GPS or visual markers such as QR codes. This means that the games can provide location-specific information and players can experience content in context. Pedagogically, given the situated nature of the learning, the students are then better prepared to solve similar problems in the future because the learning transfer is more direct (Dunleavy & Dede, 2013).

Collaborative

Freely collaborating students often fail to engage in meaningful interactions; however, INPLACE games provide the scaffolding and support necessary so that students can work together effectively. INPLACE games are designed to be collaborative, but the theory draws heavily from cooperative learning. In cooperative learning, students split the task into individual parts, work individually, and then assemble the final result, whereas collaborative learning requires that students work together (Dillenbourg, 1999). Because of the design of INPLACE games, players exhibit both cooperative and collaborative learning. In order to scaffold this type of collaborative problem solving, each player's role is differentiated. Because scientific practice in the real world is inherently social, groups of people need to put their expertise together in order to solve problems. INPLACE games provide an opportunity for such authentic, collaborative practice.

Experiential

INPLACE mobile games have a kinesthetic component that offers an immersive feel—players are up and moving around. Researchers have found that this physicality contributed to players' enjoyment (Facer et al., 2004; Dunleavy et al., 2009). However, the immersive feel comes from more than just the physical experience. In reference to the capabilities of today's new media, Chris Dede (2009) explained that "immersion is the subjective impression that one is participating in a comprehensive, realistic experience" (p. 66). Several aspects of INPLACE games seem to support this description of immersion. First, narrative plays an important role; Dunleavy and Dede (2013) summarized the literature and declared that the narrative is a critical design decision. Second, given the portability of mobile devices, the real environment can also provide an authentic context. For example, students playing *Up River* became immersed in their local watershed; throughout gameplay, players developed a detailed understanding of the environment and the conservation issues surrounding it (Wagler & Mathews, 2012). Third, the combination of the authentic context and the INPLACE game can provide an extremely robust sense of immersion. In their *EcoMOBILE* project, Kamarainen, et al. (2013) used mobile AR interfaces to contextualize information and supply just-in-time instruction as students explored a real pond; students used real probeware, which enhanced the realism because the technology was very similar to what real scientists use to take ecosystem measurements. For *EcoMOBILE*, being in the real context provided three-dimensional experiential learning, but more notably, the mobile technology augmented the experiential learning.

INPLACE: Testing the Framework

Using the INPLACE design framework outlined earlier, a game series called *School Scene Investigators* was developed to test the effectiveness of the framework for engagement and learning. The first game, *The Case of the Stolen Score Sheets*, was developed to test the framework for engagement (Bressler & Bodzin, 2013). The second game, *The Case of the Mystery Powder*, was developed to test the framework as a support for learning and collaboration (Bressler and Bodzin, 2016).

The Pilot Test

The INPLACE design framework was pilot tested with *The Case of the Stolen Score Sheets*. The game was a 30-minute activity that students played in the school's hallways. Students played in collaborative groups and explored their school to find and scan QR codes for game-related information; scanning a code could mean meeting a virtual character or discovering a piece of forensic evidence (Bressler, 2015). The game was built in ARIS (2016), which stands for Augmented Reality and Interactive Storytelling. ARIS is an open-source, Web-based programming environment that offers a free iOS application and allows anyone to freely create and play games. To authentically implement the INPLACE design framework, students selected roles: social networker, techie, photographer, or science whiz. Each role offered a unique yet integral way for the student to participate in the learning game. The roles also enabled players to participate in the narrative storyline of the game: everyone was accused of stealing the test answers and needed to prove their innocence! All players were given their own Wi-Fi–enabled iPhone. Scanning a code would access the ARIS servers and deliver role-specific information to each player.

The research intervention was conducted outside of normal school instructional time, and only one group of students played at a time. A total of 68 middle school students played through the game. Because the theory of flow (Csikszentmihalyi, 1990) was used as a measure of engagement, participants responded to a self-report survey to assess their flow experience. The survey included 10 Likert-style items that measured elements of flow as outlined by Csikszentmihalyi (1996). Possible scores ranged from 10 to 50. The survey was derived from the Short Flow State scale (S FSS-2) and the Core Flow State scale (C FSS-2) developed by Jackson, Eklund, and Martin (2010). Data sources also included field observations and group interviews. According to Bressler and Bodzin (2013), the pilot test revealed that the design framework promoted a substantive flow-like experience during a voluntary extracurricular activity. Coding of the qualitative data revealed that the participants' flow experiences seemed to stem from three experiential features: participants felt a flash of intensity, a sense of discovery, and desire for higher performance. Bressler and Bodzin (2013) attributed the experience of flow partially to the lack of technological frustration, ease of navigation, and a game design element that shifted leadership evenly between group members.

The Scale-Up Test

Because the initial pilot test was implemented as a voluntary activity, further pilot testing was conducted to test the design framework under the precedent of a mandatory school activity. It was essential to determine if the design was equally engaging when played as a required school activity with multiple groups playing simultaneously. *The Case of the Stolen Score Sheets* was used by two eighth-grade science teachers as a full class activity; over 200 students participated in the game from 10 different classes. Teachers played through the whole game with the researcher prior to using it with their classes; no special skills were required of the teachers, except the willingness to allow their students freedom to explore the school during class time. In order to accommodate a full class of students playing together, additional QR code locations were dispersed throughout the school while the game narrative stayed exactly the same.

Similar to the pilot test, all participants responded to a self-report survey to assess their flow experience immediately after gameplay. Research revealed that when the INPLACE mobile game was used as a mandatory school activity, students still experienced a substantive flow-like experience; in fact, the average voluntary participant had a similar flow experience score to the average mandatory participant (Bressler & Bodzin, 2014). Additional findings from a hierarchical regression model revealed that there were no significant differences in flow experience scores across achievement levels. When controlling for gender, group type, and teacher effects, achievement level did not predict flow experience ($t = -1.19, p = .24$); high-achieving and low-achieving students were equally engaged by the game.

The Quasi-Experimental Test

Once the INPLACE design framework proved successful as a mandatory classroom activity, a new game was developed to enhance the scientific inquiry afforded by the game and to facilitate testing players for learning gains. *The Case of the Mystery Powder* was the second game in the *School Scene Investigators* series; in this case, students investigated a mysterious powder at their school. The game narrative was set up as a forensic science mystery similar to the first game. The context of the school and the people who worked there were integrated into the narrative in order to enhance authenticity. Someone has stolen money from the cash register in the cafeteria and left behind this unknown white powder. Was it the janitor who used cornstarch as a cleaning agent? Was it the front office secretary who baked sugar cookies for the staff? Or was it the student who was making a rocket using baking soda? All had valid reasons to get supplies from the kitchen, but who stole the cash?

Both games were designed according to the INPLACE design framework; however, there were some differences between how *The Case of the Stolen Score Sheets* and *The Case of the Mystery Powder* were implemented. First, *The Case of the Mystery Powder* lasted several class periods. Most players took four to five class periods to play through the whole game. Second, in the middle of *The Case of the Mystery Powder*, a character from the game required players to conduct a real hands-on experiment and use the results in order to continue in the game. This part of the game required extensive teacher facilitation because students were asked to conduct their own tests on an actual mystery powder using iodine, vinegar, heat, and pH paper. Third, in order to document learning outcomes, students playing *The Case of the Mystery Powder* were required to write up an incident report. The game included all the prompts necessary to complete the incident report, which was a written report with all open-ended explanations.

There was an intrinsic link between the game and curricular content. First, the game was aligned to a district curriculum unit. During gameplay, students utilized the scientific method, learned about acids and bases, and conducted some basic physical and chemical tests to analyze evidence and solve the mystery. Second, the game was designed to promote Next Generation Science Standards (Achieve

Inc., 2013). In order to complete the game, students utilized scientific practices such as asking questions, carrying out an investigation, data analysis, constructing explanations, and engaging in argument from evidence.

To determine whether an INPLACE game afforded higher flow, better collaboration, and more scientific inquiry than the usual curriculum activity, *The Case of the Mystery Powder* was implemented with a quasi-experimental group and compared to a control group. Total sample size was 179 students; all students participated in both post-treatment measures, including a post-survey to empirically assess flow and the written report to measure scientific inquiry. In addition to the full sample, a small subsample of 6 student teams, comprising 23 students, participated in the qualitative data collection, which included audio recordings of collaborative discourse, photographs, and field notes.

A MANCOVA analysis was used to evaluate the amount of variance in flow experience and report score as explained by treatment group; standardized math scores were used as the covariate to eliminate systematic bias and make group comparisons fairer. Similar to the results of the pilot testing, this quasi-experimental study confirmed that students ($M = 40.50$, $SD = .72$) playing the INPLACE mobile game reported higher levels of flow than students ($M = 38.62$, $SD = .50$) in the control activity (Bressler & Bodzin, 2016). This result is best explained by game teams experiencing these specific elements of flow: challenge–skill balance, sense of control, total concentration on the task, clear goals, and an intrinsically rewarding experience. In general, control teams spoke off-topic more than game teams and were uncertain how to organize themselves in order to accomplish their tasks. Higher levels of teacher involvement also gave control teams a lower sense of control.

In comparison to business as usual, the INPLACE mobile game supported higher scientific practices in both written reports and in team conversations (Bressler, 2014; Bressler & Bodzin, 2016). It seems that INPLACE mobile games support students' newer, neomillenial learning styles, which consist of preferences for learning environments that support communal learning, active learning, and nonlinear learning (Dieterle, 2009). Specifically, INPLACE mobile games lead to deeper scientific inquiry by supporting situated learning, social construction of knowledge, and autonomy. In game teams, teachers had less direct involvement, thus students were more intrinsically motivated; they took initiative and felt in control. Emergent analysis revealed that game teams used teachers for support when necessary, which shifted the teacher's role away from direct instruction to a guide-on-the-side. Therefore, game teams developed and exhibited a *deeper* understanding of scientific inquiry and needed *less* teacher intervention to stay on task (Bressler, 2014; Bressler & Bodzin, 2016).

Given the nature of the INPLACE design framework, the big difference between treatment groups was that the game environment provided more collaborative scaffolding than the control activity. Results confirmed that conversations within game teams had more engaged responses and higher levels of communal language, whereas control teams has less engaged responses and higher levels of command language (Bressler, 2014). Because the interdependent roles that were designed into the game promoted positive communication patterns and a shared leadership structure, games teams were more successful collaborators. Unfortunately, without such scaffolding to facilitate productive communication and collaboration, control teams were less successful collaborators. In this quasi-experiment, game teams were able to share decision making and co-construct solutions because of the interdependency of their collaboration as dictated by the INPLACE design framework.

FUTURE RESEARCH DIRECTIONS

The next step in the research progression of INPLACE mobile games is to grow their presence in the classroom. Using games in the classroom has been around since the early 1980s, and there is a growing body of knowledge about the positive impacts of game-based learning, yet schools struggle to incorporate games. Why? First, schools have had to adhere to the standards set by the No Child Left Behind Act of 2001. As Klopfer (2008) stated, "The final nail in the coffin of learning games was the No Child Left Behind Act of 2001" (p. 26). Teachers and parents express concern over using games in the classroom because the skills developed during gameplay do not align to skills assessed on national, high-stakes tests (Kirriemuir & McFarlane, 2004). The other big obstacle to using games for learning is the ever-present criticism against game-based learning in schools.

Critics of game-based learning highlight problems associated with integrating games into the formal school environment, including costs, teacher role, and logistics. First, in the past, the costs of licenses, software products, and video game equipment had to be considered and were thus a barrier to implementation (Kirriemuir & McFarlane, 2004). Second, the role of the teacher is complicated; students are not necessarily used to playing a game in a classroom learning environment (Squire, 2005), so teachers need to consistently work at keeping kids on track toward the educational goal (Kirriemuir & McFarlane, 2004). Third, the constraints of the school day do not afford players several hours to immerse themselves in the gameplay and derive the maximum benefit (Young et al., 2012). The bottom line for why games do not appear more in classrooms was succinctly stated by Nicholas Negroponte

(1995): "Most adults fail to see how children learn with electronic games" (p. 204). For all these reasons, the research on games for learning needs to be convincing and concrete.

The INPLACE design framework has been tested for engagement under both voluntary and mandatory situations. It has also been investigated to determine whether an INPLACE mobile game supports collaboration and learning more effectively than a control nongame setting. Essentially, the INPLACE design framework has shown promise and potential, even in light of the criticism against game-based learning. INPLACE mobile games offer something different—they capitalize on the affordances of the device, the best practices of game-based learning, and they try to help students learn content in authentic, participatory, collaborative ways. Specifically, INPLACE games hold promise for K-12 education because:

- **INPLACE Games Are a Hybrid of Entertainment and Edutainment:** Klopfer, Osterweil, and Salen (2009) explained that there are two camps for educational gaming: in one camp, the game is pinnacle and the education is lost, whereas in the other camp, the learning goals are the focus and the game design is lost. INPLACE mobile games, with their focus on participatory learning and collaborative design, bring together the best of both camps—losing neither the game design nor the learning goals. The content is the game.
- **INPLACE Games Are Kinesthetic:** When researchers have considered the problems with educational games, they have expressed concerned for the lack of kinesthetic activities to support the virtual learning, expressing concern that there is a loss of transfer from virtual reality to physical reality (Young et al. 2012). As expressed in the definition of INPLACE games, they offer an immersive experience within a real, authentic context. Kinesthetic activity is built right into the game design.
- **INPLACE Games Require Minimal Costs:** Critics have complained that schools cannot afford the software or the equipment necessary to play games at school. The software necessary to play INPLACE games is mostly free. The platforms that are used to design INPLACE games are mostly freeware as well. Platforms such as ARIS (2016) are free to make games *and* free to play games. Critics may complain that the graphical nature of these platforms is not as realistic as what students are accustomed to outside of school. Although that is certainly the case, Vogel et al. (2006) found no significant learning differences between games of higher and lower levels of realism. As far as equipment costs, INPLACE games are mostly designed for devices that students are already carrying with them: their mobile phones. The culture of school just needs to shift the paradigm from the phone as a "threat to

learning" to the phone as a "support for learning." In fact, letting students use their own devices can contribute to an even deeper personal learning experience.

- **INPLACE Games Are Designed for Curriculum Support:** Although teachers might complain that commercial, off-the-shelf games require a lot of time and support to align the game with curriculum goals, INPLACE mobile games are designed around content. Some games are even designed with interdisciplinary learning goals such as *Alien Contact* (Dunleavy et al. 2009).

Now with AR development platforms such as ARIS (2016) that enable nonprogrammers to create their own INPLACE mobile games, the ability to make mobile games no longer resides within the commercial sector but with educators and researchers. Essentially, INPLACE provides a design framework that teachers and researchers can use for building mobile games that heightens engagement and increases learning. Incorporating more INPLACE mobile games into classrooms could better prepare students for the collaborative inquiry-based culture of their future STEM careers and provide ample opportunities for future research.

CONCLUSION

Today's students have the opportunity to learn in different ways, largely because of the technologies that surround them outside of school (Ito et al., 2010). One medium that dominates a young student's out-of-school time is gaming, and it starts from an early age. In fact, "gaming represents the central form of early computer experience for kids" (Ito et al. 2010, p. 196). By middle school, students are high-frequency users of video games and Web-based games (Spires, Lee, Turner, & Johnson, 2008). More importantly, students value the use of games for learning (Goyal, 2012; Project Tomorrow, 2010). Nikal Goyal (2012), who published his student perspective on the state of education, stated, "I have learned so much more about math by playing games…than by hearing lectures and doing textbook problems" (loc. 5400); he goes on to clearly state that he wants to see games used in the classroom. He is not the only student voicing this opinion. Students in the *Speak Up* study—approximately 300,000 students—also advocated for games in the classroom because games can get them more engaged with the subject matter, help them work in teams, and provide an easier way to understand difficult concepts (Project Tomorrow, 2010).

Games are not only what students want in their classroom, but mobile games—specifically INPLACE mobile games—offer learning advantages in STEM inquiry over traditional approaches. One of the main reasons that meaningful learning takes place during INPLACE mobile games is due to the situated, immersive context. By role playing within the game, students live and learn scientific practice. Specifically, students do not learn science by simply acquiring facts. Rather, they are immersed in a context through which science becomes a process, and more importantly, the students are eager to participate in that process.

Are INPLACE mobile games the perfect panacea to solve all of the issues currently facing our educational system? No, because nothing can do that. However, there is a strong case that they can improve engagement in science learning and STEM inquiry, which means they can address two profound issues facing education today. Ultimately, by serving the needs of administrators, educators, and parents along with the desires of the students, INPLACE mobile games are a win-win. In other words, they are the perfect "win state" for educational reform.

REFERENCES

Achieve, Inc. (2013). *Next Generation Science Standards*. Washington, DC: Achieve, Inc. Retrieved from http://www.nextgenscience.org/

Annetta, L. A., & Cheng, M. (2008). Why educational video games? In L. A. Annetta (Ed.), *Serious educational games: From theory to practice* (pp. 1–11). Rotterdam, The Netherlands: Sense Publishers.

ARIS. (2016). *Mobile learning experiences*. Retrieved July 15, 2016, from http://arisgames.org/

Aronson, E., & Patnoe, S. (2011). *Cooperation in the classroom: The jigsaw method* (3rd ed.). London, UK: Pinter & Martin Ltd.

Asbell-Clarke, J., Edwards, T., Larsen, J., Rowe, E., Sylvan, E., & Hewitt, J. (2012). Martian Boneyards: Scientific inquiry in an MMO game. *International Journal of Game-Based Learning*, *2*(1), 52–76. doi:10.4018/ijgbl.2012010104

Barab, S., Sadler, T., Heiselt, C., Hickey, D., & Zuiker, S. (2007). Relating narrative, inquiry, and inscriptions: Supporting consequential play. *Journal of Science Education and Technology*, *16*(1), 59–82. doi:10.1007/s10956-006-9033-3

Barab, S., Scott, B., Siyahhan, S., Goldstone, R., Ingram-Goble, A., Zuiker, S., & Warren, S. (2009). Transformational play as a curricular scaffold: Using videogames to support science education. *Journal of Science Education and Technology*, *18*(4), 305–320. doi:10.1007/s10956-009-9171-5

Bennani, F., & Novak, K. (2016). Project Outbreak: Take biology out of the classroom and into the world. In H. Crompton & J. Traxler (Eds.), *Mobile learning and STEM: Case studies in practice* (pp. 125–138). New York, NY: Routledge.

Bransford, J. D., Brown, A. L., & Cocking, R. R. (Eds.). (2000). *How people learn: Brain, mind, experience, and school.* Washington, DC: National Academy.

Bressler, D., & Bodzin, A. (2016). Investigating flow experience and scientific practices during a mobile Serious Educational Game. *Journal of Science Education and Technology*, *25*(5), 795–805. doi:10.1007/s10956-016-9639-z

Bressler, D. M. (2014). Better than business-as-usual: Improving the scientific practices during discourse and writing by playing a collaborative mystery game. *Double Helix: A Journal of Critical Thinking and Writing, 2*, 1-13.

Bressler, D. M. (2015). School Scene Investigators: A collaborative AR game for middle school science inquiry. In C. Holden, S. Dikkers, J. Martin, & B. Litts (Eds.), *Mobile Media Learning: Innovation and Inspiration* (pp. 87–100). Pittsburgh, PA: ETC Press.

Bressler, D. M., & Bodzin, A. M. (2013). A mixed methods assessment of students flow experience during a mobile augmented reality science game. *Journal of Computer Assisted Learning*, *29*(6), 505–517. doi:10.1111/jcal.12008

Bressler, D. M., & Bodzin, A. M. (2014, April). *Continuing to flow: Student experience during a scaled-up INPLACE mobile game.* Paper presented at the National Association for Research in Science Teaching (NARST), Pittsburgh, PA.

Chang, C.-S., Chen, T.-S., & Hsu, W.-H. (2011). The study on integrating WebQuest with mobile learning for environmental education. *Computers & Education*, *57*(1), 1228–1239. doi:10.1016/j.compedu.2010.12.005

Cheng, K., & Tsai, C. (2013). Affordances of augmented reality in science learning: Suggestions for future research. *Journal of Science Education and Technology*, *22*(4), 449–462. doi:10.1007/s10956-012-9405-9

Christensen, C. M., Horn, M. B., & Johnson, C. W. (2008). *Disrupting class: How disruptive innovation will change the way the world learns.* New York, NY: McGraw-Hill.

Crompton, H. (2013). Mobile learning: New approach, new theory. In Z. Berge & L. Muilenburg (Eds.), *Handbook of Mobile Learning* (pp. 47–57). New York, NY: Routledge.

Csikszentmihalyi, M. (1990). *Flow: The psychology of optimal experience*. New York, NY: Harper and Row.

Csikszentmihalyi, M. (1996). *Creativity: Flow and the psychology of discovery and invention*. New York: Harper Collins.

Dede, C. (2009). Immersive interfaces for engagement and learning. *Science*, *323*(5910), 66–69. doi:10.1126/science.1167311 PMID:19119219

Dieterle, E. (2009). Neomillennial learning styles and River City. *Children, Youth and Environments*, *19*(1), 245–278.

Dillenbourg, P. (1999). What do you mean by "collaborative learning"? In P. Dillenbourg (Ed.), *Collaborative learning: Cognitive and computational approaches* (pp. 1–16). Amsterdam, The Netherlands: Pergamon, Elsevier Science.

Dunleavy, M., & Dede, C. (2013). Augmented reality teaching and learning. In J. M. Spector, M. D. Merrill, J. Elen, & M. J. Bishop (Eds.), *Handbook of research on educational communications and technology* (4th ed.; Vol. 2). New York, NY: Springer.

Dunleavy, M., Dede, C., & Mitchell, R. (2009). Affordances and limitations of immersive participatory augmented reality simulations for teaching and learning. *Journal of Science Education and Technology*, *18*(1), 7–22. doi:10.1007/s10956-008-9119-1

Facer, K., Joiner, R., Stanton, D., Reid, J., Hull, R., & Kirk, D. (2004). Savannah: Mobile gaming and learning? *Journal of Computer Assisted Learning*, *20*(6), 399–409. doi:10.1111/j.1365-2729.2004.00105.x

Gatto, J. T. (2009). *Weapons for mass instruction: A schoolteacher's journey through the dark world of compulsory schooling*. Gabriola Island, BC, Canada: New Society Publishers.

Gee, J. P. (2003). *What video games have to teach us about learning and literacy*. New York, NY: Palgrave Macmillan.

Gee, J. P. (2010). *New digital media and learning as an emerging area and "worked examples" as one way forward*. Cambridge, MA: The MIT Press.

Gopnick, A., Meltzoff, A. N., & Kuhl, P. K. (1999). *The scientist in the crib: What early learning tells us about the mind*. New York: HarperCollins Publishers.

Goyal, N. (2012). *One size does not fit all: A student's assessment of school* [Kindle version]. Retrieved from Amazon.com

Ito, M., Baumer, S., Bittanti, M., Boyd, D., Cody, R., Herr-Stephenson, B., & Tripp, L. (2010). *Hanging out, messing around, geeking out: Kids living and learning with new media*. Cambridge, MA: MIT Press.

Jackson, S., Eklund, R., & Martin, A. (2010). *The FLOW manual*. Queensland, Australia: Mind Garden, Inc.

Johnson, D. W., Johnson, R. T., & Holubec, E. J. (1993). *Circles of learning: Cooperation in the classroom* (4th ed.). Edina, MN: Interaction Book Company.

Kamarainen, A. M., Metcalf, S., Grotzer, T., Browne, A., Mazzuca, D., Tutwiler, M. S., & Dede, C. (2013). EcoMOBILE: Integrating augmented reality and probeware with environmental education field trips. *Computers & Education*, *68*, 545–556. doi:10.1016/j.compedu.2013.02.018

Kanter, R. M. (2001). *Evolve! Succeeding in the digital culture of tomorrow*. Cambridge, MA: Harvard Business School Press.

Katz, J. (2006). *Magic in the air: Mobile communication and the transformation of social life*. New Brunswick, NJ: Transaction Publishers.

Kirriemuir, J., & McFarlane, A. (2004). *Report 8: Literature review in games and learning*. Bristol, UK: NESTA Futurelab.

Klopfer, E. (2008). *Augmented learning: Research and design of mobile educational games*. Cambridge, MA: The MIT Press. doi:10.7551/mitpress/9780262113151.001.0001

Klopfer, E., Osterweil, S., & Salen, K. (2009). Moving learning games forward. Cambridge, MA: Education Arcade. Retrieved from http://education.mit.edu/papers/MovingLearningGamesForward_EdArcade.pdf

Klopfer, E., Perry, J., Squire, K., Jan, M.-F., & Steinkuehler, C. (2005). *Mystery at the Museum: A collaborative game for museum education*. Paper presented at the annual conference of Computer Supported Collaborative Learning, Taipei, Taiwan. doi:10.3115/1149293.1149334

Klopfer, E., & Sheldon, J. (2010). Augmenting your own reality: Student authoring of science-based augmented reality games. *New Directions for Youth Development*, *2010*(128), 85–94. doi:10.1002/yd.378 PMID:21240956

Klopfer, E., Squire, K., & Jenkins, H. (2002). *Environmental detectives: PDAs as a window into a virtual simulated world*. Paper presented at the International Conference on Wireless Mobile Technologies in Education, Växjö, Sweden. doi:10.1109/WMTE.2002.1039227

Koutromanos, G., Sofos, A., & Avraamidou, L. (2015). The use of augmented reality games in education: A review of the literature. *Educational Media International, 52*(4), 253–271. doi:10.1080/09523987.2015.1125988

Lave, J., & Wenger, E. (1991). *Situated learning: Legitimate peripheral participation*. New York, NY: Cambridge University Press. doi:10.1017/CBO9780511815355

Looi, C. K., Zhang, B., Chen, W., Seow, P., Chia, G., Norris, C., & Soloway, E. (2011). 1:1 mobile inquiry learning experience for primary science students: a study of learning effectiveness. *Journal of Computer Assisted Learning, 27*(3), 269–287. doi:10.1111/j.1365-2729.2010.00390.x

National Research Council. (2007). *Rising above the gathering storm: Energizing and employing America for a brighter economic future*. Washington, DC: The National Academies Press.

National Research Council. (2011). *Learning science through computer games and simulations*. Washington, DC: The National Academies Press.

National Research Council. (2012). *A framework for K-12 science education: Practices, cross-cutting concepts, and core ideas*. Washington, DC: The National Academies Press.

Negroponte, N. (1995). *Being digital*. New York, NY: Alfred A. Knopf.

Norris, C. A., & Soloway, E. (2011). Learning and schooling in the age of mobilism. *Educational Technology, 51*(6), 3–12.

Project Tomorrow. (2010). *Creating our future: Students speak up about their vision for 21st century learning*. Irvine, CA: Project Tomorrow. Retrieved from http://www.tomorrow.org/speakup/pdfs/SU09NationalFindingsStudents&Parents.pdf

Ravitch, D. (2011). *The death and life of the great American school system: How testing and choice are undermining education* [Kindle version]. Retrieved from Amazon.com

Reigeluth, C. M. (1994). Introduction: The imperative for systemic change. In C. M. Reigeluth & R. J. Garfinkle (Eds.), *Systemic change in education* (pp. 3–14). Englewood Cliffs, NJ: Educational Technology Publications, Inc.

Rosenbaum, E., Klopfer, E., & Perry, J. (2007). On location learning: Authentic applied science with networked augmented realities. *Journal of Science Education and Technology, 16*(1), 31–45. doi:10.1007/s10956-006-9036-0

Shaffer, D. W., Hatfield, D., Svarovsky, G. N., Nash, P., Nulty, A., Bagley, E., & Mislevy, R. et al. (2009). Epistemic Network Analysis: A prototype for 21st century assessment of learning. *International Journal of Learning and Media*, *1*(2), 1–21. doi:10.1162/ijlm.2009.0013

Spires, H. A., Lee, J. K., Turner, K. A., & Johnson, J. (2008). Having our say: Middle grade student perspectives on school, technologies, and academic engagement. *Journal of Research on Technology in Education*, *40*(4), 497–515. doi:10.1080/15391523.2008.10782518

Squire, K. (2005). Changing the game: What happens when video games enter the classroom? *Innovate: Journal of Online Education, 1*(6). Retrieved from http://www.innovateonline.info/index.php?view=article&id=82

Squire, K. (2010). From information to experience: Place-based augmented reality games as a model for learning in a globally networked society. *Teachers College Record*, *112*(10), 2565–2602.

Squire, K., & Jan, M. (2007). Mad City Mystery: Developing scientific argumentation skills with a place-based augmented reality game on handheld computers. *Journal of Science Education and Technology*, *16*(1), 5–29. doi:10.1007/s10956-006-9037-z

Squire, K., & Klopfer, E. (2007). Augmented reality simulations on handheld computers. *Journal of the Learning Sciences*, *16*(3), 371–413. doi:10.1080/10508400701413435

Steinkuehler, C., & Duncan, S. (2008). Scientific habits of mind in virtual worlds. *Journal of Science Education and Technology*, *17*(6), 530–543. doi:10.1007/s10956-008-9120-8

Steinkuehler, C., & Squire, K. (2014). Video games and learning. In R. K. Sawyer (Ed.), *The Cambridge handbook of the learning sciences* (2nd ed.; pp. 377–396). New York, NY: Cambridge University Press. doi:10.1017/CBO9781139519526.023

Tapscott, D., & Williams, A. D. (2006). *Wikinomics: How mass collaboration changes everything*. New York, NY: Penguin Group.

Thomas, D., & Brown, J. S. (2011). *A new culture of learning: Cultivating the imagination for a world of constant change*. Lexington, KY: CreateSpace Independent Publishing Platform.

Vogel, J. J., Vogel, D. S., Cannon-Bowers, J. A. N., Bowers, C. A., Muse, K., & Wright, M. (2006). Computer gaming and interactive simulations for learning: A meta-analysis. *Journal of Educational Computing Research*, *34*(3), 229–243. doi:10.2190/FLHV-K4WA-WPVQ-H0YM

Wagler, M., & Mathews, J. (2012). Up River: Place, ethnography, and design in the St. Louis River estuary. In S. Dikkers, J. Martin, & B. Coulter (Eds.), *Mobile media learning: Amazing uses of mobile devices for teaching and learning* (pp. 23–40). Pittsburg, PA: ETC Press.

Willingham, D. T. (2010). *Why don't students like school: A cognitive scientist answers questions about how the mind works and what it means for the classroom.* San Francisco, CA: Jossey-Bass. doi:10.1002/9781118269527

Young, M. F., Slota, S., Cutter, A. B., Jalette, G., Mullin, G., Lai, B., & Yukhymenko, M. et al. (2012). Our princess is in another castle: A review of trends in serious gaming for education. *Review of Educational Research*, *82*(1), 61–89. doi:10.3102/0034654312436980

KEY TERMS AND DEFINITIONS

ARIS: The acronym stands for Augmented Reality and Interactive Storytelling. ARIS is an open-source, Web-based software platform that enables users to create mobile learning activities.

Augmented Reality (AR): Augmented reality is when the physical environment is enhanced with digital information. Specifically, in this chapter, mobile devices enabled the players to access the digital information that was embedded in the real environment.

Engagement: Student engagement occurs when students are attentive, curious, interested, and possibly passionate about their learning. Students may show high levels of motivation and optimism as well.

Flow: Flow is a positive psychological state that is challenging, intrinsically rewarding, and enjoyable (Csikszentmihalyi, 1990). Csikszentmihalyi has posited that there are nine elements of the flow experience, which include the right challenge–skill balance, sense of control, and enjoyment.

Game-Based Learning: Simply put, games used for educational purposes could be categorized as game-based learning; this is one definition used within this study. However, the term game-based learning is also used to describe the effective learning experienced during gameplay.

INPLACE: This is an acronym invented by this chapter's author that stands for Interdependent, Networked, Participatory Learning, Augmented, Collaborative Experience; it is used to describe the type of mobile games discussed in this chapter.

Quick-Response (QR) Code: A visual marker that looks like a two-dimensional barcode that can be scanned with special software. Once scanned, the special software reveals digital information embedded in the barcode such as Web site address.

Chapter 7
Learning about Complex Systems from the Bottom Up:
Role-Playing Together in a Participatory Simulation

Sharona T. Levy
University of Haifa, Israel

ABSTRACT

The topic of learning through collaborative role-playing in computer-based participatory simulations of complex systems in STEM is presented. Participatory simulations are networked classroom activities aimed at learning about complex systems. In the process of learning, students query its underlying structure and explore its spatial, temporal and mathematical patterns in various conditions. The importance of understanding complex systems is highlighted, driving the main question in this chapter: How can we design learning experiences that support students' deep learning of emergent systems? The motivations behind using participatory simulations and their various designs are described as well as some of the more central learning research, cumulating with five studies into designs for such activities in science. Based on this research, eight design principles are introduced and future research directions are proposed.

DOI: 10.4018/978-1-5225-2528-8.ch007

INTRODUCTION

Participatory simulations (hence, partSims) are networked classroom activities that immerse students in collaborative role-playing in a social experiment into complex systems. As a design for learning, students experience the system as its *entities, orchestrators* and *explorers*. They are each an *individual agentic entity* acting and interacting locally. The class itself is the investigated system made up of many such entities, where students *orchestrate experiments*. Through these experiments, they *explore emergent processes*. The unique aspects of learning with such a design lies in traversing between these three perspectives, which provides for a more robust conceptual understanding of the system. Several simulations of this type have been developed over the years, targeting the learning of systems in STEM and social systems, approaching a wide range of topics: epidemics, evolution, bee-hives, global warming, structure of matter, traffic, topics in probability and statistics, functions, geometry, sharing resources, economic markets and more. One of the goals of this chapter is to pull out a set of design principles for such simulations, both out of previous research and out of a series of studies conducted by the author and her students. Such an abstraction of a variety of designs could light the way to further design, use in schools, pedagogical supports and research into this unique learning environment.

Complex systems are prevalent in our world, spanning several orders of magnitude. Students learn about systems in science whose micro-level entities range from sub-atomic particles, electrons, molecules, cells, individual organisms in an ecosystem, all the way up to the stars and planets that comprise our universe. Reasoning about such systems has become an essential part of STEM education (Chen & Stroup, 1993) and recently, of computational literacy (Weintrop et al., 2016). Providing students with access to learning about systems has become of vital concern.

This chapter presents participatory simulations, a design for learning about complex systems that has students role-play individual elements within a system, becoming an integral part of it. In the process, they strategize and conduct experiments, analyze and reflect upon their results. The teacher's role is central in designing the activity, connecting it with other learning taking place, eliciting students' descriptions of experiences and their ideas, supporting conversations that involve local and global perspectives of the simulation that took place, and helping students concretize experiments they would like to conduct. This design can be viewed as an extension of computer models that represent dynamic multiple entities (ABMs, or agent-based models), where some of the "computation" is shifted from the computer to people. For example, in the Dancing with Molecules partSim people role-play gas molecules

(Levy et al., 2015). They determine which way and when to move. However, their speed is determined by the computer simulation based upon energy considerations. In the TrafficJams partSim (Figure 1), students drive cars (right hand interface) and decide whether to shift lanes or slow down (Levy, Rotem & Linyevsky, 2015). However, the computer simulation sets up the roads which constrain the motion and provides collective information, such as the average speed (left hand interface). For example, in the average speed plot, one can see the speed first oscillate and then take a downward dip as a traffic jam forms. On the students' computer, they can see the cars around them but not beyond. They can speed up or down and shift lanes, limited by the cars up ahead and the road itself. As a group, they can observe the teacher's computer that provides the full view of the road from a birds'-eye view, and notice how the traffic jams – high density of cars – gradually spread through the system. They can design experiments aimed at smoothing the traffic flow or speeding it up, using a variety of strategies, such as limiting the speed or the distance between cars. This distribution of computation between a group of people and the simulation provides for enough constraints from the computer so that the conceptual structure of relationships is retained and can be explored and discovered through people's actions and collaborations.

Complex systems are made up of many entities that interact among themselves. The individual actions and local interactions among these entities emerge into global behaviors (Bar-Yam, 2003; Vicsek, 2002). Global patterns are described in ways that are distinct from those used to describe the individuals. Understanding systems presents a challenge for learners as it requires a host of coordinations: between local and global perspectives, among individual elements, between static and dynamic qualities of the system. In fact, one of the more demanding coordinations involves noticing, integrating and patterning parallel processes related to several entities as

Figure 1. TrafficJams class simulation conducted in a computer lab. The simulation (left) sends and receives information from individual clients operated by the students (right)

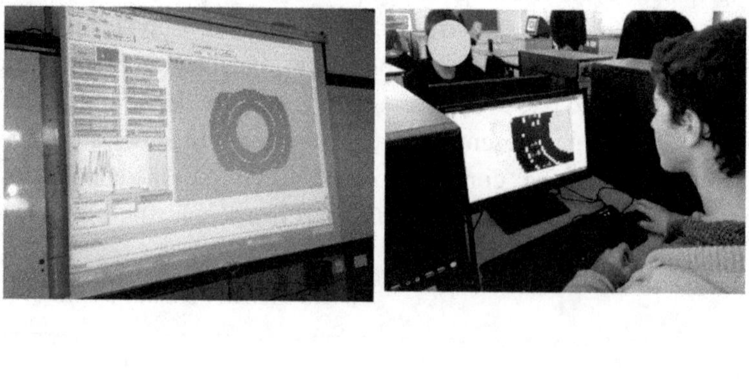

they act and interact in a shared landscape. Assisting people in learning to think in parallel ways about a collection of interacting objects clearly needs a distinct set of supports.

Simulations of epidemics have been the drosophila fly[1] of research into partSims conducted with several platforms, as will be described in the next section (Colella, 2000; Klopfer, Yoon & Perry, 2005; Stroup & Wilensky, 2014; Neulight, Kafai, Kao, Foley, & Galas, 2007). When modeling the spread of disease in a population, people are represented as micro-level individual entities. These people move about, some cross longer distances, some stay closer to home; a few move more rapidly, most are more sedate; some are friendlier than others and tend to congregate. If an individual contracts a contagious disease, their subsequent meetings with people might infect them in turn. This micro-level account of the system includes descriptors such as location, speed, meetings and the local interaction of passing on a disease. Descriptions of the whole population over time and space might include the number of sick people and the rates at which the disease spreads or how it moves between specific locations. Containing the disease may be the goal of the group investigation. Investigating how individual behaviors impact whether or not the disease is contained, helps relate the micro- and macro-levels, is one of the main goals in learning about complex systems. Noticing such relations provides clues and cues into capturing the emergent processes mobilizing and stabilizing the system.

In this chapter, research and several studies and designs of partSims are described. The motivating question in the chapter is: "How can we design learning experiences that support students' deep learning of emergent systems?"

BACKGROUND

Emergence is one of the fascinating and challenging ideas in reasoning about systems: how do many distributed simple interactions produce persistent and stable global patterns? Several designs for learning address this issue: constructing, exploring and participating in models. In *constructing computer models* people represent a system's entities and their properties. They construct the model with code or code-blocks, using simple actions (such as move, turn) and interactions (such as change direction upon collision) at the micro-level (e.g. Papert, 1983; Wilkerson-Jerde, Gravel & Macrander, 2015; Guo et al., 2016) to describe these entities' behaviors. Upon running the model, they view the emergent behaviors in a display that includes other representations, such as graphs and monitors. They can then change the model and explore it, informally or formally constructing a space of behaviors, the system's landscape. A second design involves *exploring a given model* (e.g. Pallant & Tinker, 2004; de Jong & van Joolingen, 1998; Stieff, 2011). This form is most popular in schools, as there are no coding requirements. Most models display a visualization of the computer simulation and opportunities to manipulate global variables. Students change the variables, run the model, observe it and any other associated representations, such as monitors and graphs. PartSims offer a third way of accessing the relationship between local and global patterns by *participating in a social simulation*. In partSims, people are role-play entities in a system on a social platform. The interactions between participants' actions and other objects in the simulation within the constraints built into the simulation result in the system's overall behavior. These experiences – as individuals, and as a collective - become an object of question, manipulation and discussion, transforming the classroom itself into the explored system. In this chapter, partSims are described and several studies into their instantiation are presented.

PartSims have been implemented with a variety of platforms, different from each other in whether they are portable, have a public display of the overall system, whether there is a central computational simulation, and whether or not running the simulation is part of additional activities. Three kinds of platforms are described: portable clients, computer-based clients and simulations set within a MUVE (multi user virtual environment).

Portable Clients. One early platform involved electronic badges (Borovoy, McDonald, Marin & Resnick, 1996). People wore the badges on their clothes while walking around a room in a conference. When two people met, their badges communicated the number of questions out of five they both answered the same. People looked at each other's badges and could use this information to start conversations. While it is certainly a game and fun ice-breaker, this is also a simulation of how using simple information about the relationship between people could create stronger and weaker alliances, clusters, groups and networks. The first version held information only locally between individuals. Understanding the process for the whole group over time involved conversations and analytical activities of representing information to portray the group's evolution. In a later version, Borovoy et al. (1998) extended the use of these electronic badges by including global information. Displays of "Community Mirrors" provided visualizations of the community dynamics. In one version, they simulated the passing of memes via such tags. People could copy memes they liked from other people's badges or recommend a particular meme to someone they met. As a group, this simulated the replication and obliteration of ideas in a society, so that some ideas become anchored and clustered, while others become extinct. These badges were later used to simulate epidemics and other complex phenomena and were named Thinking Tags, small wearable microprocessor-driven tags (Colella, 2000). Klopfer, Yoon and Perry (2005) further extended this work using handheld computers instead of badges, so that more intricate information could be used and displayed on the individual hardware. In their use of the handheld computers, the display turned inwards to the students, different from the badges that showed information about the person outwardly, to other people. They developed a broad range of simulations – predator-prey systems, games of economics, genetics, viruses and more.

Computer-based clients were developed by Wilensky and Stroup (1999) in their creation of HubNet, a component of NetLogo (Wilensky, 1999) which supports the creation of partSims. Considering the question of portability, they formed two versions of HubNet. One involved using graphical calculators, a portable client. The second version used computers as the clients in the system. This system has been used to create a broad range of simulations such as controlling traffic lights in a grid of driving cars and gas particles (Wilensky & Stroup, 2002; Stroup & Wilensky, 2014). It is also used in the studies described hence.

Simulations set within a MUVE. Neulight, Kafai, Kao, Foley & Galas (2006) present a third way of conducting partSims. Their investigation of the spread of disease took place in WhyVille, which is a MUVE (multi user virtual environment) that engages many students with various kinds of activities. The disease simulation is just a part of a range of events and topics participants could engage with. The larger scale of the simulation and its more dispersed focus present a form of learning through partSims that is more akin to everyday life.

Classroom research into learning with partSims is quite sparse. Colella (2000) explored how high-school biology students used Thinking Tags to learn about the spread of disease. During three weeks, the students conducted several investigations. These investigations began with the challenge of meeting as many people as possible without getting sick. There were no public displays in this study, and students accessed global data only at critical points to confirm their hypotheses. Several aspects of the simulation were explored: students' engagement as seen in their excitement and sense of truly experiencing the simulation; the process that took place as the students stepped back to figure out confusing situations; the students' design of several experiments varying their behaviors to find out about the dynamics of the system; the inclusive nature of these activities as the students were all interdependent in conducting their experiments; and how experimental design of investigations into the structure of the system happened only in the later simulations. Ares, Stroup & Schademan (2009) analyzed high-school students' discussions and learning of mathematics by having them each control an elevator among a series of elevators in a HubNet simulation. During the lesson, students were highly active both in participating in the activity and in contributions to the class conversation. These discussions primarily involved observing and describing the simulation, but also predicting, asking for clarification and explaining. The teacher's role was creating very open invitations: invitations to predict, but also to elaborate and explain. Klopfer, Yoon & Perry (2005) explored teachers' perceptions of partSims and found that they saw affordances regarding motivation, engagement and self-directed learning. Stroup & Wilensky (2014) describe sixth-grade and high-school students' reasoning about epidemics and traffic flow in the context of partSims. They find complementarity between two forms of reasoning: the agent-based account of the system, that begins with individuals and builds up; and the aggregate collective account that is framed in terms of populations, rates and flows.

These studies also present a set of *design principles* for partSims. Some of the design principles suggested by Colella (2000) include a high level of *flexibility and iterativity* to accommodate many kinds of experiments that include varying the participants' behavior patterns, and including a *clear goal for the first activity*. In using Thinking Tags, students wore the badges on their clothes so that *the technology was kept unobtrusive* to *support communication* among the students, a design principle implemented in some of the platforms but not others, *simple displays* for the individual student client, *coherent regularities* in the simulation that can be gradually uncovered through their investigation, *make the focus phenomenon happen in the inter-personal space*, facilitate similar but non-identical experiences for participants so that comparing them helps capture distributions of behaviors and associations between conditions and related results, and enabling students' invention of exploration strategies. In describing the design of the Elevators activity, Ares et al. (2009) stress the importance of *having multiple solutions to problems* posed to the students; and the *combination of private spaces* on the students' displays *and public displays* for group deliberations. Articulating these design principles is important to further develop, test and question this learning design. The next section offers additional principles to help build up an understanding of the kinds of learning that can take place.

RESEARCH INTO LEARNING WITH PARTICPATORY SIMULATIONS

This chapter presents several studies into social partSims: a collaboration among a group of people and a computer simulation in enacting the behaviors of a system. In our lab, we have developed several partSims, geared at learning systemic phenomena. These are presented as five short studies. These studies are described as *designs and principles* of the learning environment and as experimental results regarding *learning and processes*. Threaded throughout the studies is an accumulating set of design principles guiding the creation of each environments as well as the designs themselves. Design principles that have been described in previous research are not included. Each study also describes the process of learning and its outcomes. The topics learned include chemistry (global warming, phases of matter), traffic flow and epidemics. The audiences are diverse, ranging from elementary, middle-school to high-school among regular and special education students. In fact, one intention of this presentation is to show the diversity of users and uses for this learning design.

Dancing With Molecules to Learning About Global Warming

Design principles:

1. Everyday experiences as a resource for students' actions.
2. Riding then Driving inanimate objects.
3. Scaffolding models for simple component processes.

The simulation "Dancing with Particles" was developed in response to a middle-school science teacher's enduring difficulties in teaching about global warming (Levy et al., 2011; 2015). Upon surveying existing teaching materials, several drawbacks were detected. Some materials were verbal and visual-spatial but not dynamic, so that the many different processes taking place at the same time are difficult to relate (Groves & Pugh, 1999). In some model-based learning materials, the basic entities in the system are moving packets of energy, which are a less familiar and non-intuitive, preventing learning of the system through the micro-level entities' behaviors (NetLogo Climate Change model, Tinker & Wilensky, 2007; Varma & Linn, 2012).

We wanted to approach the topic *by using everyday experiences as a resource for students' actions*, the first design principle. We selected the experience of moving through world at different speeds, as speed is the individual molecule's "contribution" to the gas temperature. This led us to re-frame global warming in terms of air molecules' *speed.* Since speed and energy are interchangeable for heat conduction, the main mechanism by which atmospheric air is heated, this seemed a fair exchange[2].

However, even after this transformation, we could not avoid the many parallel processes taking place in global warming, one of the main challenges in forming a coherent explanation of the phenomenon. Our solution was to design a partSim that would have students be a molecule in the system. For a single molecule, there is only one process happening at a given time. The molecule moves in a straight line unless other forces act on it, it can change direction and speed upon collision and gravity pulls the molecule towards earth. If it is an emission gas molecule, such as carbon dioxide, then upon absorbing infrared radiation it can transform the absorbed radiation into another molecule's speed by colliding with it.

Our initial conjecture was that starting with a single entity and experiencing its history through the different transformations would be easier to follow and provide for deeper learning of how global warming transpires. Participants would be able to note changes and what led to these changes. Relating these causes and effects would support a more mechanistic understanding. By placing several students-as-molecules together, they could then broaden their scope to the space of the class and the simulation. By comparing their experiences within the shared space, they could see the similarities in the underlying rules and notice the distribution of properties such as speed. This could enable students to gradually understand and integrate both the sequence for a single molecule with the parallel processes taking place for many molecules.

The Dancing with Molecules simulation was designed so that participants can control their molecule's direction of motion, but not their speed (Figure 2). They decide which way to turn and when to move forward. However, their speed is determined by the simulation code based on energy and momentum conversation. Gravity is automatically added as a bit of extra downwards motion at each time segment. Molecules that reach the top of the screen disappear. Molecules that hit the ground bounce like a basketball.

What is important is that the main outcome of global warming – increasing temperatures – is also the main change experienced by the students at the micro-level. They experience speed as their own increased or decreased motion across the screen: for each forward mouse tap, they advance further (fast) or closer (slow). This form of representing speed is quite common in gaming environments. As a group, some are faster and some are slower at a given time; however, on average, they all become faster when the temperatures rise. After the initial experiences of moving as molecules in the system, they could also follow the collective changes on the temperature plot. Finally, they also experienced how emission gases (such as carbon dioxide) transform radiation into increased molecular speed. This would be a crucial step in understanding the paradox of how such a small proportion of

Figure 2. Dancing with molecules simulation

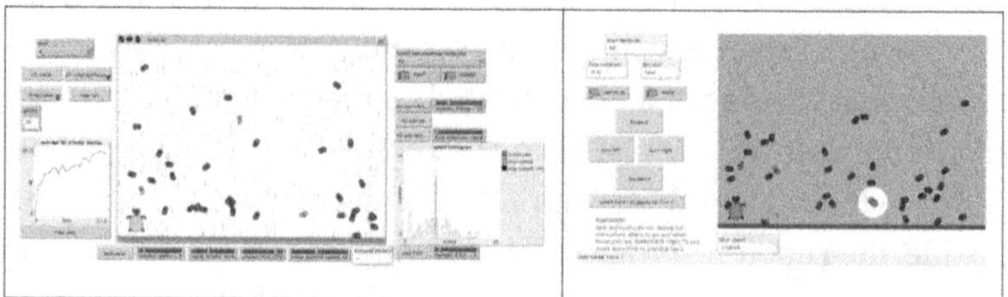

Dancing with Particles class simulation (server) for learning about global warming.	Dancing with Particles personal interface (client)

emission gases could impact the whole atmosphere, making all of earth's gaseous molecules faster.

Two rounds of classroom studies were enacted. The first round (Almog, 2011) was with 18 middle-school students. The second round (Zoubi, 2015) took place with 14 high-school students, majoring in science. In both cases, the learning sequence was accompanied by paper or web-based worksheets and involved both class-wide and individual learning. The unit began with a motivating scenario describing the impact of global warming on polar bears, followed by a general presentation of global warming. When shifting to the simulation, one of the central concerns was that the students would develop a perception of molecules as living decision-making entities. Here, the second design principle – *Ride then Drive inanimate objects* – comes into play. Build-up introductory activities had students "ride" the molecules before they could actually "drive" them, to ensure they begin with a scientific understanding of molecular motion. The unit ended with a teacher-led integration and consolidation of the topic.

In the earlier round of research, middle-school students were challenged with saving all living creatures in the world (Almog, 2011). They played nitrogen molecules in the atmosphere, first riding then driving their molecule. A turtle representing all living things could be either healthy, sick or dead depending on the temperature. The temperature was displayed in a plot. They could do this by coordinating their own motion and behavior as nitrogen molecules, such as running away from carbon-dioxide molecules. They could also adjust general features in the model, such as radiation flux, the amounts of different substances in the air, adding and removing factories.

In practice the activity was quite confusing. The first activities involved getting used to being a nitrogen molecule and finding themselves in their personal and class view. Only later did they control their motion and follow the temperature plot. In the class discussion, they suggested a few strategies that could control the temperature. These ideas included figuring out the impact of emission gases by adding a factory and adding water to the simulated world. The designs were written up on the class display, discussed and voted upon. After clarifying what it meant to run a particular strategy as an individual, they tried some of these out. However, the goal of the activity was not clear. Some of the processes taking place were not even noticed. At the end, we found that they had secretly used a background chat tool, mainly in banter and jest. However, interspersed among these jokes, they also asked each other advice as to what they should be doing in the simulation and what some of the representations meant.

Questionnaires used to assess their understanding of global warming showed no significant differences with a pair t-test as the class means regressed (!) from pretest to posttest: 25 (10) to 19 (11). A typical answer in the posttest shows a superficial understanding that did not include any mechanistic information: *Global warming is a process by which all of the world warms up by a few degrees. Because of the increasing amounts of emission gases. Less sun rays can leave the atmosphere.* The interviews and enactment were analyzed in depth to find the disconnections in reasoning about the system, which processes were not mentioned or were not tied into the whole explanation. Based on these results, the simulation was re-designed.

Several modifications were made. One involves the third design principle: *including scaffolding models for simple component processes.* Activities were designed to clarify some of the component processes and relationships, such as the impact of gravity on molecules' motion, and the transfer of energy upon collision. Some of these activities took place outside of the partSim, using simple models. Planting trees was added to the simulation to broaden the scope of solution strategies for counteracting the effect of emission gases. The questionnaires were also re-designed to test for each of the component processes and concepts, to clarify which concepts were learned and which weren't and how they were integrated into an explanation of global warming.

Research with high-school students in the second round (Zoubi, 2015) showed a rise in questionnaire scores from 46 (20) to 86 (17), with a paired t of 6.477 ($p<0.01$). They demonstrated a deeper understanding in the interviews and questionnaires regarding the components of gravity, solar radiation, absorption of radiation by earth, molecule speed and collisions, temperature. They could also combine these components into a mechanistic explanation of global warming, e.g. *The small amount of CO2 influences global warming by absorbing heat radiation and transferring this energy to other molecules through collisions, which raises the molecules' speed, which is expressed as a rise in temperature.* Several runs of the simulation were conducted, experimenting with the simulation, such as changing gravity. During the simulation students called out to each other, e.g. *Come on… Hurry up!* Class discussions of the simulation were marked by gestures and body motion, that seemed to emulate either the molecules, such as moving their arms to mimic the shape of the carbon-dioxide molecule before and after absorbing radiation; or, a view of an individual molecule, for example tracing the parabolic path of the molecules.

The following discussion of whether the molecules' speed changes shows the generative role of the social arena as the students' parallel experiences are compared:

S1: *Changes, changes… [S1's molecule's speed]*
S2: *For me, it's constant!*
S3: *For me, it changed twice!*
S4: *Also for me, it's constant.*
S2: *You're right! It does change.*
S5: *It doesn't collide (laughter).*
S6: *That means that each molecule has its own motion!*

S7: *Here! Look, this time it collided.*

In comparing experiences of their molecule's speed, some students claim it doesn't change, while others declare it changes. Notice how several students participate in the discussion, a noted feature of partSims. They share their varied personal experiences of speed. This helps S2 shift in her perception of her own experience: she now notices her own speed change. Immediately following this shift, an explanatory idea about collisions is introduced. S6 then comes up with a global concept of diversity or distribution of the molecules' histories "each molecule has its own motion". This helps S7 shift the talk about speed from a general perspective to a moment-by-moment perspective, noticing a collision, an event crucial to speed change. To summarize, in this short exchange we can see how having students be air molecules together in the Dancing with Molecules simulation results in animated discussion ranging from micro-level molecular descriptions to macro-level distributions and back to a more mechanistic micro-level description.

Students gradually coordinated their personal experiences as a molecule to an integrated view of the system. In order to follow this process of coordination, the students' utterances during several simulation runs were coded as (a) being about their personal molecule alone, e.g. *My molecule is moving slowly* or *Here, I'm starting to move;* as (b) focusing on their own molecule but also noticing its local surroundings, e.g. *I'm moving faster than you!* or *Don't collide into me*"; or as (c) taking a global view of the system *Because there are more collisions between the molecules, they change their speed accordingly, which influences the average temperature.* Figure 3 shows how these perspectives evolved during nine runs of the simulation. Earlier, an individual personal view of the system was dominant. Gradually a local-interactive view became more apparent (runs 2-5). Once perception of the local surroundings of the molecule was reached (run 5), it was then integrated with a global view as well so that in the next (run 6) and final (run 9) run we see all three perspectives used in the class conversations. The gradual shift from an individual to a local view, and only later to include a global view has been previously demonstrated by Stroup and Wilensky (2002).

One aspect of running such simulations in a classroom involves noise. These activities captivate most students and involve intrigues and excited exchanges. While the classroom will not be silent, it will be engaged. In measuring the level of noise in the class during the nine runs, we could see an increase between the first half and the second half of the activities. Rather than petering out, the activity continues to engage them throughout, as seen previously by Colella (2000), Ares et al. (2009) and Klopfer et al. (2005). The conclusion of analyzing the videos for on-task versus

Figure 3. Students' views of the system along successive simulation runs

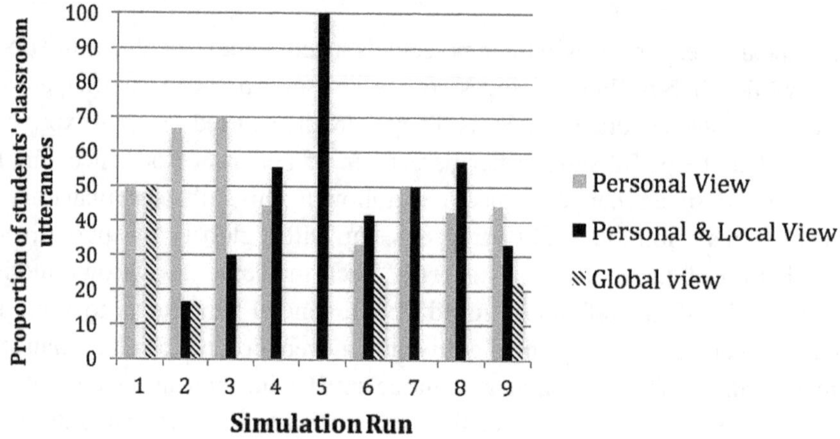

Figure 4. Level of noise in class during activities with the participatory simulation, on-task and off-task activities

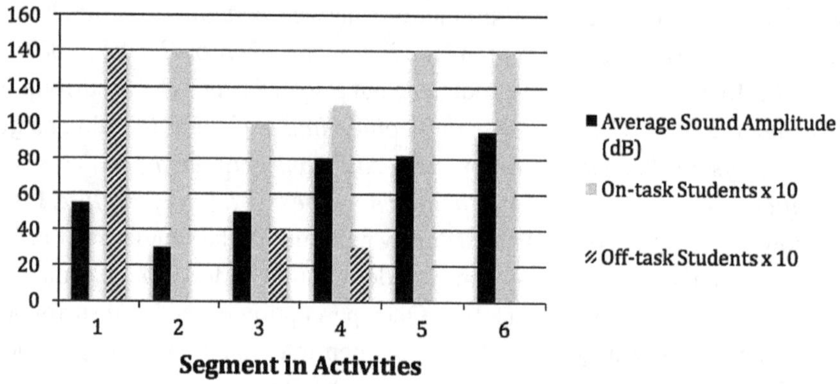

off-task noise-generating activities for each student (Figure 4) we can see a strong reduction in off-task activities and the activities progress.

To conclude this section, we have seen students' engagement sustain during the activity, as their local and global perspectives developed, changed and coordinated. We have also seen the modifications needed to make the simulation accessible to learning that included, broadening the range of activities, creating clear goals and supporting learning of the separate processes when the studied phenomenon involves several combined processes.

MeParticle WeMatter: Learning About the Structure of Matter and Phase Change

Design principles:

1. Seemingly off-task talk during the simulation reflects processing and learning in a convivial way.
2. Design for local micro-level understanding in the initial partSim runs.

After successfully simulating a gas in Dancing with Molecules, the next step was simulating a liquid and its transitions to the gaseous phase. Including inter-particulate forces was the ensuing challenge. As others before us, we were inspired by the Nobel-prize physicist Richard Feynman's summary of the most basic principles of scientific knowledge: *If, in some cataclysm, all scientific knowledge were to be destroyed, and only one sentence passed on to the next generation of creatures, what statement would contain the most information in the fewest words? I believe it is the atomic hypothesis (or atomic fact, or whatever you wish to call it) that all things are made of atoms — little particles that move around in perpetual motion, attracting each other when they are a little distance apart, but repelling upon being squeezed into one another. In that one sentence you will see an enormous amount of information about the world, if just a little imagination and thinking are applied.* (Feynman, 1964). We wanted to transform this minimal set of assumptions –particles' perpetual motion and inter-particulate repulsions and attractions - into students' experiences, forming a basis for understanding a variety of chemical and physical phenomena (Langebeheim & Levy, 2015; 2016).

Let's imagine water in a glass as many particles moving about and then focus on two neighboring particles. When these two particles come close together, pulls attract them together and pushes keep them apart. When the pushes and pulls are balanced, we have reached equilibrium. However, this equilibrium is dynamic. When the water is hotter, this means the particles can move faster and "jiggle" about more frequently. We could have the students move closer and farther away from each other, once again constrained by energy conservation. However, here's the big leap - let us apply this to many particles at once. Each particle can "feel" a number of neighbors, attract and repel with each – the sum of these forces operating right back at the particle. The MeParticle WeMatter partSim (Levy, Langbeheim et al., 2014) that was designed for learning about phases of matter, using a complex systems approach. As in the Dancing with Molecules partSim, the students first ride the molecules and then drive them. They determine their direction of motion and when it takes place. Lennard-Jones forces between the closer particles in the fluid are computed and these in turn, determine the particle's speed (Verlet 1967).

Imagine you were the brown particle (Figure 5) and decided to approach a gang of green particles. You might tread carefully. As you come closer, you are "captured" by sticky forces pulling you towards two particles. Because one particle is closer than the other, its force is also stronger. You can see this as thicker or thinner bubble gum connecting the particles. If you try to detach yourself, this may not be so easy; although with the right angle, you might pull apart one of the bubble gum threads, probably the one with the weaker force.

You then dive *into* the gang of moving green particles. You are gradually surrounded by more and more neighbors and the web of forces gets stickier to maneuver. When you move around, you can see that some particles detach from your back, those in the middle keep you stable and you attach to new particles further up. You are always keeping some of your webby forces, detaching from some in the back, attaching to some in the front.

These experiences provide students with a particle's perspective on what it means to be a liquid or a gas, or what it means to turn from a gas particle to a liquid particle. Positioning students as one-particle-among-many focuses them on the interactions the particle undergoes. Moreover, as action-based perception helps them attend to the local environment, they may notice the formation and dissipation of flexible "clumps" of particles. From a global perspective, the social aspect of the activity of conducting an experiment together may encourage students to notice larger emergent patterns in the system, such as clustering, temporary ordering within a cluster or evaporation mechanisms.

Research into learning vaporization was conducted with two classes of ninth-grade students where one class participated in the MeParticle WeMatter simulation while the second class only observed the simulation as it was projected on a screen (Langhebeheim & Levy, 2015). Both classes completed pre- and post-test questionnaires. Comparison of their questionnaire results shows that the participatory group explained heat conduction significantly ($p<0.05$) better than the observing

Figure 5. MeParticle WeMatter participatory simulation

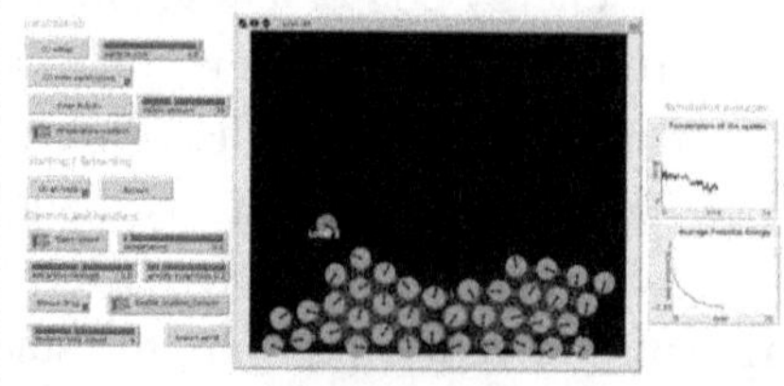	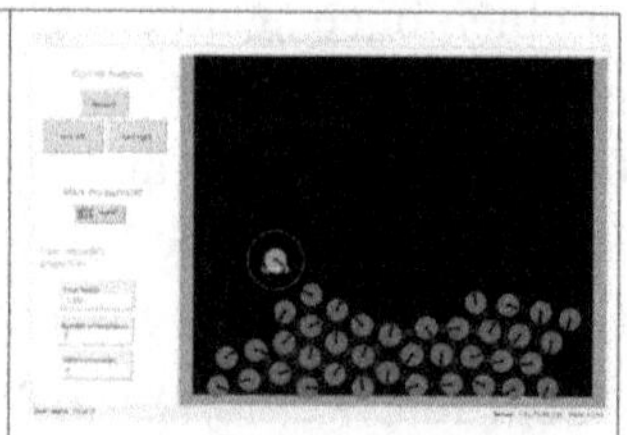
MeParticle WeMatter class simulation (server) for learning about the structure of matter	MeParticle WeMatter personal interface (client)

group in the post-test. For example, in the post-test responses to the question: "Explain how the hot plate causes the fluid's temperature to increase", the participatory group introduced more ideas related to heat conduction as due to energetic interactions between particles and intensified motion of the particles in the liquid when the temperature goes up. At the same time, in the observers group we found a greater increase in students' correct drawings of the global configuration of particles during phase transition from liquid to gas. Figure 6 shows two responses when students are asked to draw the state of the system during phase change. The frequency of correct responses to both heat conduction mechanism and phase transition global configuration questions is shown in Figure 7. We can see that in the question requiring a micro-level understanding of how heat is conducted, observers did not change their responses from pre- to posttest, while the participatory group increased by about 30%. For the question requiring a global view of system, both groups increased in their perception of phase change; however, the observers' increase was much greater (43% versus 25%).

Among the students participating in the simulation, we see a focus on individual particles. During the simulation run, personal identities and interactions are in the limelight:

Figure 6. Correct (left) and incorrect (right) students' drawings of the liquid-gas phase transition

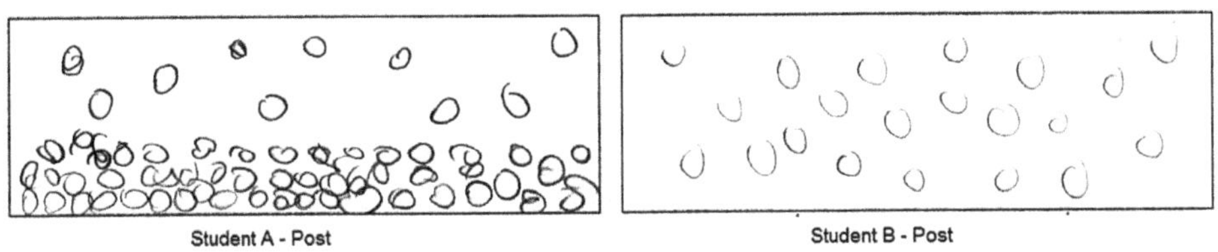

Figure 7. Questionnaire results regarding phase transition drawings (left) and mechanism of heat transfer (right). Participants, n=23; Observers, n=18

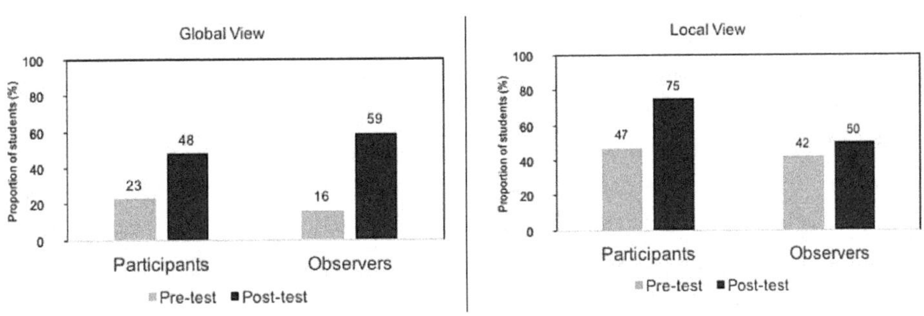

Student 1: *... I am "Safed cheese"* [they could label their particles]
Other Students: *Booom! We got one out. Yeah!*

This exchange demonstrates the fourth design principle: *Seemingly off-task talk during the simulation reflects processing and learning in a convivial way.* Similarly, for the students who observed the simulation the focus is on the complete aggregation of particles, as they argue what its shape is like:

Student 1: *It looks like a boat, a man lying in a boat ...*
Student 2: *No, it is a dead bird! .... Ahh...it IS a boat... but really small.*
Student 3: *But you aren't paying attention to the motion of the particles!*
Student 2: *No, it is actually a plate*
Student 1: *It's a boat, like in Venice*

In this brief case study, we can see that one of the main advantages to using partSims is in noticing interactions and building up a mechanistic explanation of change in the system and does not necessarily support a perception of global patterns. However, this probably results from the brevity of the study. In lengthier interventions (Colella, 2000; Stroup & Wilensky, 2014) and in the study into students' learning about global warming, it was found that students gradually decentered from the individual view to include and coordinate it with the global view. This demonstrates the fifth design principle: *Design for local micro-level understanding in the initial partSim runs.*

Keep Me Healthy!

Design principles:

1. Adaptation with respect to population and current goals.
2. Reruns for reflection.

The study engages students with moderate cognitive-impairment in learning about the spread of contagious diseases and applying this learning to improving their own health. Among cognitively impaired individuals, there is a much higher rate of contagion, resulting mainly from behaviors such as using someone else's cutlery (Krahn, Hammond & Turner, 2006). We have replicated the study in three schools in the north of Israel and found a three-fold increase in morbidity with respect to other students in the same school (Okal, 2015). Normally, these students are taught about health-preserving behaviors via instructions with no explanations.

The goal of the design was to support the students in creating a rich explanatory model of disease transmission they could adapt to with health-preserving behaviors. The overall design of the learning unit is based upon a well-established constructivist framework for special education students, The Circle of Internalization Method (Reiter, 2003)[3] that includes students' successive problem-solving and discussions to establish explanatory knowledge. We adapted the Disease epidemic partSim (Wilensky & Stroup, 1999), demonstrating the seventh principle: *Adaptation with respect to population and current goals*. Students move their avatars in a common space. An individual is infected and can further infect nearby individuals upon contact. Adaptations included (Figure 8): having movies created of each run, so they could be viewed repeatedly and discussed in detail; creating spaces where the sick individuals could go rest or receive care so they could simulate health-seeking in real-life situations; replacing text by icons; personalization of the avatars by using human figures colored with an individually-meaningful color; increasing the size of the widgets and reducing their number (Levy & Okal, 2015). The partSim was used several times along a sequence of nine activities and was a central focus of the unit. The activities were grounded in students' story-telling stories of their own experiences being ill and included alternating individual work and group problem-solving and reflective activities. Reflective discussions related to the simulation runs, using their movies to re-enact the experience, demonstrated the eighth principle: *Reruns for reflection*. Each lesson included at least three such group activities, discussing the experiences.

Figure 8. Keep me Healthy! Participatory simulation

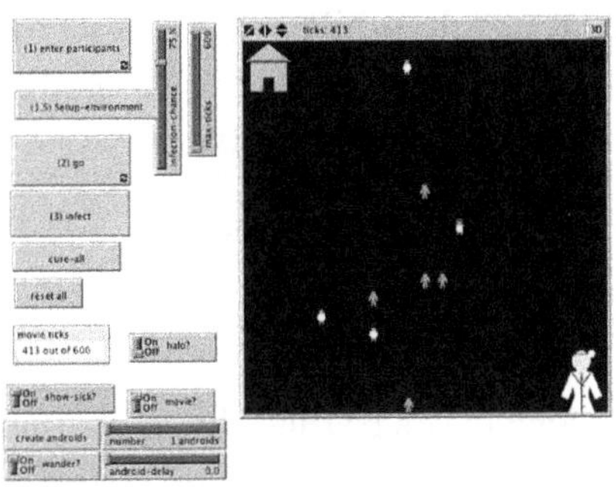	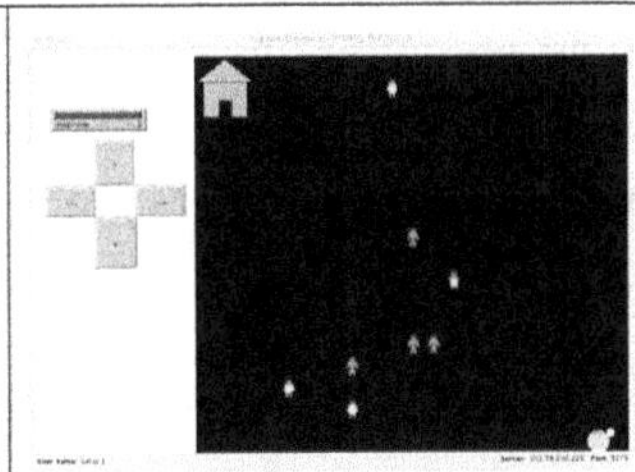
Keep me Healthy! class simulation (server) for learning about epidemics among students who are cognitively-impaired	Keep me Healthy! personal interface (client)

Four 9-11 years-old students (two males, two females) with moderate cognitive-impairment and no other impairments participated in the study (Okal, 2015). Two additional students participated in the learning activities, but were not part of the data collection. A pre- and posttest questionnaire regarding perceived health locus of control (Wallston, 2007) was adapted and reviewed by experts to ensure suitability. Another pre- and posttest engaged the students as a group in a physical simulation of contagion. Activities with the simulation were videotaped at each computer. Analysis of students' actions in the simulations and their comparison provided a measure of health-seeking behaviors and their progression.

The pre- and posttest questionnaire did not provide us with results, as the students did not understand the questions, even with additional supports advised by an expert researcher in the field. The pre- and posttest physical simulation showed an increase in the mean number of health preserving behaviors from M1=20 (SD=2) to M2=34 (SD=4), a 66% increase. When analyzed in greater detail, the increase was due mainly (two-thirds) to self-preserving behaviors (e.g. not using another child's towel) rather than others-preserving (e.g. staying at home when you're sick so that you won't pass the disease to other children).

During the activities and discussion, students grew to articulate an understanding of why specific behaviors were important to preserving their health. In comparing four consecutive runs of the simulation (Figure 9), the number of steps the students made until they got sick, until they reached the doctor and until they went home was recorded. Across time, we can see an increased "survival" rate in the simulation as seen in the rising number of steps taken before succumbing to disease, and a greater swiftness in adopting self-preserving behaviors (getting home or to a doctor).

To conclude, the intensive intervention that involved constructivist pedagogies emphasizing students' reasoning, problem solving and reflection together with several simulations of contagion were effective in supporting both related behaviors and understanding within the time span of the research. While enacted as a group or a system, students were focused on "saving" themselves, rather than on others, and noticed how different aspects of the system impacted them personally.

SOLUTIONS AND RECOMMENDATIONS

The set of studies presented above shows the potential for learning through participating in social experiments set within computer-based simulations. These studies demonstrated students' experiences with partSims as its entities, as orchestrators and as explorers of the system. Through personal deliberation and action, they gained an understanding of interactions and patterns in their local environment. Through sustained activities, they came to grasp the global emergent phenomena, and related

Figure 9. Participants' moves in the simulation across several runs

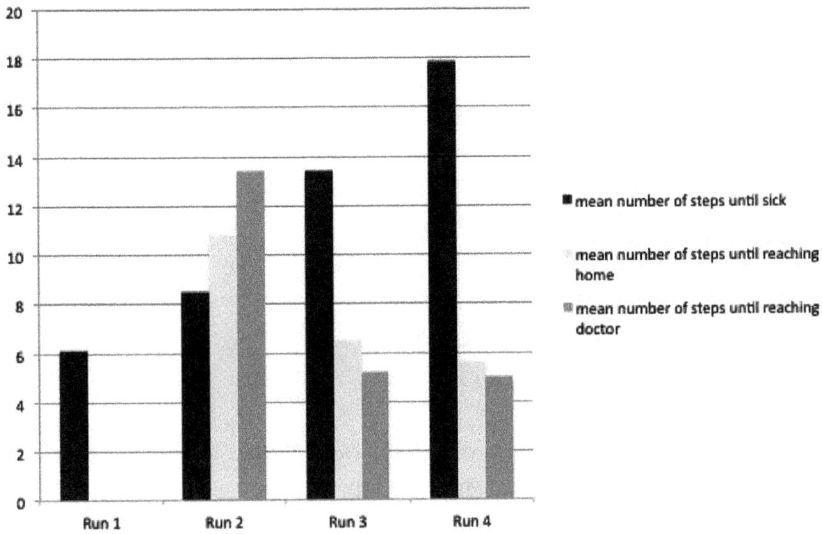

them to their own personal action. In terms of learning about complex systems, we have seen the growth of understanding in some of these studies, that is related to greater attention to interactions as connecting micro- and macro-levels. In terms of conceptual learning in science, this understanding of complex systems is tightly woven with a deep understanding of systems in science.

Several challenges currently limit the use of simulations in classrooms.

One challenge is that the use of such an intricate architecture for class activity requires much planning and organization. Another challenge is the lack of pedagogical guidance for teachers in running partSims in their classrooms.

Currently, teachers may not see the value of expending this effort, given the relative lack of research with partSims. Further research is needed in collaboration with teachers to develop, pedagogically support and study learning through partSims. This research needs to be able to compare learning gains with other learning environment so that the contribution of such learning is clarified. Moreover, once the advantages and disadvantages are delineated, further design can improve upon these results – within the partSims, or by combining with other forms of learning. Pedagogical support needs to be developed, tested and refined, providing examples and vignettes that can help teachers use these systems.

Previous work that was described in the introduction has proposed a number of design principles or guidelines for constructing partSims:

1. Create a combination of private spaces on the students' displays and public displays for group deliberations.
2. Simple displays for the individual student client.
3. Have the focus phenomenon happen in the inter-personal space.
4. Coherent regularities in the simulation that can be gradually uncovered through their investigation.
5. Facilitate similar but non-identical experiences for participants.
6. Have multiple solutions to problems.
7. High level of flexibility and interactivity.
8. Create a clear goal for the first activity.
9. Support communication among the students.
10. Enable students' invention of exploration strategies.

Several additional design principles have been added to this set of guidelines through presentation of the studies in this chapter:

1. Everyday experiences as a resource for students' actions.
2. Riding then driving inanimate objects.
3. Scaffolding models for simple component processes.
4. Seemingly off-task talk during the simulation reflects processing and learning in a convivial way.
5. Adaptation with respect to population and current goals.
6. Reruns for reflection.

This set of principles can be further categorized into four groups: design of the architecture, the simulation itself, the activities with the simulation and the teacher's support, as seen in Figure 10.

Existing partSims may benefit from additional design and research to improve the simulation, activities and teacher support based on these ideas. Future development of new simulations can be supported with such a framework.

FUTURE RESEARCH DIRECTIONS

Conspicuously missing from the design of partSims is the pedagogical aspect. The classroom culture is unique when students are conducting these social experiments. The whole class collaboration led by the students themselves increases their engagement across the board, forming a more equitable classroom. Teachers who experience these activities see their central affordances in engaging students' understanding of their own decisions and actions to comprehend complex phenomena (Klopfer, Yoon

Figure 10. Design principles for creating and using participatory simulations

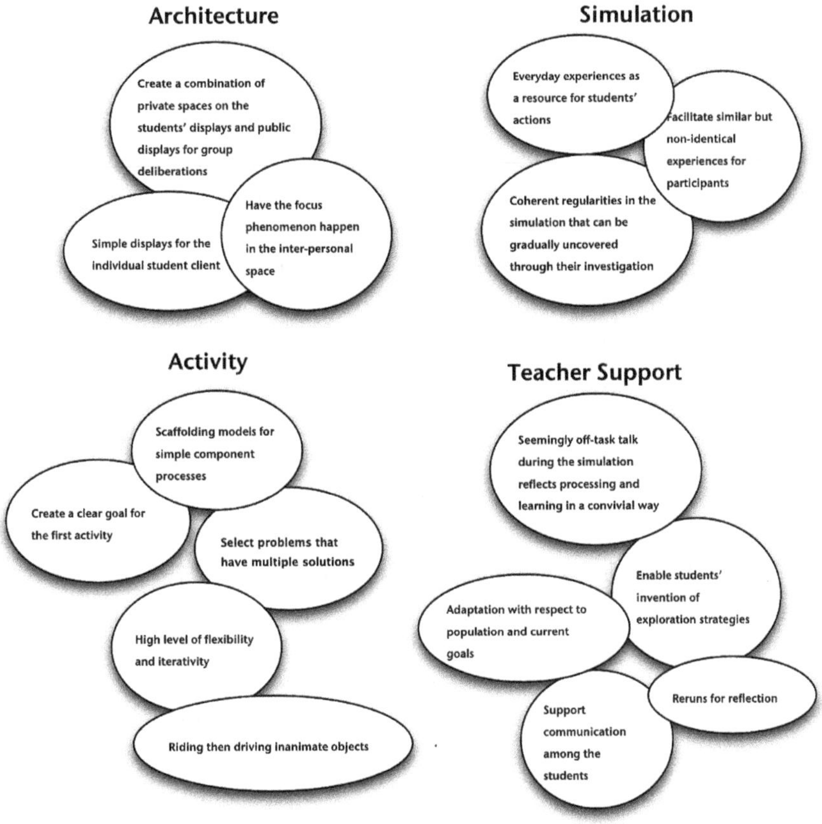

& Perry, 2005). However, few studies explore the teachers' role in class. Notable is the Ares et al. (2009) study that devoted a section in their analysis to the topic, observing the different kinds of invitations the teacher introduced during the lesson. A more comprehensive exploration of the various ways to conduct such activities within classrooms is of essence.

CONCLUSION

Participatory simulations, networked class-wide simulations of complex systems in science have been presented. This chapter has offered a map for the design of participatory simulations that relates to the architecture, simulation and activity design and teacher supports. Being able to separate these categories and see what is missing within each one opens the way for further development of learning in such settings.

Regarding the design of participatory simulations, we have seen several dimensions that relate to their architecture, which help distinguish between several applications. The advantage of seeing these dimensions is that architectures can then be broadened or used to enable a variety of interaction forms.

For example, having portable clients enables participants' movement in space and direct interactions with other participants. Considering whether a simulation will be used with computers or portable devices is much easier today, as various applications are in the process of creating web-based versions.

Using a public display of the whole system creates an emergent and collective object-to-think-with, one individuals and the group can relate to and discuss. However, it is also possible that in some simulations we purposefully want to hide the global picture and have students focus only on the local interactions. Thus, the use of such a public display becomes a question, rather than a given.

While most uses of partSims focus on the simulation alone, we have also seen a case where such a simulation was planted into a virtual world and was only one experience out of many (Neulight et al., 2006). This idea opens up the design of partSims, helping us consider embedding such activities as part of a larger set of activities.

In designing the simulations, it was noted that having simple client helped students focus on the central phenomenon. Harnessing students' everyday experiences as guides for designing their actions as individuals in the simulation was productive for learning to use the simulation quickly and noticing changes at the individual level. Moreover, alignment between the enabled actions and explored phenomenon made some of the more sophisticated concepts accessible. Helping students see the simpler phenomena and then engage with their combination was important in disentangling some the threads of several related phenomena happening at the same time. Planning for multiple investigations and activities is important in making this a rich environment, where the students can invent and study different questions.

Research into classroom learning with partSims has repeatedly shown the deep sustained engagement of students participating in such activities and the inclusive nature of such participation. It also shows students' gradual adoption of a multi-faceted and integrated view of systems, connecting micro-level behaviors, local interaction and global properties of the system at hand. It is important to note that one or two runs would not be enough to achieve such an integrated view, so that using partSims is an extended activity. Students' conversations capture the relationship between their individual experiences and the system's behavior. The richness of this platform is seen in the many and various experiments students design and explore. The nature of such activities involves going back and forth between immersed action during the simulation and stepping back for discussion and reflection when confusing events occur, or after a run is completed. This process in fact supports learning to shift between micro-level and macro-level phenomena and gradually integrate them, as the core of a deep understanding of systems. Being able to experience how local interactions emerge into group behaviors is accompanied by a more global view of the system's behaviors evolving over time. During the activity, students' disparate local and global views of the system increasingly focused on interactions with the local environment, later combining once again with the global and individual perspectives. Finally, research has shown that partSims can be adapted and designed for a variety of populations, including special education students, making even small group partSims a viable avenue to learning.

Teachers' roles were central in setting up the activity itself, designing for multiple entry points, inviting students' predictions, elaborations and explanations. It is important to note that having the students play around in the earlier parts of the activities is necessary for them gain a sense of their virtual body and to go beyond the initial fun-factor of meeting their friends in such a virtual world. Once these needs are met, one can proceed to the more experimental activities. However, even then playful talk is not necessarily a hurdle as it often reflects processing of the system's behavior and can in fact, be tapped onto by the teacher as a source of students' perspectives.

To conclude, it would seem that a combination of playful interactions with partSims and a teacher-supported set of discussions to make sense of and plan the following interactions provide fertile ground for learning about complex systems.

REFERENCES

Almog, H. A. (2011). *Learning about complex systems through exploring models and taking part in participatory simulation in the domain of ecology* (Unpublished master's thesis). Faculty of Education, University of Haifa.

Ares, N., Stroup, W. M., & Schademan, A. R. (2009). The power of mediating artifacts in group-level development of mathematical discourses. *Cognition and Instruction*, *27*(1), 1–24. doi:10.1080/07370000802584497

Bar-Yam, Y. (2003). *Dynamics of complex systems*. New York: Perseus Publishing.

Borovoy, R., Martin, F., Vemuri, S., Resnick, M., Silverman, B., & Hancock, C. (1998). Meme tags and community mirrors: moving from conferences to collaboration. In *Proceedings of the 1998 ACM conference on Computer supported cooperative work* (pp. 159-168). ACM. doi:10.1145/289444.289490

Borovoy, R., McDonald, M., Martin, F., & Resnick, M. (1996). Things that blink: Computationally augmented name tags. *IBM Systems Journal*, *35*(3.4), 488-495.

Chen, D., & Stroup, W. (1993). General system theory: Toward a conceptual framework for science and technology education for all. *Journal of Science Education and Technology*, *2*(3), 447–459. doi:10.1007/BF00694427

Colella, V. (2000). Participatory simulations: Building collaborative understanding through immersive dynamic modeling. *Journal of the Learning Sciences*, *9*(4), 471–500. doi:10.1207/S15327809JLS0904_4

de Jong, T., & van Joolingen, W. R. (1998). Scientific discovery learning with computer simulations of conceptual domains. *Review of Educational Research*, *63*(2), 179–201. doi:10.3102/00346543068002179

Feynman, R., Leighton, R., & Sands, M. (1964). Atoms in Motion. In *The Feynman Lectures on Physics, Volume I*. Retrieved from the Feynman Lectures Website: http://www.feynmanlectures.caltech.edu/I_01.html

Groves, F. H., & Pugh, A. F. (1999). Elementary pre-service teacher perceptions of the greenhouse effect. *Journal of Science Education and Technology*, *8*(1), 75–81. doi:10.1023/A:1009433705790

Guo, Y., Wagh, A., Brady, C., Author, S. T., Horn, M. S., & Wilensky, U. (2016, June). Frogs to Think with: Improving Students' Computational Thinking and Understanding of Evolution in A Code-First Learning Environment. In *Proceedings of the 15th International Conference on Interaction Design and Children* (pp. 246-254). ACM. doi:10.1145/2930674.2930724

Klopfer, E., Yoon, S., & Perry, J. (2005). Using palm technology in participatory simulations of complex systems: A new take on ubiquitous and accessible mobile computing. *Journal of Science Education and Technology*, *14*(3), 285–297. doi:10.1007/s10956-005-7194-0

Krahn, G. L., Hammond, L., & Turner, A. (2006). A cascade of disparities: Health and health care access for people with intellectual disabilities. *Mental Retardation and Developmental Disabilities Research Reviews*, *12*(1), 70–82. doi:10.1002/mrdd.20098 PMID:16435327

Langbeheim, E., & Author, S. T. (2015). *Learning Affordances related to Participation-in and Observation-of Particle Simulations: Hints from Seemingly Off-task Talk*. Paper presented at the National Association for Research in Science Teaching (NARST) annual conference, Chicago, IL.

Langbeheim, E., & Author, S. T. (2016). ConfChem Conference on Interactive Visualizations for Chemistry Teaching and Learning: Learning by Being: Playing Particles in the MeParticle–WeMatter Simulation. *Journal of Chemical Education, 93*(6), 1145–1147. doi:10.1021/acs.jchemed.5b00443

Levy, S. T., Langbeheim, Roller, E., Peri, G., Grady Drozd, Y., & Azzam, Y., E. (2014). MeParticle WeMatter. Participatory simulation for role-playing particles in fluids. Developed at the Systems Learning & Development Lab (SLDL), University of Haifa.

Levy, S. T., Almog, H., Izkovich, A., Zlotnick, L., Eldar, S., & Ofeck, E. (2011; 2015). Dancing with Particles. Participatory simulation for social explorations of global warming. Developed at the Systems Learning & Development Lab (SLDL), University of Haifa.

Levy, S. T., & Okal, Z. (2012). *Keep me Healthy! Participatory simulation for health education among students who are cognitively-impaired*. Retrieved from http://ccl.northwestern.edu/netlogo/models/HubNetDiseaseHubNet

Levy, S. T., Rotem, S., & Linyevsky, A. (2015). *TrafficJams. Participatory simulation for collaborative experiments in driving. Developed at the Systems Learning & Development Lab (SLDL)*. University of Haifa.

Neulight, N., Kafai, Y. B., Kao, L., Foley, B., & Galas, C. (2007). Childrens participation in a virtual epidemic in the science classroom: Making connections to natural infectious diseases. *Journal of Science Education and Technology, 16*(1), 47–58. doi:10.1007/s10956-006-9029-z

Okal, Z. (2015). *Learning about contagion and health in human systems among students with retardation* (Unpublished master's thesis). Faculty of Education, University of Haifa.

Pallant, A., & Tinker, R. F. (2004). Reasoning with atomic-scale molecular dynamic models. *Journal of Science Education and Technology, 13*(1), 51–66. doi:10.1023/B:JOST.0000019638.01800.d0

Papert, S. (1980). *Mindstorms: children, computers, and powerful ideas*. New York, NY: Basic Books.

Reiter, S. (2003). Life skills versus the skill of living: The Circle of Internalization Method for the enhancement of the "Skill of Living". *Proceedings for the International Conference on Inclusive Education ICIE 2003 Inclusive Education: A Framework for Reform*.

Stieff, W. (2011). When is a molecule three dimensional? A task-specific role for imagistic reasoning in advanced chemistry. *Science Education*, *95*(2), 310–336. doi:10.1002/sce.20427

Stroup, W. M., & Wilensky, U. (2014). On the embedded complementarity of agent-based and aggregate reasoning in students developing understanding of dynamic systems. *Technology. Knowledge and Learning*, *19*(1-2), 19–52. doi:10.1007/s10758-014-9218-4

Tinker, R., & Wilensky, U. (2007). *NetLogo Climate Change model.* Retrieved from http://ccl.northwestern.edu/netlogo/models/ClimateChange

Varma, K., & Linn, M. C. (2012). Using interactive technology to support students understanding of the greenhouse effect and global warming. *Journal of Science Education and Technology*, *21*(4), 453–464. doi:10.1007/s10956-011-9337-9

Verlet, L. (1967). Computer experiments on classical fluids I. Thermodynamical properties of Lennard-Jones molecules. *Physical Review*, *159*(1), 98–102. doi:10.1103/PhysRev.159.98

Vicsek, T. (2002). Complexity: The bigger picture. *Nature*, *418*(6894), 131–131. doi:10.1038/418131a PMID:12110869

Weintrop, D., Beheshti, E., Horn, M., Orton, K., Jona, K., Trouille, L., & Wilensky, U. (2016). Defining computational thinking for mathematics and science classrooms. *Journal of Science Education and Technology*, *25*(1), 127–147. doi:10.1007/s10956-015-9581-5

Wilensky, U. (1999). *NetLogo*. Retrieved from http://ccl.northwestern.edu/netlogo/

Wilensky, U., & Stroup, W. (1999). Learning through participatory simulations: Network-based design for systems learning in classrooms. *Proceedings of Computer Supported Collaborative Learning (CSCL'99).*

Wilkerson-Jerde, M. H., Gravel, B. E., & Macrander, C. A. (2015). Exploring shifts in middle school learners modeling activity while generating drawings, animations, and simulations of molecular diffusion. *Journal of Science Education and Technology*, *24*(2-3), 396–415. doi:10.1007/s10956-014-9497-5

Zoubi, H. (2015). *Studying complex systems by participation in a collaborative simulation study unit on the topic of global warming* (Unpublished master's thesis). Faculty of Education, University of Haifa.

KEY TERMS AND DEFINITIONS

Agent: An entity that takes up an active role in creating an effect.

Agent-Based Models: A class of computational models that simulate agents actions and interactions to build up the system's behavior.

Client: An application that runs on a computer that accesses a service provided by a server.

Complex System: Complex systems are made up of many entities that interact among themselves. The individual actions and local interactions among these entities emerge into global behaviors.

Computer Model: A program that is designed to represent and simulate a system or phenomenon.

Design Principle: A formalism that relates a design feature to a rationale.

Emergence: The process by which many distributed simple interactions produce persistent and stable global patterns.

Participatory Simulation: A design for learning about complex systems that has students role-play individual elements within a system, becoming an integral part of it.

Role-Play: Act out a particular role while assuming its attitudes, actions and discourse.

Simulation: Running the representation of the operation and functioning of a system, process or phenomenon to observe its outcomes.

ENDNOTES

1 The *drosophila melanogaster*, colloquially known as the fruit fly, became a favorite model for genetics (and later, developmental biology) research from its earliest days.

2 Consultation with an atmospheric chemistry and climate change expert, Yinon Rudich from the Weizmann Institute, one of the leading scientists in the field was an integral part of this development.

3 Shunit Reiter (University of Haifa) herself advised us in the educational design process and reviewed the study's design.

Chapter 8
Simulations in Chemistry for Conceptual Understanding and Assessment of Student Knowledge

Tanya Gupta
South Dakota State University, USA

Gregory Albing
South Dakota State University, USA

Zachary P. Ziolkowski
South Dakota State University, USA

Akash Mehta
Independent Researcher, USA

ABSTRACT

Simulations are dynamic resources that have been found useful for communicating abstract fundamental ideas such as stoichiometry and several other concepts. In this chapter the authors present their recent work on designing and implementing an interactive simulation called Combustion Lab based on reaction stoichiometry - a topic that has continually been a challenge for chemistry learners. Several researchers have reported persistent student misconceptions in stoichiometry. In order to address this challenge, a novel computer simulation was developed to assess student understandings of stoichiometry based on student problem solving performance, and also to promote student conceptual understanding. The Combustion lab was particularly focused on the stoichiometry of these reactions, problem solving, and the relevance of stoichiometry for its everyday applications. Results of this sequential exploratory study show that the simulation was effective in revealing student understanding and student treatment of stoichiometry problems based on analysis of various data collected.

DOI: 10.4018/978-1-5225-2528-8.ch008

INTRODUCTION

The use of interactive simulations in chemistry has seen a rapid surge in the past decade. Simulations are digital environments that allow users to interact with models of various concepts and processes. A simulation uses a mathematical or a logical model to illustrate real world phenomenon with a goal of providing opportunity to students to understand the underlying concept or scientific model (Suits & Sanger, 2013). Prior researchers have found simulations to be effective in helping students visualize relative motion and in improving problem solving and critical thinking by supplementing sensory experiences of learners (Falvo, 2008; Gerjets & Hesse, 2004; Stieff & Wilensky, 2003).

Simulations have been found effective in engaging students; improving student exam scores and their representational competence, and fostering student understanding of specific concepts such as the particulate nature of matter and electrochemistry (Barak 2013; Falvo, Urban, & Suits, 2011). Simulations allow students to explore phenomena and their representations while manipulating variables. Simulations thus provide a certain degree of control to learners through their interactive features such as play and pause buttons, adjusting variables such as mass, volume etc. using buttons and sliders, and stop, review, and exit functions. Some researchers have focused on design features of simulations. These researchers suggest that simulation developers should design simulations that are simple to use; display a balance of information, have audio and visual components in design; allow user control; reduce extraneous cognitive load on learners; and accurately portray the concepts and processes in chemistry. Simulation designers should also consider the prior knowledge of learners, which plays an important role in student use of simulations (Landriscina, 2013; Plass, Homer, & Hayward, 2009)

Despite numerous studies on simulations in chemistry, there is a lack of evidence of simulations being effective in assessing student understanding based on problem solving abilities of students with specific problems incorporated within the simulation. This chapter highlights the development and implementation of a stoichiometry based simulation called "Simulation Lab" and its implementation in organic chemistry laboratory with students taking first semester of a two-semester sequence of organic chemistry laboratory course in a mid-western university. The goal of simulation was to assess student understanding based on their performance on stoichiometry problems. The study also determined student approach to problem solving based on qualitative analysis of simulation lab data and student - written laboratory reports.

Rationale for Developing Simulation on Stoichiometry

Chemistry is a unique subject that requires an understanding of submicroscopic processes through macroscopic observations and symbolic depiction of various chemical processes. Many concepts studied in chemistry are abstract and are inexplicable without the use of analogies or models (Bou Jaoude & Barakat, 2003). Hence students perceive chemistry to be a difficult subject. A reason behind this perception of difficulty for novice learners could be that chemical processes are physically observable at macroscopic level, but can only be conceptually explained at the particulate (submicroscopic) level. A deeper and correct conceptual understanding of chemistry requires connecting the three representational levels coherently – the macroscopic, particulate level, and symbolic level of representations. Novice students struggle with integrating these three levels and treat each level as independent of the other (Gulacar, Eilks, & Bowman, 2014; Marais & Combrinck, 2009).

Stoichiometry is a foundational concept in chemistry. It provides a quantitative relation between reactants and products for chemical reactions that occur in nature or are conducted in laboratory (Finley, Stewart, & Yarroch, 1982; Hessen, Ågron, Anders, Elser, & Ruiler, 2004). Understanding of stoichiometry is important to determine the amount of a certain chemical needed or to predict the theoretical amount of product that can be generated during a chemical reaction. It also helps determine the efficiency of a chemical reaction by comparing the yield of product to theoretical or predicted amounts for a certain reaction. Though stoichiometry is a fundamental concept behind every chemical reaction, students seem to have difficulties in making conceptual connections based on correct understanding of various representations of these chemical reactions. Prior research studies have shown that students are able to provide correct responses to stoichiometry problems without applying the law of conservation. For example, studies conducted by Nurrenbern and Pickering (1987), Gabel and Bunce (1984), and Yarroch (1985) demonstrate that students often rely on the use of algorithms when solving chemistry problems that involve proportional reasoning. The problem solving approaches of a majority of students rely on rote memorization and regurgitation of formulas based on the surface features of stoichiometric problems. Research studies on student misconceptions in stoichiometry (Huddle & Pillay, 1996) have demonstrated that students:

1. Often confuse subscripts and coefficients in a balanced chemical equation
2. Are able to solve direct ratios such as 1:1 stoichiometry when the reactions involve two reactants leading to simple products, but are lost when the reaction schemes become complicated and ratios of reactants are changed from simple 1:1 to 1:3 or 3:5 leading to different product ratios

3. Do not correctly understand and apply the law of conservation of mass to stoichiometry-based problems often changing the number of atoms before and after the reaction.
4. Tend to mistake mass with moles and vice versa due to incorrect understanding of mole concept or applications of Avogadro's number.
5. Lack an understanding of the limiting or excess reagents in reactions and the role of a limiting reagent in determining the reaction yield.
6. Incorrectly display various states of reactants and products (solids, liquids or gas) and lack a fundamental understanding that the states of matter especially of reactants play an important role in determining reaction conditions and product formations.
7. Do not see a relation between symbolic representation and macroscopic phenomena and the conceptual connection at particulate level.

Reasons behind this lack of student understanding could be attributed to weak foundational knowledge of chemistry. However, a challenge that often remains unattended is that most students experience an instructional emphasis on computational problem solving in stoichiometry irrespective of the context being lecture or laboratory. The difficulty of the subject matter and the ease of reducing stoichiometry to manipulation of mathematical symbols and use of algorithms offer a glimpse of what most curriculum materials and instructional interventions offer to students in various settings. Understanding of Stoichiometry is crucial for students pursuing various chemistry courses and aspiring for majors that require basic understanding of chemical reactions. Despite research on the importance of stoichiometry, there have been no interventions reported to date that are helpful in assessing student understanding of stoichiometry online. Particularly research is lacking on the use of simulations for the assessment of student understanding. This study addresses a significant gap in current literature on the use of simulations for assessment of student knowledge in stoichiometry.

BACKGROUND

Simulations and Their Effectiveness in Chemistry: Summary of Literature

Simulations can serve as a scaffold for students to make connections by allowing students to manipulate variables, and having them engage in self-explanations of the phenomena experienced through the simulated models (Kearney, 2004; Keeney-Kennicutt & Merchant; 2013). Experts can easily construct mental models of phenomena by connecting various visuals and conceptual information deeply ingrained in their minds through years of experience. However, novice learners struggle to construct such models due to their undeveloped understandings and lack of sufficient prior knowledge of a given topic (Fiore & Schooler, 2002; Gupta, PhD Dissertation, 2012; Gabel, Sherwood, & Enochs, 1984; Monaghan & Clement, 2000).

An expert chemist solves complex problems in chemistry by drawing on his/ her prior understandings and by using various mental operations and models to arrive at the solution. Drawing from this observation, the conceptual understanding of students can be measured by studying the correctness of their representations, self-explanations, and expressions of their mental models for solving various complex chemistry problems (Chi, Feltovich, & Glaser, 1981; Kozma & Russell, 1997). Conceptual understanding in chemistry depends on the acquisition and integration of several ideas and representations at the macroscopic level (observable or concrete level such as laboratory experiments and demonstrations or phenomena occurring in real world outside laboratory), submicroscopic level (involves interactions of atoms, molecules or ions) and symbolic level (equations, graphs or depiction of a chemical process through mathematical and chemical formulas). Submicroscopic level representations provide an opportunity to learners to convert their observations of phenomena to the particulate level of interactions. Such particulate behavior though not directly observable can be physically abstracted through simulations. Symbolic representations are at the far right of this continuum that reduces a chemical process or a phenomenon to a set of formulas or equations. These various representations offer challenge to chemistry students who often compartmentalize representations independent of each other instead of connecting the dots between the three and mentally integrating representations as coherent mental model of the chemical process (Boo, 1998; Bou Jaoude & Barakat, 2003; Monaghan & Clement, 1999). A coherent mental model will encompass all three levels of representations as overlapping triangles at the circle as shown in Figure 1.

Several researchers have developed and studied the effectiveness of chemistry simulations. For example; Lancaster, Moore, Parson, and Perkins (2013) have developed PhET simulations which are a library of about 125 interactive simulations in physics and chemistry. PhET simulations aim at helping students learn chemistry through inquiry and include linked multiple-representations. These simulations were initially developed using Java programming language and are now being converted to HTML due to the limitation of these simulations to run on windows, macs and Linux computers. PhET simulations have been studied extensively till date comparing them with traditional and inquiry based approaches and have been founds effective by various researchers with mixed outcomes on student conceptual understanding (Wieman, Adams, & Perkins, 2008).

Figure 1. A coherent mental model interconnecting three levels of representations in chemistry

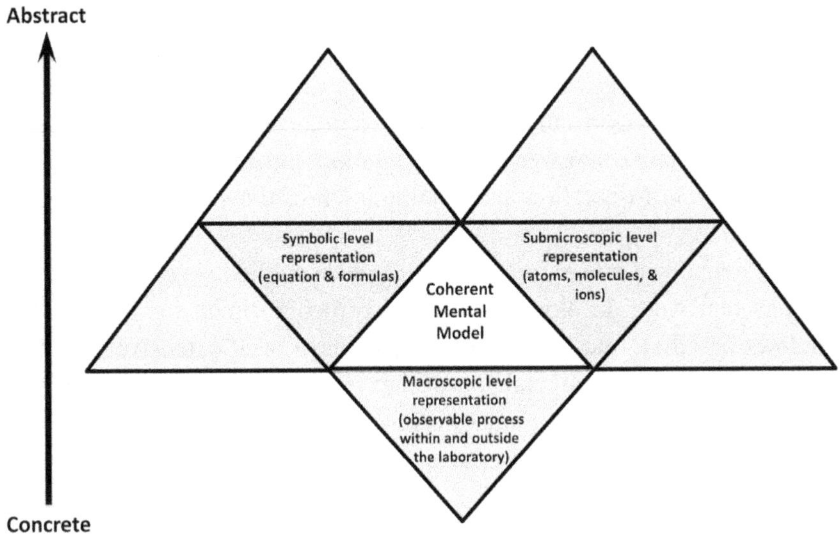

Ashe and Yaron (2013) developed and studied the effectiveness of analogy-based simulations that use familiar objects such as balls and staircase as a concrete system to depict molecular level interactions for physical chemistry concepts. Analogy based simulations are helpful in providing a correct conceptual understanding via a concrete dynamic representation that draws parallels between familiar concrete objects as visuals and unfamiliar abstract chemical process at particle level (Yaron, Karabinos, Lange, Greeno, & Leinhardt, 2010) Some developers have embedded simulations into virtual worlds as virtual laboratories or as web-based assignments (Cole & Todd, 2003; Laughlin, 2001). Virtual laboratories provide opportunities to users to interact with one another in a three-dimensional virtual world, and with objects available in virtual reality using second life or avatars. Students in a virtual world can perform experiments and make observations. These virtual experiments have led to improvement in procedural knowledge and conceptual knowledge of students. Some researchers have integrated virtual laboratories in their classroom and studied their effectiveness. Such studies have demonstrated a better understanding of students for chemical structures when compared to 2-D representations (Keeney-Kennicutt & Merchant, 2013).

Various software such as, Jmol, Chemsketch, and ChemDraw are used as representational tools to help students explore, and interpret three-dimensional molecular geometries and stereochemistry (Barnea & Dori, 1999; Charistos, Tsipis, & Sigalas, 2005, Hanson, 2010). Jmol is an open-source Java based visualization tool that can be used to observe and manipulate large and small molecules and several crystal structures in both two and three dimensions. Such software help students understand the reaction mechanisms with respect to changes in the three-dimensional geometry of molecules during various steps of reactions and through reaction profile diagrams. Students can explore the binding sites of proteins and configurational changes occurring during reactions (Herraez, 2006). These tools help students internalize the external representations into their conceptual framework at a concrete level of operation and manipulation of the software (Hegarty, Kriz, & Cate, 2003).

Burke, Greenbowe and Windshitl (1998) have studied several animations and simulations that were developed using macromedia director and flash and run on windows based system. These simulations have been extensively studied and compared with other pedagogical approaches by various researchers. The findings of these studies indicate that simulations have been helpful for students in terms of understanding acid-base reactions, current flow, and thermochemistry.

RESEARCH ON DESIGN FEATURES OF STUDENT-CENTERED SIMULATIONS

Development of simulations has captured the interest of several researchers who have examined these pedagogical interventions at various levels of chemistry instruction. The overarching goal of simulation designers is to enable students to visualize a concept on the computer screen and construct mental models akin to the visuals and models that are mentally envisioned by experts. Few researchers have focused on the design featured that foster development of coherent mental models of students (Falvo & Suits, 2009; Falvo, Urban, & Suits, 2011).

Of particular importance for simulation designers is an emphasis on essentials, which is the chemical process itself, and not the peripheral features such as color, brightness or other visual displays that distract learners. The fundamental belief of researchers engaged in this process is that simulations should focus on providing an interactive learning environment across a spectrum of learning goals. Such simulations should clearly state the learning objectives and take in account the prior knowledge of students (end users) for a given topic. Such a simulation will culminate in assessment that goes beyond mere recall or facts or algorithms while encouraging students to develop their own conceptual or mental models that are consistent with accepted scientific models (Reiber, 2005; Reigeluth & Schwartz, 1989). The assessment can take place through an instructional sequence that requires continuous interaction between the teacher and instructor and also includes interaction with simulation.

Researchers recommend simulation designers to first identify the key concepts and then the specific features of a chemical system. The next step should involve establishing a correspondence between the features of interest of the chemical system and that of concrete system that is being used as a visual for simulation development. Such a design would help students visually connect the concrete system with an abstract idea behind the chemical process via observational matching. Further, simulations should be simple enough to reduce any extraneous cognitive load that might negatively impact the connection and transfer of short-term experience of manipulating the simulation and forming of long-term memory via practice and repetition through simulation (Hasler, Kersten, & Sweller, 2007; Schnotz & Kurschner, 2008; Sweller, Van Merrienboer, & Plaas, 1998).

Another recommendation is that simulations should be simple yet realistic to an extent (Solimondau, 2009). A realistic simulation can be too complicated to convey a phenomenon as simple as dissolution of salt in water. A complete dynamic visual representation of dissolving salt would require the simulation developer to consider hydrogen bonds between water molecules, formation of charged species Na^+ and Cl^-, representation of polar water molecules, attraction of Na^+ ions to oppositely charged end of water molecules, vibration of particles in NaCl crystal lattice. Such a complete realistic representation of dissolving would be very complicated and pose a challenge for students. It will be hard for students to determine which specific aspect of visual is important and to make sense of several changes or sequences occurring one after another for a given concept. Hence the simulations should focus on a few interrelated sequences to ensure that leaners can focus on key aspects of the concept such as hydrogen bonding or vibrations in crystal lattice (Akaygun & Jones, 2013).

Simulations can be continuous or segmented depending on the learning goals. In general scientists contemplate various chemical reactions using verbal descriptions of various structures and functions of the reaction components. It is recommended that simulations be segmented or chunked into small steps or episodes. When chemical processes are depicted as a continuous visualization, they require processing of large amounts of information by learners (Suits & Diack, 2002). Such a simulation includes thousands of frames of changes in structures accompanied by changes on the screen in terms of shapes and colors of the structures. These rapid transitions can lead to cognitive load on learner making it difficult for students to interpret all the information that is being presented on screen. The advantage of segmenting is that it provides opportunities to students to focus on each aspect of simulation individually and see the connection of new structure with the preceding structure in the simulation. Segmented or distinctive episodes presented through simulations can contribute to the meaningful learning of a chemical process. Such design requires careful planning of each sequence in terms of the importance of the aspects of phenomena being presented, and can be achieved by including interactive features such as pause or play, rewind, or slide functions within the simulation.

Simulations can also be accompanied by audio or verbal input (Falvo, 2008). This can provide learners with information regarding the concept that they are studying through simulation or directions for exploration of the chemical phenomena. Audio needs to flow along with each sequence and should not be too lengthy for learners to loose engagement. Lastly it is important to draw on learner's knowledge before, during and after their experience with dynamic visualizations. A large number of simulations allow users to interact and use their knowledge while they are constructing their mental models (Monaghan & Clement, 2000). Yet simulations developed thus far lack in-built assessments and ability to engage learners in authentic inquiry experiences that build on a comprehensive research-based model of inquiry such as the learning cycle approach.

LEARNING CYCLE APPROACH TO INQUIRY TEACHING

Inquiry based teaching extends to classroom and laboratory setting. The role of teachers in an inquiry-based laboratory translates from being the sage-on-stage to being a guide-on-side. Instructor thus actively facilitates student learning while students engage in hands on inquiry activities. The goal of inquiry-based instruction is to establish the understanding of scientific concepts and processes and provide skills to students that are consistent with scientific practices of the experts in the discipline (Gupta, Burke, Mehta, & Greenbowe, 2015). Learning Cycle (LC) is a student-centered inquiry-based approach derived from constructivist theory of teaching that emphasizes knowledge construction through individual experiences or activities that involve group work or collaborative learning (Abraham, 1986).

The LC approach is divided into distinct phases of instruction that are conducted in a specific sequence (Lawson, Abraham, & Renner, 1989). These phases include exploration, concept invention or concept introduction, concept application, and concept evaluation. The goal of concept exploration is to provide opportunities to learners to explore relations and patterns in data by engaging in hands-on experiences. Such experiences can include laboratory experiments, demonstrations, videos, and simulations. In this phase students focus on what they see and observe, and emphasis is on gathering data. During the concept introduction phase, the goal is to have students develop and understand concepts with teachers and peers through a brief lecture or class discussion. Students make sense of their experiences of the exploration phase from data gathered by asking questions such as-"what does it mean" and then proceed to explaining patterns and anomalies in the data. The goal of concept application stage is to ensure that students apply, reinforce, review and extend their understanding of concepts beyond the first two phases of exploration and introduction. This can be achieved by reading further, solving new problems, or engaging in verification based laboratory activity tied to the main concept. The questions during this phase focus on asking students "what they know", and to apply their knowledge to answer or address specific questions or problems that extends the concept to the next level of application. The last stage of concept evaluation involves assessment of student knowledge acquired through the first three phases through, quizzes and examinations. The LC model is depicted in Figure 2. The simulation designed by the authors is based on this LC model of inquiry with in-built assessment for assessing the prior knowledge and student understanding (knowledge evaluation component).

A NEW SIMULATION ON STOICHIOMETRY: COMBUSTION LAB

Description of Simulation

The Combustion Lab simulation is a comprehensive unit on stoichiometry of combustion reactions and is based on LC approach to inquiry-based instruction.

Figure 2. The learning cycle model

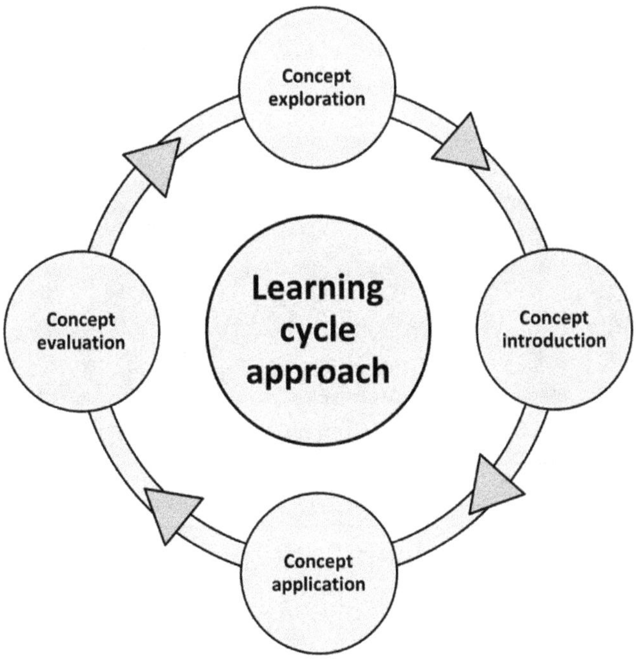

Combustion Lab is a 2-D simulation focused on student conceptual understanding and assessment of student knowledge. It accomplishes this goal by providing various experiences of concept exploration, concept introduction, and concept application and knowledge evaluation as four buttons in the opening screen (Figure 3). In addition to these four phases of LC, the simulation incorporates a key element of student knowledge assessment at various stages beginning from prior knowledge activation, to concept exploration, concept introduction, and concept application. Also, the researcher or instructor has access to behind-the scenes data for various clicks and moves of students when they do the simulation activity. Knowledge evaluation is embedded within each component of simulation. The simulation is segmented to cover different combustion reactions. Students can start with any of the component by clicking on any four buttons, however it is recommended that students start with each component and follow the sequence of buttons from top to bottom for an authentic LC based inquiry experience.

The Prior Knowledge Activation (PKA) segment of simulation includes multiple-choice type question bank that was partly developed by first author. A few questions in this question bank are adapted from Chemical Concept Inventory (CCI) developed and validated by Mulford and Robinson (2002) with prior written permission from the author. This segment of Simulation examines student prior knowledge based on an inbuilt assessment of student understanding prior to engaging in the exploration phase. Figure 4 provides a screen-image of a question from this segment of combustion lab.

Figure 3. Opening screen of Combustion Lab Simulation displaying the four stages of LC approach embedded in simulation

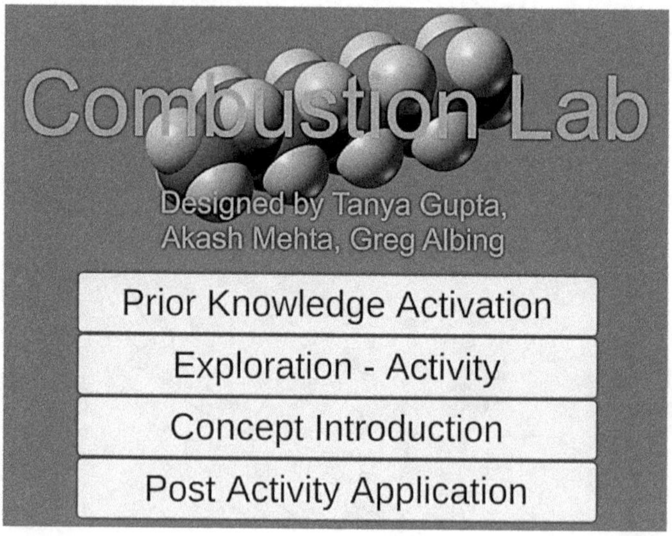

Figure 4. A screen-shot of the PKA segment of combustion lab simulation

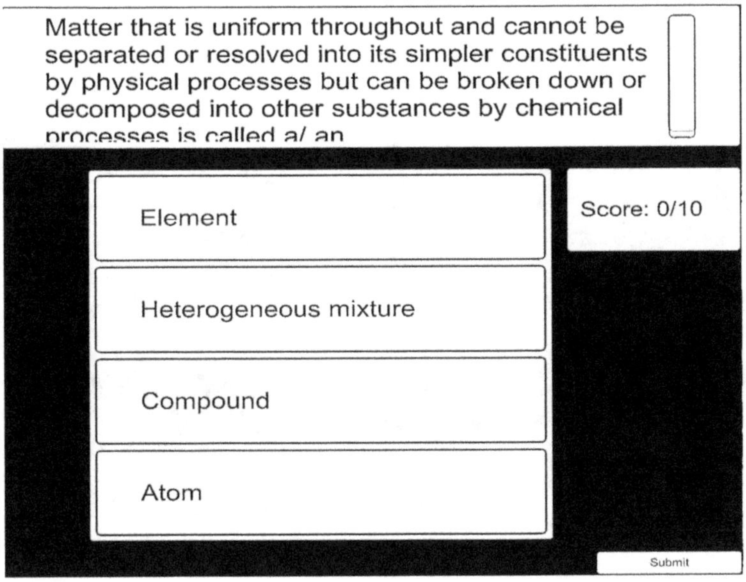

In PKA segment students are presented with 10 multiple-choice questions. In order to proceed from one question to next students must select an answer choice. Each correct response increases score by a point and overall 10 points can be achieved in the PKA segment.

The Concept Exploration (CE) segment of simulation is titled Exploration-Activity or (EA). On clicking this button students are presented individual combustions reactions that are further segmented into series of macroscopic, submicroscopic and symbolic representations. Students can use arrows to explore a given reaction and move back and forth. Figure 5 depicts the combustion reaction at macroscopic level. Figure 6 and 7 represent the symbolic and molecular representations of these reactions during the concept exploration. Only one reaction of the four different combustion reactions is depicted here to give an idea of each sub-segment of exploration-activity. For example when a student explores an activity that involves combustion of octane – the student is presented with sliders to adjust the ratio of air to fuel. Once the student arrives at the correct ratio (25:2) there is an ignition observed in the simulation at the moving piston that indicates that car engine has started. As the fuel ignites, a forward arrow appears on the bottom right corner of the screen indicating user to proceed to the next sub-segment which is a blackboard (see figure 6 for digital chalkboard with symbolic representation or reaction equation). On this digital blackboard the user can balance the reaction equation for the reaction explored in previous segment, by entering coefficients for reactants and products.

Figure 5. Concept Exploration (CE) segment of combustion lab simulation depicting reaction at macroscopic level

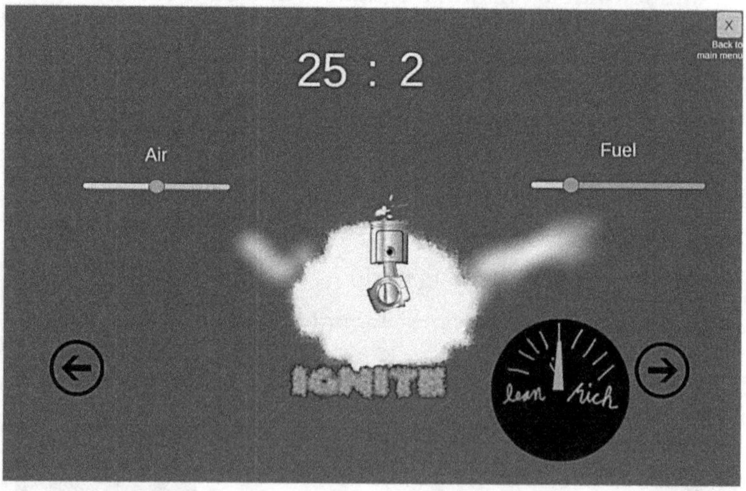

Figure 6. Concept Exploration (CE) segment of combustion lab simulation: symbolic representation of reaction with digital board for balancing reaction equation

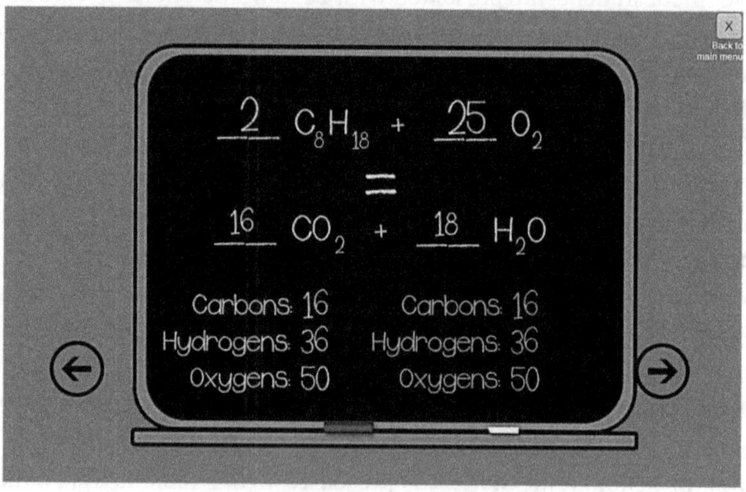

A forward arrow at the bottom right of simulation window leads users to the next screen or sub-segment of simulation wherein the conversion factors for various amounts of reactants and products are provided along with the space-filled models of each reactant and product (Figure 7-molecular level representation with units). In exploration students can use sliders and set the amount of reactants and products

Figure 7. Concept Exploration (CE) segment of combustion lab simulation: molecular level representation of reaction with units included for conversions

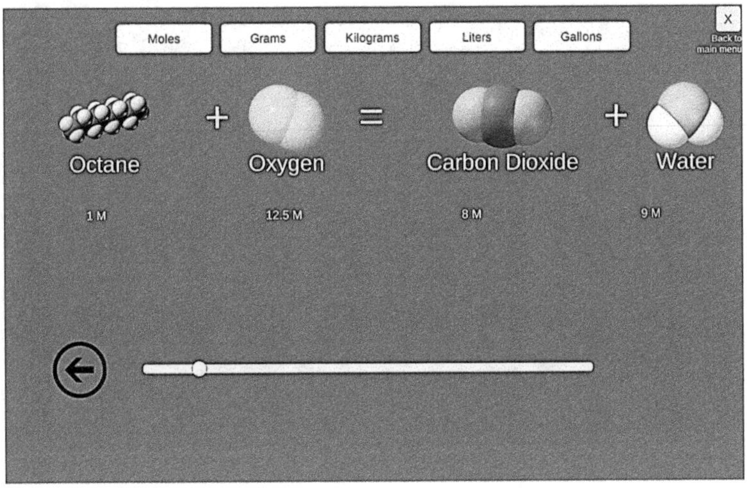

in units of moles, grams, kilograms, liters and gallons. When students change the amount on the slider the amount of reactants and products displayed underneath molecular representation also changes. The EA involves inbuilt trials and errors for students, which is consistent with the learning cycle approach. Students can play with various ratios for each reaction and try different coefficients and explore conversions for different amounts of reactants and products.

The Concept Introduction (CI) segment of the simulation introduces students to concept on a screen with essential information (Figure 8-Concept Introduction segment of combustion lab) and two buttons. In CI phase students are presented brief information on coefficients, law of conservation of mass, combustion reactions and the impact of the use of fossil fuels on global warming. First button, which is "go back", allows students to revisit any of the four combustion activities. The second button "visit EPA website links the EPA (Environmental Protection Agency) website from within the simulation (Figure 9- users can visit EPA website from within the simulation by clicking button). This inbuilt feature provides students an opportunity to understand the relevance of the stoichiometry of combustion reactions in light of the effects of these fuels on global warming leading to climate change.

The concept application segment of simulation or Post Activity Application (PAA) has specific questions on problem solving approaches for stoichiometry problems including questions on limiting reagent, particulate or submicroscopic representations, and applications, for example, calculating the gas mileage and cost for a round-trip to Disneyland, and the resultant quantity of a greenhouse gas

Figure 8. Concept Introduction (CI) segment of combustion lab simulation

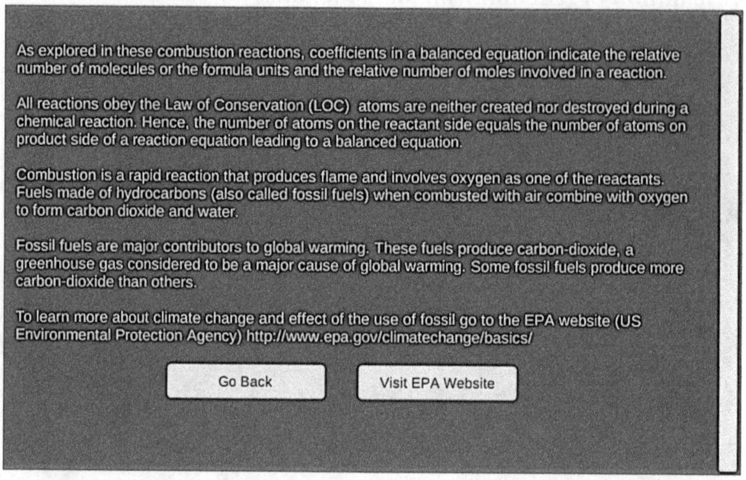

Figure 9. Users can visit EPA website using a button within the simulation

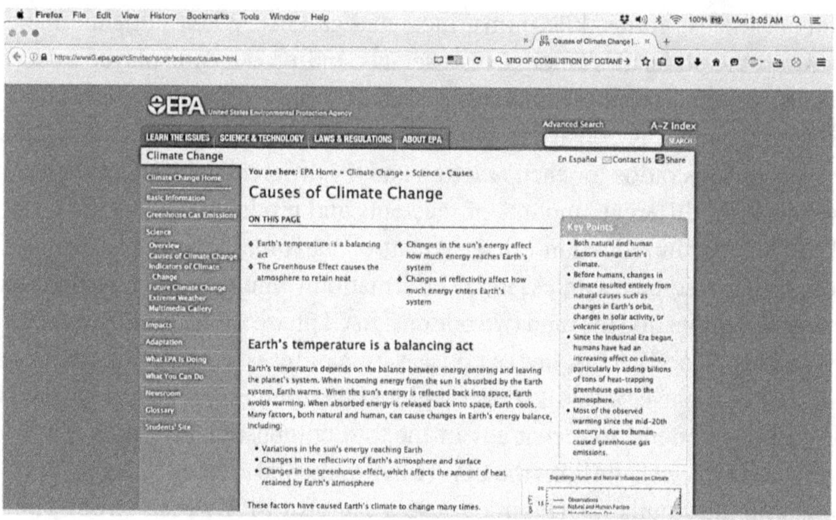

(carbon-dioxide) emitted by car as a result of the trip (Figure 10- Example of an open-ended PKA problem). These problems apply concepts from basic to advanced level on stoichiometry of combustion reactions. Students can calculate their answers and provide response to these open-ended problems in a box for student response within simulation problem. The simulation provides an inbuilt feedback as red flash based on the student final answer being correct or wrong.

Figure 10. Post Activity Application (PAA) segment of combustion lab simulation depicting a problem based on real-life application on stoichiometry

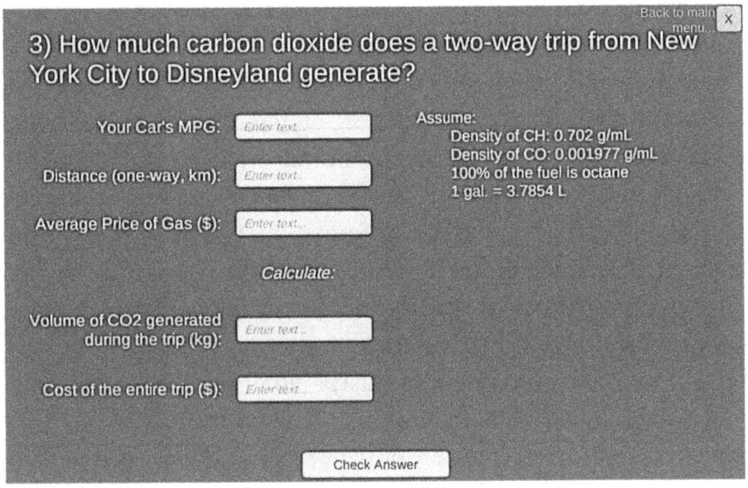

RESEARCH STUDY ON EFFECTIVENESS OF OUR NEW SIMULATION

A key objective of authors was to implement combustion lab and to study its effectiveness. Authors wanted to examine if the combustion lab met the goal of assessing and aiding student conceptual understanding of stoichiometry of combustion reactions through analysis of student problem solving. In light of this objective- the implementation of simulation and the student use of combustion lab in an undergraduate level organic chemistry course was examined with a focus on two specific research questions as below:

1. What does combustion lab simulation reveal about student understanding of stoichiometry based on student performance on stoichiometry based problems?
2. What sequence of steps (paths) do students take for solving stoichiometry problems?

Theoretical Framework for Study

The theoretical frameworks of constructivism, symbolic interactionism and model-based learning and teaching informed this study on effectiveness of combustion lab. The basis of constructivism is the building of knowledge. According to Gabel

(1999), chemistry concepts are abstract. There needs to be something in the long-term memory of learners to which a new concept can be related for it to hold on to learner's conceptual framework. Prior experiences of individuals influence construction of knowledge. New ideas or experiences are interpreted through a process of assimilation that makes new knowledge and experiences fit into our existing mental structures. Accommodation, on the other hand, takes into account the processing of this new information by an individual and subsequently changes learners existing mental framework. Any new knowledge is solidified when an individual actively accommodates new information. In addition to assimilation and accommodation, disequilibrium is important to examine. All learning is motivated by a desire to maintain a state of equilibrium. Disequilibrium occurs when a learner encounters new information that contradicts previously learned ideas. Cognitive conflicts that cause a shift in equilibrium must be either assimilated or accommodated. Through this struggle back to equilibrium, an individual reaches an understanding that deepens and augments one's view of the world. The construction of knowledge should be coherent, viable and consistent with the established principles of the discipline (Herron, 1975).

Symbolic interactionism is a based on the need for symbols and their importance in knowledge construction. This framework focuses on the different meanings people place on objects, interactions, and people, and the corresponding behaviors that reflect this range of interpretations (Blumer, 1969) Symbolic interactionism reflects an individual's desire to interpret events and knowledge of their surrounding environment. This is especially relevant for chemistry as the development of chemistry knowledge to a large extent relies on the use of symbols as a language (formulas and equations) and microscopic and macroscopic representation of phenomena as static 2D-diagrams and images or dynamic simulations (2D or 3D).

Models are central to scientific reasoning and comprehension. Scientists often construct and use models to explore and explain phenomena, to arrive at correct explanations and to build evidence and arguments. Model-based teaching and learning is an important aspect of chemistry instruction. Conceptual models can promote deep understanding of science at a level that goes beyond rote memorization of facts, equations and procedures (Campbell, Oh, & Nielson, 2013). Model-based instruction can improve students' ability to make sense of scientific phenomena by enabling them to construct causal explanations. However, it is challenging to engage in models-based learning when teachers lack experience in developing and using models for instruction and struggle with conceptualizing models-based teaching. It is also difficult to bring forth hidden-mental models of students and connecting these student models to abstract scientific models (Clement, 2000). To address this challenge of developing scientifically coherent conceptual models, one needs to consider target model that is a qualitative, simplified analogous model of a specific phenomena. Such a model takes into account learner's preconceptions, their initial models, and their natural reasoning skills and uses these as building blocks through one or more intermediate models to reach target models. Simulations can play an important role in this process through mediating sequence or intermediate models. Various models in the form of pictorial representations, animations and simulations can positively impact student conceptual understanding.

Constructivism and symbolic interactionism tie into the teaching and learning of stoichiometry. For example, the teaching of chemistry involves manipulation of various physical models such as ball-and-stick models. However, once chemical symbols and equations are introduced, individuals begin to associate these equations as symbols for stoichiometry. This, in turn, perhaps leads to algorithmic thinking which hinders the growth of the initial conceptual teachings. Wolfer (2000) expressed concerns with use of equations stating that algorithms (equations) can be viewed as a structure or framework that students use in solving problems, but algorithms can also be used as a shortcut when understanding is limited or as a way to avoid expending cognitive energy in a solution. For students who seek equations, it does not remain necessary to connect their observations of macroscopic behavior with the symbols and/ equations or to explain the process at atomic or molecular level.

A reason behind poor comprehension of stoichiometry could be a lack of reinforcement of conceptual ideas during instruction. In a study by Wolfer and Lederman (2000) the instructor teaching an undergraduate-level introductory chemistry course emphasized the importance of the conceptual ideas and microscopic level of stoichiometry, but on final assessment the instructor asked algorithmic questions. Researchers also suggest that in order to fully understand stoichiometry, analogies must be used to relate complex and abstract concepts to ideas already known by the learner. These analogies become symbols to the individual in order to further make sense of the concept. However, analogies are known for eventually breaking down under some circumstances. This breakdown of analogies leads to gaps in understanding. In order to be fully effective, individuals need to comprehend the analogy and notice the connection between the new concept and his or her preexisting idea. Researchers have also shown that students approach stoichiometry problems differently. Gulacar, Eilks, and Bowman (2014) noticed that high-achieving and low-achieving students approach stoichiometry problems differently. High achieving students organize their approach to a various problems around the underlying chemistry principles, whereas low-achieving students rely on memorized formulas only. An objective of the use of simulations is to provide further insight into the steps of students for solving stoichiometry problems.

Research Methods

This study was conducted using a sequential explanatory mixed-method (Creswell & Plano Clark, 2011). A sequential explanatory study is characterized by collection of both quantitative and qualitative data. Quantitative data was collected and analyzed first followed by qualitative data to support or explain quantitative findings. For this study, the sequence of administration of quizzes, simulation as intervention and data collection through student written laboratory reports is summarized in Figure 11. The qualitative data in the form of student written laboratory reports was integral in explaining and interpreting the results of the quantitative data of the tests.

Students enrolled in an undergraduate first semester organic chemistry laboratory course (of a two-semester course sequence) were invited to participate in this study after obtaining Institutional Review Board's (IRB) approval for project. The simulation was implemented as an alternate online laboratory activity. The combustion lab simulation was posted on the D2L (Desire to Learn) learning management system (LMS) and students were provided instructions for downloading and conducting the combustion lab and writing laboratory report. The rationale behind selecting the first semester of a two-semester organic chemistry laboratory sequence course was that all students enrolled in this course had already experienced stoichiometry during the two-semester sequence of general chemistry. It was hypothesized that students had sufficient prior knowledge to engage in a stoichiometry based online lab with some minimal directions on downloading and doing the laboratory activity. The online laboratory activity was assigned during the first few weeks before students engaged in any wet laboratory on organic synthesis that would involve stoichiometry based calculations (mole ratios and limiting reagents). The D2L was set for students to download and run the combustion lab during the laboratory section meeting for the same amount of time (2 hours and fifty minutes) as their regular in laboratory session meeting.

Data Collection and Analysis

Overall 261 students were enrolled for the one credit organic chemistry laboratory course out of which 215 students (82.4%) consented to participate in this study. Students enrolled in the organic chemistry laboratory were mainly pre-med, pre-pharmacy, pre-dental or allied health science majors. The male to female ratio in the laboratory course was about 43% males and 57% females. Students enrolled in a laboratory were also enrolled in organic lecture course, which is a co-requisite. It is important to note that the implementation of combustion lab simulation was done in the laboratory and the data is collected only from the laboratory component of the course. The sequence for simulation implementation and data collection is presented in Figure 11.

Prior to the implementation of combustion lab, the students were administered a pre-test that had problems similar to the assessment within stoichiometry combustion lab simulation. Out of the possible list of twenty questions, five questions were randomly selected for each student. The questions were in multiple-choice format but were carefully designed including distractors with different units which meant that students could not simply guess correct answers, but had to calculate before submitting a response to each quiz question. D2L folder was conditionally set such that students could access simulation lab on D2L only after completing the pre-quiz. Once the intervention was completed, students were given a post-quiz which became available only after students completed the simulation and submitted a screen-shot of the PKA problem in D2L dropbox. The post-test also included five randomly selected questions from the same list of twenty questions as the pre-test. Students had to write a laboratory report for each component (segment) of simulation and show their calculations for their responses to the PAA questions for full credit for the laboratory report component. Following Simulation on Combustion Lab week students began laboratory experiments (in the actual lab) wherein they performed experiments on synthesis of organic compounds. These laboratory activities required calculation of reaction stoichiometry and percent efficiency (yield) of the reaction. Final assessment of students was based on the laboratory practical exam at the end of the semester. The exam included questions based on the concept of combustion reactions and organic synthesis reaction stoichiometry.

At the end of semester data was downloaded from D2L website as Microsoft Excel files. The files were then cleaned and organized- students who did not consent to participation in the research study were omitted from the data analysis. Further students with incomplete data (missing a quiz or laboratory report) were also eliminated which left with sample size of 193 students. Upon examining the data, a pre- and post-quiz question sequence was developed for each student to match randomized questions for pre- and post-quiz. The user data generated from the simulation use by students was also converted to Microsoft Excel spreadsheets. These files were then organized into the different sections of the simulation (PKA, PAA and CI) for generating response frequencies and studying user behavior. A

Figure 11. Sequence of simulation lab implementation and data collection

two-tailed matched-pair t-test was performed (Howell, 2007) to compare student performance on pre- and post-quiz and for PKA and PAA combustion lab questions. For this chapter the authors will only focus on use of simulations for assessment of student understanding based on student problem solving for stoichiometry based questions and student approach to problem solving based on analysis of quantitative and qualitative data generated during simulation implementation.

Lab reports written by the students were retrieved from the Dropbox on D2L learning management system. The lab reports provided a detailed description on the steps each student took in order to answer the questions within the exploration-activity and PAA sections of the simulation. Student responses to PAA Question 1 (which detailed the typical order that each student takes when solving stoichiometric problems) were then compared to the second PAA question, and to the detailed problem-solving steps that they took when answering other questions within the simulation. These reports were analyzed using the ATLAS.ti software through open coding followed by thematic analysis of laboratory reports for the combustion lab simulation. The reports had an inter-rater agreement of 87% between two raters who independently coded various laboratory reports for students who took the pre- and post-quiz, completed all components of simulation, and completed the laboratory reports.

Quantitative Results and Discussion

Research Question 1: Student Performance on Stoichiometry Problems

The data for the pre- and post-quiz, PKA and PAA problems within simulations and specific questions on laboratory final exam are displayed in Table 1. The results of two-tailed t-test for pre- and post-quiz indicate a statistically significant difference ($p=0.008$; $\alpha=0.05$, DF, 192) between student scores from pre to post-quiz. It is important to note that the quiz was based on stoichiometry and the randomized questions were similar in levels of difficulty from pre-to post-quiz but essentially not the same (wording or numbers in questions could have differed due to random assignment). Students also performed significantly better on stoichiometry questions in laboratory final exam (two-tailed t-test; $p=0.0001$; $\alpha=0.05$, DF, 192). However, they performed poorly on PAA questions as is evident from student average for this segment of simulations (Table 1). These results indicate that students struggle with transferring their understanding to complex problems of real world significance as evident from very low mean scores (0.57) on PAA questions in simulation lab.

Table 1. Comparison of student performance on stoichiometry problems in pre- and post-quiz; PKA & PAA problems within simulation, and specific questions on final laboratory practical examination

(N=193)	Mean (Out of 10)	Standard Deviation	Paired T-Test
Pre-Quiz	6.90	2.14	p=0.0008*; DF=192; 95% confidence
Post-Quiz	7.60	2.25	
PKA	5.84	0.39	0.1000; DF=192; 95% confidence
PAA	0.57	.07	NA
Stoichiometry questions in lab practical exam	6.39	1.65	p=0.001*; DF=192; 95% confidence

* Significant at α=0.05

Research Question 2: Sequence of Steps (Paths) of Students for Solving Stoichiometry Problems

The student sequence for specific steps of problem solving for a combustion reaction of methanol were also analyzed based on student responses on PAA and the corresponding end-user data collected in excel format. We first used excel data for identifying the steps that were used by students by reading through the excel spreadsheet and determining the frequencies of steps. We then used most frequently used steps or actions of students for solving stoichiometry problems as qualitative descriptors or codes. Based on excel data we found that six steps or problem solving actions were most common. These six steps or actions used during simulation exercises corresponded directly to the actual steps taken by most students for PAA question on the combustion of methanol in water with different starting amounts of the reactants.

1. Calculate the molecular masses of methanol and oxygen.
2. Identify the type of reaction based on knowledge of the reactions involving oxygen.
3. Set up the equation to calculate the amount of products formed using the factor-label method of dimensional analysis.
4. Write the molecular formula for methanol and oxygen.
5. Find the limiting or excess reagent.
6. Write the balanced chemical equation for the reaction of methanol and oxygen

Student steps were organized by permutation and combination method after thematically coding student problem solving approach from sequence of steps generated by excel spread-sheet and on qualitative analyses of student work on the laboratory report for this particular question. These are interesting results with 35.61% students using a somewhat conceptual approach to solving problems (steps 421635) starting with molecular formulas of reactants→identifying reaction type (conceptual framework)→calculating molecular mass of reactants→writing balanced equation for reaction→setting up equation for calculation using factor label method→finding the limiting or excess reagent. Majority of students used algorithmic approach to solve this problem with the second largest and almost equivalent percent of 35.93% students using sequence 146352 that begins with first step being calculation of molar mass of reactants→writing molecular formula of reactants→setting up balanced equation for reaction→set up equation to calculate products formed using factor label method→finding limiting reagent and→identifying the type of reaction.

Lastly, the steps outlined by each student in the first post-activity application question were compared to the steps they took to arrive at their final answer for the second question of the same section. For this analysis, the sample size was 142 because only the students who completed the pre- and post-simulation quizzes, post-activity application, and laboratory report were considered. Only 142 students attempted to solve the PAA question on a real word stoichiometry problem to a certain extent before they gave up. Out of these only 31 students (21.83%) correctly solved the problem of cost of trip and the amount for carbon dioxide generate in the reaction during a round trip to California (see Figure 10 for example of this question). The data show a positive outcome of simulation implementation. Though the means for the entire class sample were statistically significant for pre- to post quiz and students showed similar trend for PKA questions that were inbuilt in simulation, the PAA question mean scores showed that students were not able to completely transfer their understanding of stoichiometry from one type of question to another (Table 1).

Table 2. Student path showing numbered sequence of steps or problem solving actions for Stoichiometry problems in PKA segment of simulation

Sequence of Problem Solving Path	Percentage of Students Choosing Similar Approach
4, 1, 3, 6, 2, 5	11.50%
4, 1, 3, 2, 5, 6	15.96%
4, 2, 1, 6, 3, 5	35.61%
1, 4, 6, 3, 5, 2	36.93%

Qualitative Results and Discussion

Upon examining the laboratory reports of students, three different levels of understanding of stoichiometry were observed for student approach to solving PAA questions. Only 35% of students had a conceptual approach that involved considering the law of conservation and interactions at the particulate level (Figure 12). Majority of students (65%) were algorithmic and tried to solve problem based on direct mathematical relation. These results are consistent with sequence of steps of students who had a conceptual approach to solving problems as shown in Table 2 and highlight student difficulties with transfer of knowledge to application based questions wherein only 21.83% of students managed to correctly solve an application based problem.

Only a few students seemed to have a grasp on the conceptual idea of stoichiometry. Student written reports further provide insights into this challenge of knowledge transfer to higher order stoichiometry problems. For example one student conceptually explained stoichiometry to be:

Figure 12. Student approach to stoichiometry problem solving PAA question as analyzed from student written laboratory reports

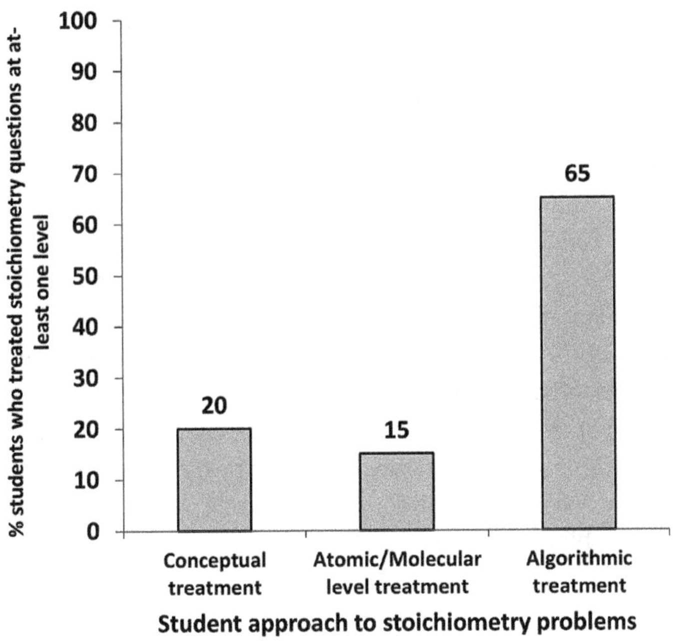

Stoichiometry is the balancing of molecules on both the reactants and the products sides to ensure that the number of atoms in the reactants equals the number of atoms in the products. Stoichiometry is important to learn because the necessary input and the probable output of a reaction can be calculated without the reaction-taking place. It is a form of planning ahead.

Other students understood that stoichiometry was important at the molecular level but failed to actually make that connection with macroscopic observation they experienced during the exploration of combustion reactions and applying it to PAA questions. One student wrote-

This is an important concept to learn because it helps make sure that there is the correct ratio of atoms so that the reaction reacts completely and so that the correct makeup of the atoms can be determined.

Though this student appears to have an understanding of the concept, on further analysis it was found that this student missed the question that focused on the microscopic level of stoichiometry on both the pre- and post-quizzes. Laboratory reports further ascertained that many students in this study treated the stoichiometry problems algorithmically. For instance, a student reported that-

Stoichiometry can be used to convert between metric units and imperial units.

Qualitative analysis of student reports showed that student's understanding of stoichiometry was filled with gaps and students had a difficult time relating the symbolic, submicroscopic and macroscopic levels of the simulation for a complete understanding of stoichiometry. Students do show some degree of conceptual understanding of stoichiometry and were able to think in terms of atoms or molecules but most students had these representations as a set of disconnected ideas. This indicates that student mental models of stoichiometry are fragmented. Considering that participants in this study had experienced two semester of general chemistry before engaging in simulation laboratory this is an alarming picture revealed by our simulation. Without simulation it would not have been possible to see the sequence of exact steps (actions) taken by students to solve post activity questions and match these to student work in the laboratory reports to develop step sequence (specific student path to problem solving) as codes.

FUTURE RESEARCH DIRECTIONS

The study had few limitations to begin with. One limitation of this study was that simulation was implemented online. It will be worthwhile to have students experience these simulations in a classroom and/or laboratory setting and to actually structure a unit around simulations to promote conceptual understanding through small group, in-class discussions and then assess student knowledge of stoichiometry. Further, due to time constraints qualitative interviews of students in this particular group could not be conducted to generate further information about student thought processes while solving problems.

For future studies an aspect worth considering would be to actually implement simulations with students in general chemistry during the stoichiometry unit as an in-class activity for combustion reactions, involving instructor facilitation for discussion of applications. This will provide further evidence on the effectiveness of simulation for assessment of student knowledge. First author is developing new simulations based on learning cycle approach to inquiry for other concepts. Implementing these inquiry-based simulations at various points during the semester will provide a rich data on the effectiveness of such simulations and provide insights on student knowledge. The study has important implications for the use of simulations as assessment tools for online and in class instruction. Instructors can use this simulation in multiple ways depending on their need. One important use for instructors would be to exactly pinpoint and address gaps in student understanding. Instead of waiting for final assessment, the student data can be used to provide remedial instruction targeted to individual students in real time based on the back end user data.

CONCLUSION

The study established the usefulness of combustion lab simulation in assessing student conceptual understanding based on problem-solving abilities of students. Quantitative findings showed a significant gain for students from pre to post-quiz with implementation of simulation and also for laboratory final exams. However, based on analysis of user data generated from simulation and qualitative laboratory reports it can be concluded that a vast majority of students who participated in this study and were college juniors aiming for a health science career do not understand stoichiometry conceptually. They are unable to connect the macroscopic observations with symbolic and submicroscopic representations. This is especially alarming considering that these students want to pursue careers that require a fundamental knowledge of chemical processes. While students show adeptness in solving problems that are similar to the concept, students struggle with applying their understanding to real world issues, for example, the release of greenhouse gases into the environment by combustion of fossil fuels. Our simulation proved to be a useful tool and allowed for an in-depth examination of how students use and apply their stoichiometry knowledge. From this study, it is evident simulations can be invaluable tools to generate a clear and comprehensive picture of student knowledge in chemistry and other STEM areas.

REFERENCES

Abraham, M. R. (1998). The learning cycle approach as a strategy for instruction in science. In K. Tobin & B. Fraser (Eds.), *International Handbook of Science Education* (pp. 513–524). Kluwer. doi:10.1007/978-94-011-4940-2_30

Akaygun, S., & Jones, L. L. (2013). Animation or Simulation: Investigating the Importance of Interactivity for Learning Solubility Equilibria. In J. P. Suits & M. J. Sanger (Eds.). *Pedagogic Role of Animations and Simulations in Chemistry Courses. ACS Symposium Series*, (*vol. 1142*, pp, 127-159). Washington, DC: American Chemical Society. doi:10.1021/bk-2013-1142.ch006

Ashe, C. A., & Yaron, D. J. (2013). Designing analogy-based simulations to teach abstractions. In J. P. Suits, & M. J. Sanger. (Eds.), *Pedagogic Roles of Animations and Simulations in Chemistry Courses. ACS Symposium Series*, (*vol. 1142*, pp, 367-388). Washington, DC: American Chemical Society. doi:10.1021/bk-2013-1142.ch015

Barak, M. (2013). Making the unseen seen: Integrating 3D molecular visualizations in elementary, high school, and higher education. In J. P. Suits, & M. J. Sanger (Eds.), *Pedagogic Roles of Animations and Simulations in Chemistry Courses. ACS Symposium Series*, (*vol. 1142*, pp. 273-291). Washington, DC: American Chemical Society. doi:10.1021/bk-2013-1142.ch011

Barnea, N., & Dori, Y. J. (1999). High school chemistry students performance and gender differences in a computerized molecular modeling learning environment. *Journal of Science Education and Technology*, *8*(4), 257–271. doi:10.1023/A:1009436509753

Blumer, H. (1969). The methodological position of symbolic interactionism. In Symbolic interactionism: Perspective and Method (pp. 1-60). Englewood Cliffs, NJ: Prentice-Hall.

Boo, H. K. (1998). Students understandings of chemical bonds and the energetics of chemical reactions. *Journal of Research in Science Teaching*, *35*(5), 569–581. doi:10.1002/(SICI)1098-2736(199805)35:5<569::AID-TEA6>3.0.CO;2-N

Bou Jaoude, S., & Barakat, H. (2003). Students' problem solving strategies on stoichiometry and their relationships to conceptual understandings and learning approaches. *Electronic Journal of Science Education*, *7*(3), 1–42.

Burke, K., Greenbowe, T. J., & Windschitl, M. (1998). Developing and using conceptual computer animations for chemistry instruction. *Journal of Chemical Education*, *75*(12), 1658–1661. doi:10.1021/ed075p1658

Campbell, T., Oh, P. S., & Neilson, D. (2013). Reification of five types of modeling pedagogies with model based inquiry (MBI) modules for high school science classrooms. In M. S. Khine & I. M. Saleh (Eds.), *Approaches and strategies in next generation science learning* (pp. 106–126). Hershey, PA: IGI Global. doi:10.4018/978-1-4666-2809-0.ch006

Charistos, N. D., Tsipis, C. A., & Sigalas, M. P. (2005). 3D molecular symmetry Shockwave: A web application for interactive visualization and three dimensional perception of molecular symmetry. *Journal of Chemical Education*, *82*(11), 1741–1742. doi:10.1021/ed082p1741.2

Chi, M. T., Feltovich, P. J., & Glaser, R. (1981). Categorization and representation of physics problems by experts and novices. *Cognitive Science*, *5*(2), 121–152. doi:10.1207/s15516709cog0502_2

Clement, J. (2000). Analysis of Clinical Interviews: Foundations and Model Viability. In A. E. Kelly & R. Lesh (Eds.), *Handbook of Research Design in Mathematics and Science Education* (pp. 547–589). Mahwah, NJ: Lawrence Erlbaum Associates.

Cole, R. S., & Todd, J. B. (2003). Effects of web-based multimedia homework with immediate rich feedback on student learning in general chemistry. *Journal of Chemical Education*, *80*(11), 1338–1343. doi:10.1021/ed080p1338

Creswell, J. W., & Plano Clark, V. L. (2011). *Designing and Conducting Mixed Methods Research*. SAGE.

Falvo, D. A. (2008). Animations and simulations for teaching and learning molecular chemistry. *International Journal of Technology in Teaching and Learning*, *4*(1), 68–77.

Falvo, D. A., & Suits, J. P. (2009). Gender and spatial ability and the use of specific labels and diagrammatic arrows in a micro-level chemistry animation. *Journal of Educational Computing Research*, *41*(1), 83–102. doi:10.2190/EC.41.1.d

Falvo, D. A., Urban, M. J., & Suits, J. P. (2011). Exploring the impact of and perceptions about interactive, self-explaining environments in molecular-level animations. *Center for Educational Policy Studies Journal, 1*(4), 45–61.

Finley, F. N., Stewart, J., & Yarroch, W. L. (1982). Teachers perceptions of important and difficult science content. *Science Education, 66*(4), 551–538. doi:10.1002/sce.3730660404

Fiore, S. M., & Schooler, J. W. (2002). How did you get here from there? Verbal overshadowing of spatial mental models. *Applied Cognitive Psychology, 16*(8), 897–910. doi:10.1002/acp.921

Gabel, D. L., Sherwood, R. D., & Enochs, L. (1984). Problem-solving skills of high school chemistry students. *Journal of Research in Science Teaching, 21*(2), 221–233. doi:10.1002/tea.3660210212

Gabel, D. S., & Bunce, D. M. (1994). Research on Problem Solving: Chemistry. In D. L. Gabel (Ed.), *Handbook of Research on Science Teaching and Learning. A Project of the National Science Teachers Association.* New York: Macmillan Publishing Company.

Gerjets, P. H., & Hesse, F. W. (2004). When are powerful learning environments effective: The role of learner activities and of students' conceptions of educational technology. *International Journal of Educational Research, 41*(6), 445–465. doi:10.1016/j.ijer.2005.08.011

Gulacar, O., Eilks, I., & Bowman, C. R. (2014). Differences in General Cognitive Abilities and Domain-Specific Skills of Higher- and Lower-Achieving Students in Stoichiometry. *Journal of Chemical Education, 91*(7), 961–968. doi:10.1021/ed400894b

Gupta, T. (2012). *Guided-inquiry based laboratory instruction: Investigation of critical thinking skills, problem solving skills, and implementing student roles in chemistry* (Unpublished doctoral dissertation). Iowa State University, Ames, IA.

Gupta, T., Burke, K. A., Mehta, A., & Greenbowe, T. J. (2015). Impact of Guided-Inquiry-Based Instruction with a Writing and Reflection Emphasis on Chemistry Students Critical Thinking Abilities. *Journal of Chemical Education, 92*(1), 32–38. doi:10.1021/ed500059r

Hanson, R.M. (2010). Jmol-a paradigm shift in crystallographic visualization. *Journal of Applied Crystallography, 43*(5-2), 1250–1260.

Hasler, B. S., Kersten, B., & Sweller, J. (2007). Learner control, cognitive load and instructional animation. *Applied Cognitive Psychology*, *21*(6), 713–729. doi:10.1002/acp.1345

Hegarty, M., Kriz, S., & Cate, C. (2003). The roles of mental animations and external animations in understanding mechanical systems. *Cognition and Instruction*, *21*(4), 325–360. doi:10.1207/s1532690xci2104_1

Herraez, A. (2006). Biomolecules in the computer: Jmol to the rescue. *Biochemistry and Molecular Biology Education*, *34*(4), 255–261. doi:10.1002/bmb.2006.494034042644 PMID:21638687

Herron, J. D. (1975). Piaget for chemists. *Journal of Chemical Education*, *52*(3), 146–150. doi:10.1021/ed052p146

Hessesn, D. O., Ågren, G. I., Anderson, T. R., Elser, J. R., & Ruiler, P. C. (2004). Carbon sequestration in ecosystems: The role of stoichiometry. *Ecology*, *97*(11), 1179–1192. doi:10.1890/02-0251

Howell, D. C. (2007). *Statistical Methods for Psychology*. SAGE.

Huddle, P. A., & Pillay, A. E. (1996). An in-depth study of misconceptions in stoichiometry and chemical equilibrium at a South African university. *Journal of Research in Science Teaching*, *33*(1), 65–77. doi:10.1002/(SICI)1098-2736(199601)33:1<65::AID-TEA4>3.0.CO;2-N

Johnstone, A. H., MacDonald, J. J., & Webb, G. (1977). Misconceptions in school thermodynamics. *Physics Education*, *12*(4), 248–251. doi:10.1088/0031-9120/12/4/011

Kearney, M. (2004). Classroom use of multimedia-supported predict-observe-explain tasks in a social constructivist learning environment. *Research in Science Education*, *34*(4), 427–453. doi:10.1007/s11165-004-8795-y

Keeney-Kennicutt, W. L., & Merchant, Z. H. (2013).Using virtual worlds in the general chemistry classroom. In J. P. Suits, & M. J. Sanger (Eds.), *Pedagogic Roles of Animations and Simulations in Chemistry Courses. ACS Symposium Series*, (*vol. 1142*, pp, 181-204). Washington, DC: American Chemical Society. doi:10.1021/bk-2013-1142.ch008

Kozma, R. B., & Russell, J. (1997). Multimedia and understanding: Expert and novice responses to different representations of chemical phenomena. *Journal of Research in Science Teaching*, *34*(9), 949–968. doi:10.1002/(SICI)1098-2736(199711)34:9<949::AID-TEA7>3.0.CO;2-U

Lancaster, K., Moore, E. B., Parson, R., & Perkins, K. (2013). Insights from using PhET's design principles for interactive chemistry simulations. In J. P. Suits, & M. J. Sanger (Eds.), *Pedagogic Roles of Animations and Simulations in Chemistry Courses. ACS Symposium Series*, (*vol. 1142*, pp, 97-126). Washington, DC: American Chemical Society. doi:10.1021/bk-2013-1142.ch005

Landriscina, F. (2013). *Simulation-Based Learning. In Simulation & Learning: A model-centered approach* (pp. 99–36). New York: Springer. doi:10.1007/978-1-4614-1954-9_5

Laughlin, J. S. (2001). Breaking out of the box: Teaching biology with web-based active learning modules. *The American Biology Teacher*, *63*(2), 110–115. doi:10.1662/0002-7685(2001)063[0110:BOOTBT]2.0.CO;2

Lawson, A. E., Abraham, M. R., & Renner, J. W. (1989). *A theory of instruction: Using the learning cycle to teach science concepts and thinking skills.* National Association for Research in Science Teaching.

Marais, F., & Combrinck, S. (2009). An Approach to Dealing with the Difficulties Undergraduate Chemistry Students Experience with Stoichiometry. *South African Journal of Chemistry. Suid-Afrikaanse Tydskrif vir Chemie*, *62*, 88–96.

Monaghan, J. M., & Clement, J. (1999). Use of computer simulation to develop mental simulations for understanding relative motion concepts. *International Journal of Science Education*, *21*(9), 921–944. doi:10.1080/095006999290237

Monaghan, J. M., & Clement, J. (2000). Algorithms, visualization, and mental models: High school students interactions with a relative motion simulation. *Journal of Science Education and Technology*, *9*(4), 311–325. doi:10.1023/A:1009480425377

Mulford, D. R., & Robinson, W. R. (2002). An Inventory for Alternate Conceptions among First-Semester General Chemistry Students. *Journal of Chemical Education*, *79*(6), 739–744. doi:10.1021/ed079p739

Nurrenbern, S. C., & Pickering, M. (1987). Concept learning versus problem solving: Is there a difference? *Journal of Chemical Education, 64*(6), 508–510. doi:10.1021/ed064p508

Plass, J. L., Homer, B. D., & Hayward, E. O. (2009). Design factors for educationally effective animations and simulations. *Journal of Computing in Higher Education, 21*(1), 31–61. doi:10.1007/s12528-009-9011-x

Reiber, L. P. (2005). Multimedia learning in games, simulations, and microworlds. In R. E. Mayer (Ed.), *The Cambridge Handbook of Multimedia Learning* (pp. 549–567). Cambridge, UK: Cambridge University Press. doi:10.1017/CBO9780511816819.034

Reigeluth, C. M., & Schwartz, E. (1989). An instructional theory for the design of computer-based simulations. *Journal of Computer-Based Instruction, 16*(1), 1–10.

Schnotz, W., & Kurschner, C. (2008). External and internal representations in the acquisition and use of knowledge: Visualization effects on mental model construction. *Instructional Science, 36*(3), 175–190. doi:10.1007/s11251-007-9029-2

Solomonidou, C. (2009). Constructivist design and evaluation of interactive educational software: A research based approach and examples. *The Journal for Open and Distance Education and Educational Technology, 5*(1), 6–24.

Stieff, M., & Wilensky, U. (2003). Connected Chemistry-Incorporating interactive simulations in chemistry classroom. *Journal of Science Education and Technology, 12*(3), 285–302. doi:10.1023/A:1025085023936

Suits, J. P., & Diack, M. (2002). Instructional design of scientific simulations and modeling software to support student construction of perceptual to conceptual bridges. *World Conference on Educational Multimedia Hypermedia and Telecommunication*, (vol. 1, pp. 1904-1909). Denver, CO: AACE.

Suits, J. P., & Sanger, M. J. (2013). Dynamic visualization in chemistry courses. In J. P. Suits, & M. J. Sanger (Eds.), *Pedagogic Roles of Animations and Simulations in Chemistry Courses. ACS Symposium Series*, (*vol. 1142*, pp, 1-13). Washington, DC: American Chemical Society. doi:10.1021/bk-2013-1142.ch001

Sweller, J., Van Merrienboer, J. J. G., & Plaas, F. (1998). Cognitive architecture and instructional design. *Educational Psychology Review, 10*(3), 251–296. doi:10.1023/A:1022193728205

Wieman, C. E., Adams, W. K., & Perkins, K. K. (2008). PhET: Simulations that enhance learning. *Science, 322*(5902), 682–683. doi:10.1126/science.1161948 PMID:18974334

Wolfer, A., & Lederman, N. (2000). *Introducing college chemistry students understanding of stoichiometry: Connections between conceptual and computational understandings and instructions*. Paper presented at the meeting of the National Association for Research in Science Teaching, New Orleans, LA.

Wolfer, A. J. (2000). *Introductory College Chemistry Students' Understanding of Stoichiometry: Connections between Conceptual and Computational Understandings and Instruction* (Unpublished doctoral dissertation). Oregon State University, Corvallis, OR.

Yaron, D., Karabinos, M., Lange, D., Greeno, J. G., & Leinhardt, G. (2010). The ChemCollective: Virtual labs for introductory chemistry courses. *Science*, *328*(5978), 584–585. doi:10.1126/science.1182435 PMID:20431007

Yarroch, W. L. (1985). Student understanding of chemical equation balancing. *Journal of Research in Science Teaching*, *22*(5), 449–459. doi:10.1002/tea.3660220507

KEY TERMS AND DEFINITIONS

Algorithms: Set of rules for problem solving.

Constructivism: Theory of how people learn based on observations.

Models: A small scare replica of a phenomena or a system.

Simulation: A representation of the behavior of a system through another system, especially a computer program designed for the purpose.

Chapter 9
Science Educator Professional Development:
Big Data and Inquiry Learning

Anna Lewis
University of South Florida – St. Petersburg, USA

George Matsumoto
Monterey Bay Aquarium Research Institute, USA

ABSTRACT

Scientific discovery, problem-solving, and hypothesis testing requires observation, data analysis and synthesis of new knowledge. In today's world, this process is highly dependent on computer-based data exploration of high volume, high velocity, and high variety data streams (3HV) However, though the power of 3HV surpasses the amount of information gathered from more familiar lab experiences, data-intensive science has not yet achieved the same impact or prominence in public education. This chapter provides an examination of the Education And Research: Testing Hypotheses (EARTH) Science Educator Professional Development (PD) which was developed to bridge this gap, bringing data-intensive science into the classroom while supporting inquiry learning practices.

INTRODUCTION

In the last ten years, how scientists think about and use data has changed. According to the latest research from Technology Consultancy IDC's 2014 Digital Universe report, the amount of data created is doubling in size every two years and will be almost

DOI: 10.4018/978-1-5225-2528-8.ch009

44 zettabytes (that's 21 zeroes!) by 2020 (Turner, Gantz, Reinsel, & Minton, 2014). This type of data is commonly characterized by three properties termed 3HV – high volume, high variety, and high velocity (the speed at which the information changes). We cannot ignore this reality. Big data is woven into our daily lives. People check the Internet for the exact location of an incoming flight or track their fitness with mobile apps. Students now graduating will be part of a workforce that must understand and develop tools that will allow them to make good use of these heterogeneous datasets. Many science educators are adept at using laboratory experiments and field studies to help students learn science; these types of activities are deeply embedded into the curricula at all levels. However, teaching students about big data and big data techniques is rarely, if ever, mentioned in science classrooms, science curricula, or informal science education. In this chapter, a unique science educator professional development (PD) model is examined. The intention is to expand the notion of how PDs are conceived and provide another possible solution that can assist educators in understanding and utilizing digital tools for inquiry-based STEM learning.

BACKGROUND

In 2000, approximately 25% of the world's information was stored in a digital format. Today, more than 98% of all stored information is digital (Cukier & Mayer-Schoenberger, 2013). This recent surge in data accumulation heralds the changes in the collection, organization, storage, and analysis of information. Scientists are now integrated into this new technical environment. In the past, data collection was costly and time-consuming. For this reason, random sampling methods were developed and used to infer something (with a margin of error) about the total population or phenomenon under investigation. This methodology works well in answering many questions, such as what food is the most likely to be eaten at a certain middle school cafeteria. By randomly selecting a group of 75 students from a population of 400 students, one could predict middle-grade food preferences fairly accurately. However, the strength of this statistical method falls apart when researchers ask about the favorite food of African American girls at that school. This random sample is too small to make any meaningful prediction in that regard. However, if a researcher collected data from *all* 400 students, the question could be answered easily. When researchers are limited by collecting only a percentage of the population (sample), they must know, before data collection begins, exactly what data to collect and what question(s) are to be answered. By having all the data (or at least significantly more data regarding any population or phenomenon) on-hand, there is no longer a need to know the questions beforehand. Thus, the task becomes how to harness and extrapolate meaning from enormous pools of information. Analysis focuses on finding patterns and relationships across data types and sources, rather than carefully curating pristine samples from information-constrained datasets. Our technological revolution has brought about this important paradigm shift.

This transition from capturing *some* data to capturing *all* data has contributed to a change in how many scientists regard the goals of their research. This transition "represents a move away from always trying to understand the deeper reasons behind *how* the world works to simply learning about an *association* among phenomena… from causation to correlation" (Cukier & Mayer-Schoenberger, 2013, p. 3). Though understanding how and why things happen the way they do is important and satisfying, it is often very difficult. For this reason, correlations can be extremely valuable, even when their underlying causes are unknown. One example of this new methodology at work is the development of a fire-prevention strategy in New York City. A team of analysts generated a database of 900,000 buildings in the city and added data collected from 19 other city agencies. Using this data, they compared five years of records on building fires. Using their findings, they were able to correlate building characteristics with an increased or decreased risk of fire. Surprisingly, several of the characteristics identified were not actually related to fire hazards. Using this program, fire inspectors increased their efficiency in identifying potentially hazardous buildings from 13% to 70% (Cukier & Mayer-Schoenberger, 2013, pp. 5-6). Here, we can see the power of big datasets; billions of data points can be used to find patterns and thus predict future occurrences, without actually knowing the underlying reasons these predictions are accurate.

When acknowledging the benefits of this new information age, it is also wise to consider the potential challenges inherent in the use of big data. Researchers must develop ways to find value in data that are imperfectly or incompletely captured, overlaps, or are repetitive because they have been taken from multiple data sources; this situation calls for new statistical methods with which to handle uncertainty, error, and data integration (Jagadish et al., 2014). The challenge of integrating datasets with accuracy and relevance, as well as creating new ways of visualizing and representing data complexities from programs and databases that run concurrently, is in its early stages of development. Another huge concern is the privacy issues inherent in the collection and distribution of data not only from the public's perspective but also considering the fact that scientists must deal with the attribution of scientific discoveries if data from multiple research projects are used (Chen & Zhang, 2014). Managing these concerns will require partners from social and governmental arenas to work alongside researchers to develop equitable solutions. It is clear that "while [the] benefits of big data are real and significant and some initial successes have already been achieved, there remain many technical challenges that must be addressed to fully realize this potential" (Jagadish et al., 2014, p. 90).

The information age is changing how scientists and citizens view and understand the world. Because big data is transforming virtually all academic disciplines, researchers must prepare students for this new reality by integrating its use into all levels of education. We must help students understand both the benefits and risks inherent in big data to better enable them to appropriately harness this tool. Students who possess the skills and experience needed to use big data will become integral in addressing many of the world's most pressing problems, such as tackling climate adaptation and mitigation, improving human health, and bringing about a more equitable global economy. It the science educator's job to help prepare educators and students to use this information to create new ways of thinking and conducting science.

Publically Available Big Datasets

As noted above, the amount of data created is enormous. More interestingly, the amount of data available to the public free of charge is also massive. The most well-known science data repositories in the United States are the National Aeronautics and Space Administration (NASA) and National Oceanic and Atmospheric Administration (NOAA). A visit to these organizations' websites shows that they not only provide raw data but also provide lessons and educational interfaces for many data sources. There are also many other online resources that provide access to a variety of publicly available datasets. *Awesome Public Datasets* (GitHub, Inc., 2016) is one such resource; here, one can find information regarding genomes, international economic data, and healthcare data. For instance, the AMPds site (the Almanac of Minutely Power dataset) "contains electricity, water, and natural gas measurements at one-minute intervals — a total of 1,051,200 readings per meter for 2 years of monitoring. Weather data from Environment Canada's YVR weather station has also been added. This hourly weather data covers the same period of time as AMPds and includes a summary of climate norms observed from the years 1981-2010. Utility billing data is also included for cost analyses" (Makonin, 2016). Many of these publically available datasets are educational treasure troves that can be used free of charge to help students begin to learn about big data collection and analysis. Helping educators to become more at ease with big data will encourage them to tap into these types of existing resources.

CORE PROFESSIONAL DEVELOPMENT (PD) FEATURES

Science educator professional development (PD) is the ongoing learning experience that builds on the initial teacher educational experience. The emphasis and intensity of learning changes as educators move from initial certification, during which the focus tends to be on basic teaching practices such as classroom management and student assessment, to understanding and refining teaching methods and deepening content knowledge (Kleickmann et al., 2013). Science educator PD is unique in that there is specialized content that "span[s] the alphabet from aerodynamics to zoology" (Gardner, 1972, p.26) and there are science process skills that cross all the content domains of science, such as the rules via which knowledge is generated and the logic and reasoning used in the process of inquiry (Gardner, 1972). Progress has been made regarding the successful planning and implementation of PD (e.g., Loucks-Horsley, Stiles, Mundry, Love, & Hewson, 2009), and while there is no definitive answer to the question of what makes science educator PD successful, there are several core features that many have agreed are important: 1) increased content knowledge, 2) active learning experiences, 3) program and curriculum coherence, 4) sufficient duration, and 5) collective participation in activities. A multitude of studies have provided evidence of the utility of these core characteristics.

Research has shown that increased teacher content knowledge positively impacts student achievement (e.g., Hill, Rowan, & Ball, 2005; Darling-Hammond, 2000; Kunter et al., 2013). PD that includes active learning experiences increases educator engagement and, in turn, student learning outcomes (e.g., Banilower & Shimkus, 2004; Heller, Daehler, Wong, Shinohara, & Miratrix, 2012). Teaching strategies and methods that adhere to required standards, district policies, and school cultures are more likely to be successfully implemented and used over time (e.g., Penuel, Fishman, Yamaguchi, & Gallagher, 2007; Saad & BouJaoude, 2012). Researchers have not been able to define a specific duration that is essential for successful PD. However, it is reasonable to assume that learning requires sufficient time to be effective; '*sufficient time*' would depend on the desired goals and outcomes. Even though there is no magic number of hours, the duration of the PD experience has proven to be an important component in evaluating successful outcomes (e.g., Desimone, 2009; Hill, Beisiegel, & Jacob, 2013). PD that provides opportunities for collective participation is more likely to create teachers who able to integrate what they have learned with other aspects of instructional content (Garet et al., 2001; Joyce & Showers, 2002), provide forums for debate, and show an improved overall understanding of both pedagogical matters and content (Ball, 1996).

The ways in which these core features are implemented are numerous and can lead to successful or unsuccessful outcomes. The subject of this chapter is how these important core research-based features are enacted in the context of the EARTH PD experience, enabling educators to understand, utilize, and disseminate big data tools, resources, and curricula.

OVERVIEW OF EARTH PROFESSIONAL DEVELOPMENT (PD)

The Education And Research: Testing Hypotheses (EARTH) Professional Development (PD) workshop (www.mbari.org/EARTH) was first developed and offered in 2003 by George Matsumoto and his team at the Monterey Bay Aquarium Research Institute (MBARI). The goal of the PD is to engage science educators to co-design lessons with "in-the-field" scientists. EARTH allows educators and researchers to connect, share stories and enthusiasm for their subjects, and facilitate the development of lesson plans based on real people and data. Workshop locations move around the country, and the topic of each PD is determined by the region and research focus of the host organization and participating scientists.

The EARTH PD is a four-and-a-half-day event. Participants gather the evening before the workshop starts for group introductions and an overview of the week (usually on a Sunday). Four full days are spent engaged in the activities of the workshop, and the last day, usually a Friday, is a half-day, during which participants present and share the lessons they have created over the previous days of the PD. All PDs are held at locations that have access to computers and high-speed Internet. All costs to participants are covered, including travel, lodging, and food expenses. During the first three days of the workshop, two to three presentations per day are offered by various scientists, who introduce and explain their research and data to the PD participants. After each presentation, there is time to investigate the datasets and ask questions. Interspersed throughout the first day, PD participants present the EARTH lessons they conducted in their own classrooms before coming to the workshop. These reports describe what worked and what they would change to improve the lesson. At the end of the third day (in the late afternoon and evening), teachers are asked to pick a dataset from which to develop a lesson. Groups are formed around each dataset to develop grade-appropriate lessons (depending on the group, there could be multiple grade levels) using the selected datasets. The entire fourth day is devoted to developing the materials needed to conduct a data-driven lesson and creating the lesson itself. During this time, PD participants work together with the scientists on hand to assist in developing their lessons. The fifth and final day of the workshop is a half-day, during which educators present their newly developed lesson plans to the rest of the group. All lessons developed are then reformatted by the EARTH facilitators and posted to the website for use by the participants and the public.

The next section describes the data collection process and the evidence indicating the success of EARTH PD in motivating teachers to conduct inquiry STEM lessons using big datasets.

METHODS AND EVIDENCE OF PD SUCCESS

As noted above, the EARTH PD is a project developed and run by the Monterey Bay Aquarium Research Institute (MBARI). The program has been funded through corporate donations, which mandates yearly reports and evaluations of the project. However, since the program's inception in 2003, the evaluation process has undergone several iterations, thus making it difficult to compare participant feedback because the surveys have been modified over time. In addition, not all of the data from 2003 to 2009 were available for analysis at the writing this chapter. Therefore, aggregated data from 2010 to 2015 surveys where questions aligned were utilized when reporting results and participant feedback.

Just as the evaluations have evolved over the years, the underlying features of the program have matured into what is used today. Through focus groups, facilitator discussions, and scientist observations, practices and characteristics that supported participants before, during and after the PD were established. The aim of this chapter is to share those features that have been developed and refined over the past 15 years. These features, as identified and expanded upon later, are what make this program unique and of interest to other developers and designers. The purpose of this study is to share the design characteristics that have succeeded in helping participants learn to utilize and disseminate big data throughout the K-12 arena.

Very little demographic information was consistently collected. The majority of participants were women, as identified by reviewing the names of the participants (approximately 85%). Participant teaching roles were collected, however, no data regarding participant education or science and technology experience. Table 1 shows the aggregate of participants from the 2010-2015 workshops. The majority of participants taught high school (49%). "Other Educator" refers to coaches, principles, and other administrators who attended the PD.

Most of the data used in this study were taken from the 2010-2015 end-of-day surveys. As the name suggests, at the end of each day, participants were asked to rate various aspects of the PD. These on-line surveys were a collection of Likert scale

Table 1. 2010-2015 participant teaching roles

Participants	%
Primary/Elementary	8%
Middle School	28%
High School	49%
College	7%
Other Educator	9%

questions and open-ended questions that allowed participants to describe what they enjoyed most and least during the day and comment on any topic of their choosing. The on-line format allows participants to easily respond and offer a quick tabulation of results. These formative assessments provide facilitators with valuable insight into the pacing and the content to be delivered the next day. Participants reported feeling respected and part of the team when facilitators responded to the end-of-day surveys by taking action to ameliorate the concerns identified by participants, such as "the room was too cold," "candy during the day would be great," and "we need more time to assimilate." The aggregate of responses from questions common to all surveys distributed in 2010-1015 EARTH PDs were examined. The 7 (categories 1 & 2; 3,4,5; and 6 & 7 were collapsed to form 3 categories) and 5 (categories 1 & 2; 3; 4 & 5 formed the 3 categories) point Likert scales were reduced to 3 point scales and the frequency of responses in each category are shown in Tables 2 & 3. This data supports the claim that participants in the EARTH PD found the workshops to be relevant, interesting, and meaningful.

In traditional PD, the goal is often to show educators how to use some activity or curriculum so that it can be replicated with accuracy in the classroom. Frequently, this involves learning a process or using a pre-packaged kit or lesson (e.g., Newman et al., 2011) developed by experts, without any real need to understand the underlying

Table 2. Aggregate of frequency of participant responses from end-of-day survey for 2010-2015 EARTH participants

Rate Your Satisfaction With Today:		
Not Satisfied	**Satisfied**	**Very Satisfied**
10%	25%	64%
Relevance to Your Job:		
Not Much	**Okay**	**Very Much**
11%	30%	59%
Appropriate for Your Students:		
Not Much	**Okay**	**Very Much**
8%	25%	67%
Interesting to You:		
Not Much	**Okay**	**Very Much**
3%	9%	89%
Useful to You:		
Not Much	**Okay**	**Very Much**
5%	16%	79%

Table 3. Aggregate of frequency of participant responses from end-of-day survey for 2010-2015 EARTH participants

New Learning/New Content			Interactions with Colleagues		
Not Enough	Just Right	Too Much	Not Enough	Just Right	Too Much
4%	91%	5%	7%	92%	0%
New Experiences			**Interactions with Scientists**		
Not Enough	Just Right	Too Much	Not Enough	Just Right	Too Much
8%	91%	1%	4%	96%	0%
New Teaching Resources			**Interactions with Workshop Staff**		
Not Enough	Just Right	Too Much	Not Enough	Just Right	Too Much
6%	92%	2%	8%	92%	0%
Enjoyment/Fun			**Interactions with Educators**		
Not Enough	Just Right	Too Much	Not Enough	Just Right	Too Much
8%	89%	3%	2%	98%	0%

content driving the activity. The EARTH PD focuses on teaching content and models inquiry pedagogy in ways that are personally meaningful to each participant. Traditional PD is geared toward "make and take" activities, which can be immediately implemented in the classroom without any real understanding of the underlying science concepts. In the EARTH PD, the focus is understanding science and data skills at a very high level and then transforming this information into something that is applicable at the primary and secondary levels. One measure of success is to see what participants have actually "learned" by reviewing the lessons they developed during the PD. It is beyond the scope of this chapter to provide a three-dimensional analysis of these lessons (as described by the NGSS); however, Table 4 provides a sample of the types of lessons developed during the EARTH PD. All of these lesson make use of big datasets and required participants to understand the underlying science concepts at the depth needed transform high-level information, data, and practices into practical lessons applicable to K-12 student.

Anecdotal evidence suggests that after participating in the EARTH PD, educators use big datasets in their classrooms, demonstrate an increased ability to use a variety of big datasets, and are willing to share what they have learned with local colleagues.

To investigate the validity of these claims, a survey was sent out a year after the 2015 summer EARTH workshop in Portland, OR. Fifteen out of the 26 attendees responded to the survey. This 58% response rate may be due to some teachers no longer having the same email addresses. They may have moved to other schools/ districts, the survey may have landed in their spam folders, or there may have been

Table 4. Example lessons created by EARTH participants

Making Air Out of Oceans
In this two-part lesson, students will create a graph using data collected by scientists (Dr. Alex Worden and Dr. Amy Zimmerman) to test their predictions about the effects of nutrients (nitrogen and phosphorous) on the abundance of phytoplankton (*Ostreococcus)* in the world's oceans. Students will then use a blank world and an interactive map database to make and analyze predictions about what parts of the world's oceans are the most biologically productive. Throughout this activity, students will answer written prompts that ask them to think critically about the data they are working with and the larger processes they are analyzing. This assignment can be done individually or in small groups.
Can You Hear Me Now?
In this lesson, students will compare sound data to the length of the animal making that sound. The students will attempt to create a correlation between the frequency and the length of the animal, which will lead to a discussion/exploration of the connection between frequency and wavelength. This lesson falls into the larger context of teaching the principals of sound and waves. This lesson can be used to introduce the concepts of the frequencies and wavelengths of sound waves or to help reinforce and build on prior knowledge of the properties of sound waves. The students will analyze data collected from hydrophones, which are shown on spectrographs. The students will explore recordings of marine mammals to gain knowledge of waves and then apply that knowledge to analyze the link between frequency and wavelength.

To view these and other lessons, go to http://www.mbari.org/products/data-repository/.

a lack of interest on the part of the teacher. Based on the data returned, it is possible to infer support for the claims made regarding PD outcomes, though the data are by no means definitive.

Among the EARTH participants responding to the survey ($N = 15$), there was a statistically significant difference between the number of participants who used the lessons (M = 9.33, SD = 1.15) and those who did not use any lessons (M=5.97, SD = 1.15), $t(4) = -3.89$ $p \geq .05$, CI= 0.6012-7.257. Cohen's d (4) =3.17 and effect-size r (4) = 0.846 were calculated; both results show a large effect size.

Table 5 breaks down the data further, showing that of those who responded to the survey, 66% of the participants used their own lesson and used a lesson developed by their peers at the EARTH PD. In addition, 53% of those who responded used additional lessons posted on the EARTH website. Lessons developed by the participants or their peers were typically used on one or more occasions. Lessons downloaded

Table 5. Post 2015 EARTH PD survey responses N=15 out of 26 participants

# of Teachers that Used Their Own Lessons	Total # of Times Teachers Used Own Lesson	# of Teachers That Used a Peer's Lesson	Total # of Used Peers' Lessons Used	Used Lessons Posted to EARTH Website	Total # of Posted EARTH Lessons Used
10	23	10	25	8	34
66%		66%		53%	

from the EARTH website were used 34 times, and some lessons were used more than once. Open-ended survey responses clarified how some of these lessons were used: "We used it in two classes for 3 days each"; "Most lessons I only used once, but the microbe one I used at least 3 times"; "Used it both semesters in six classes: three biology and three physical science"; "We don't have an oceanography class or anything close to it, but I fit it in as often as possible."

These responses support the notion that the 2015 EARTH PD successfully impacted the educational landscape of the participants and their students by allowing students to interact with datasets as a part of their routine science experience. In addition, four (26%) of the 15 respondents indicated that they had either provided a workshop or professional learning session in their school district over the 2015-2016 school year. This entailed sharing what they had learned about big datasets and developing science lessons with their peers. This is a promising indicator that big data information and skills learned in the EARTH PD are being disseminated to a wider audience.

CORE FEATURES OF THE EARTH PD

As alluded to above, several unique features of the EARTH PD format have contributed to the successful outcomes and longevity of the 15 year program:

- Participant selection,
- Scientist interactions,
- Extended time for learning and planning,
- Extended peer discussion and peer review,
- Expectation of professional accomplishment.

All these characteristics are closely related, overlap, and are interdependent. However, each pertains to unique attributes conceptualized by the PD developers as important qualities needed to help educators learn how to use and become comfortable with big data resources. Data from online survey responses collected at the end of each day from the 2010 to 2015 EARTH PDs are used to illuminate and describe each of these features in greater depth.

EARTH PD Core Feature 1: Participant Selection

Participants in the EARTH PDs are selected based on geographic location. Preferably, a minimum of ~50% are local to the region in which the PD is held. This allows educators to form relationships with one another and allows local scientists to support future collaborations. Workshop coordinators see the ideal group of participants as a collection of four to six cohorts, with each cohort being composed of an administrator and four to six teachers from a single school who teach various grades and subjects. The workshop is open to all K-12 and post-secondary educators. The total number of participants per workshop is approximately 20-30 educators. This cohort model has been shown to strengthen educator support systems so that new innovations are more likely to be used and successfully maintained (Smith et al., 2007). A unique feature of this PD model is that past participants are encouraged to apply each year. These returning educators play a vital role in the success of this model. They demonstrate engaged participation, share their experiences using data-rich science lessons, support first-time participants, and continue to build their professional repertoires. Fourteen years of workshops have resulted in 212 unique participants (135 educators and 77 researchers). Table 6 indicates the number of participants between 2010 and 2015 and the percentage of past participants accepted to each PD.

Another essential feature is the selection and application process for EARTH participants. As part of the application process, educators must conduct at least one lesson posted on the EARTH website *and* complete an evaluation of the lesson. This prerequisite provides a preview, or model, of the quality of work to be developed *during* the workshops, and it is a strong indicator that the applicant will be motivated to grapple with new content and skills. By providing an evaluation of the lesson, these educators indicate their willingness to think about their teaching experiences and share their expertise. This increases the quality of lessons posted to the EARTH website by providing feedback for their improvement or refinement and assisting in lesson dissemination. After the applications are submitted, the facilitation team invites educators who have written cogent and complete evaluations who fit the desirable demographic to participate in the EARTH PD (before the 2016 workshop, this was approximately 50% locals and 50% past participants; in 2016, this ratio was adjusted to approximately 30% past participants in order to involve a greater number of first-time participants).

Table 6. EARTH PD participant #s over time

	2010	2012	2013	2014	2015
# of Participants	30	21	21	23	26
% of Past Participants	52%	43%	38%	48%	50%

EARTH PD Core Feature 2: Scientist Interactions

Scientist Interactions deliver three core research-based aspects of successful PD implementation: 1) increased content knowledge, 2) active learning experiences, and 3) collective participation in activities. Over the first three and a half days of the PD, two to three scientists present their research each day (a total of six to eight scientists present over the course of the PD). They share personal stories about what led them to study the topic and why it was of interest to them. Scientists tell how they went about developing questions and collecting data to answer those questions and reveal how their results often lead to even more questions. After each talk, participants are given the opportunity to explore the scientist's data, ask questions, and clarify concepts. In today's complex world, it often takes multiple experts from many domains to truly understand an issue. EARTH participants develop a sense of this through interacting with a variety of scientists and datasets. They enact the process of inquiry and discovery to learn science content at a very deep level and are required to "stretch" themselves to grasp the concepts under discussion.

At times, all this information was a little overwhelming, but all [the] information was relevant. I do feel like I am going through "smoltification" [transformation]. What and how I have taught science vs. starting now... I'm figuring out what I can use to teach so that my students can better understand what they are learning and use tools that bring their learning to another level. --2012 participant

Interacting with the scientists working on the data provided was probably the most attractive [aspect]. It really reminded me why I love science, and it was fascinating to learn about the different aspects of the research going on. The exposure to data and the activities developed to help bring that data and real science into the classroom was the most practical and useful. This week really helped rejuvenate me and re-spark that love of teaching and reaffirm the decision I made to become a science teacher. --2013 participant

New content on Ocean Acidification provided new info and ideas I can use in my classroom. Katy's methods of communicating the science will provide me with new ways to motivate my students to translate their science discoveries. --2014 participant

...gaining a better understanding of how data is used in modeling is very helpful to my instruction. --2012 participant

Having time to work with the scientists and their data was so great in [terms of] being able to understand it well enough to use it in lessons! ...having access to the scientists during lesson plan developments made it so easy to incorporate the data into lessons that we can actually use! For me, being in my home area was especially helpful. I learned about a ton of organizations and resources I can use that I didn't know about before! --2013 participant

Another important feature embedded in each week-long PD is a field experience, during which participants experience collecting data with local scientists or explore local research centers. These fieldtrips provide bonding time for the participants, put local educators in touch with local science resources, and allow all participants to explore how data is collected first-hand. These are opportunities for local educators to create relationships with research scientists who are willing to be available even after the workshop to answer questions and talk directly with students. These experiences further deepen science content learning and bring greater meaning to the big datasets that they are exposed to during the in-class presentations.

For example, in the 2010 workshop, participants learned and explored the *Phoebe* Glider datasets from the Center for Coastal Margin Observation and Prediction (CMOP). Gliders are autonomous, unmanned underwater vehicles used for collecting data in remote locations. The *Phoebe* glider collected data on a variety of variables, such as tides, winds, and currents, and then relayed the information back to scientists via satellite. After exploring the datasets, participants actually built models of drifters onsite. This type of process-oriented guided inquiry is designed to enable participants to develop important concepts and refine their understandings in an environment structured to support the development of process skills. The scientists do not simply deliver content but also model how to link data collection activities with data synthesis and analysis. Feedback from 2010 PD participants, as shown below, reflects that this type of learning was effective and engaging.

Great day! I enjoyed both the data interpretation and drifter assembly – both very interesting. It's interesting to see how instrument failure or problems can affect data.

Seeing the simplicity of the drifter was very useful because after seeing it, I think I can modify and create [such equipment] in my classroom.

I really enjoyed seeing the parts of the gliders and how they worked together to produce the results. It made the data more meaningful.

Meeting the scientists who actually work on this was very inspiring. It was really helpful to see them in one of their environments and talk to them about the work they actually do. Charle's presentation was really helpful for understanding how to use the data in a more meaningful way. I also think seeing the drifters and actually trying to put them together/seeing them operating was interesting. [It] made me think of trying to do similar/simpler things with my students.

I think putting together the drifters really gave me a better understanding of the components. I will be able to describe the equipment to my students [more easily]. I also enjoyed working with the data on the CMOP site. I think that if I play with it more, I will be able to pose questions to my students and have them investigate [these questions] using the data.

These types of comments are common in all EARTH workshops. Learning deep science content connected to real-world experiences is an effective and exciting way to learn science. Intellectually, most educators know this, but still, they may not create these types of learning opportunities for their students. However, the data indicate that after having been involved in this type of meaningful learning, the likelihood of enacting what they know to be inherently desirable becomes more probable.

EARTH PD Core Feature 3: Extended Time for Learning and Planning

Extended Time for Learning and Planning provides opportunities for all of the researched-based core characteristics of a successful PD: 1) increased content knowledge, 2) active learning experiences, 3) program and curriculum coherence, 4) sufficient duration, and 5) collective participation in activities. Almost all the recent literature on teacher learning and professional development calls for professional development that is sustained over time (Hill, Beisiegel, & Jacob, 2013). The activities in this PD begin when an educator first downloads and conducts a lesson from the EARTH website to submit an application and continues long after the PD through social networking and satellite workshops. Sustained engagement by participating educators after the event is ensured by employing strategies such as 1) personal interactions with scientists so that teachers are comfortable with science content and datasets, 2) collaborative group support that assists teachers in developing their own unique lessons that can be used without delay in the classroom, and 3) additional lessons that have been vetted, explored, and explained by peers at the PD and are directly available for use in the classroom. All these strategies support the national-science-standard goals of systemically incorporating real data usage into educational environments.

Over four and a half days, EARTH PD facilitators provide large blocks of unstructured time, allowing participants to assimilate and engage with the science content and pedagogy under examination. After the scientists present their data, participants are given time to explore the data and ask questions. This type of scaffolding allows educators to interact with new content and practice data skills with assistance from colleagues, scientists, and facilitators. Providing this amount of time with a variety of datasets and data tools creates a foundation for understanding how to approach and manipulate the information, which builds confidence in using big data and integrating it into the classroom.

Fluidity is another important characteristic of this PD. With some major exceptions (such as field trips), the schedule is not rigidly adhered to. Unscheduled discussions may arise regarding a topic of particular interest to the group, such as how to help students not be 'afraid' of data or how to adapt a concept to a particular grade level. By allowing enough time for discussions to run their course and prioritizing interactions and collaborative work over schedules, the PD promotes a safe environment in which to explore new concepts and ways of thinking and teaching.

Perhaps, the most the important aspect of this PD characteristic is the amount of time educators have to collaborate in developing their own lessons based on the science content and data skills developed over the first previous three days. The evening of the third day and all of the fourth day, educators create and develop materials in collaboration with their peers. Instead of being provided with a lesson associated with a dataset, they spend time translating the concepts learned into integrated lessons. This is a simple but often neglected component of PD: time to synthesize and internalize learning. In every EARTH PD, teachers express their appreciation at having time, time to learn, time to discuss, time to think, and time to simply absorb the information.

What I loved best was... time to work on lessons because I need some time to process and plan and decide what... to use for my classroom. --2012 participant

With the data given, we are able to spend time thinking about it and talking about how it works in our personal classroom. There is so much out there, and we were allowed time to synthesize... it. --2015 participant

I love the time working with colleagues, looking over the data and talking about the possibilities. It's pretty awesome having that time built in. --2013 participant

I appreciate time to work with peers to apply what I have learned. --2010 participant

In this PD, duration is conceptualized as not only the length of time spent participating in the EARTH program overall but also includes the length of the *unstructured time available* *within* the workshop. This important and often overlooked aspect assists participants in getting the most out of the PD by enhancing collaborations and learning experiences.

EARTH PD Core Feature 4: Extended Peer Discussion and Review

Extended peer discussion and review provides opportunities to enact four of the researched-based core characteristics of successful PD: 1) increased content knowledge, 2) active learning experiences, 3) program and curriculum coherence, and 4) collective participation in activities. This activity is closely related to the last characteristic discussed: *extended time*. Extended time is essential in developing partnerships; however, *extended peer discussion and peer review* is conceptualized separately because effective collaborations require much more than simply time. Successful collaborations require 1) a focus, a need, or a clearly stated purpose for working together; 2) a safe environment in which collaborators are able to speak openly about what they believe, feel, and think; 3) multiple means of communication including regular face-to-face communication – talking; 4) the sharing of authority and credit; and 5) respect for each person's talents and contributions (Smallen & Leach, 1999). These important attributes are intentionally embedded within the EARTH PD in a variety of ways.

Participants are encouraged to arrive the evening before the first full day of the PD to enjoy a meal together and begin introductions. During this time, an overview of the workshop is provided, and each person shares what he or she hopes to learn over the next few days. The schedule and intentions for the day are clearly outlined each morning before any of the activities are begun, and at the end of each day, participants are asked to share what worked and did not work for them. These activities ensure that participants understand the purpose of the PD and remain focused on the desired objectives and outcomes.

Before participants arrive, they are required to conduct and evaluate a lesson from the EARTH website. When they are selected to participate in the PD, they are asked to prepare an 8-11 min presentation describing their experiences and sharing their lesson evaluation. These presentations are interspersed throughout the first day of the PD. This format serves to provide fodder for rich discussions and pedagogical learning. It unites the new and experienced participants because they all share common classroom interactions, and it allays some of the fears of the new participants about creating and conducting a 'less than perfect' lesson. All participants are afforded this opportunity to share their ideas and experiences, which cultivates an intimate and safe environment for discussion and learning early in the PD. It is during these presentations that participants begin to grapple with the pedagogical issues related to using big data in the classroom. Rich discussions enable educators to pinpoint impediments to implementation and ways to overcome them, describe successful strategies and lesson modifications, identify how to integrate national and local science standards into lessons that are not perfectly aligned with the datasets at hand, and determine a variety of approaches to meet the needs of diverse learners.

It is wonderful to interact with people from all over the country. We all have similar problems and concerns. We are more connected than we think. --2010 participant

Listening to teacher presentations of lessons they have tried was really helpful to understand the different ways teachers approach lessons, the different strategies they incorporated into the same lessons, etc. Jennifer's presentation on the oil spills was also really useful because it is SO relevant. Katie's presentation was also really cool. [I] can't wait to try voice threads. --2010 participant

I loved how a roomful of strangers quickly become so comfortable with each other, which fosters a creative collaboration environment. --2014 participant

I think I create good lessons, but they can always be improved, and another perspective invariably sees something that I missed... I'm grateful for that insight so that I can impact more students. It is imperative for teachers to be able to bounce ideas off one another, share what they learn, and evaluate/edit work. PD needs to be dynamic in this way to be successful and relevant to teachers no matter the content. --2013 participant

After three days of learning subject matter content and pedagogical skills from scientists and peers, participants are ready to take on a project that requires them to collaborate on developing a lesson that incorporates a big dataset. This project is the meat of the PD. Here, educators must synthesize all they have learned and translate it into a cogent and cohesive lesson for their own classrooms. The activity models a process-oriented guided inquiry approach, in which students work cooperatively in self-managed groups, taking the lead in creating and developing a lesson. The datasets to be incorporated are selected by personal preference, and groups are formed around the selected datasets. These groups may contain educators from a single grade or multiple grades – and a variety of lessons may be produced or a single large unit. Each group decides what to do and how to accomplish the task. On the last day of the PD, each group presents its lesson. To meet this challenge, participants must respect their team members, make use of everyone's unique talents and expertise, and share in the decisions and final presentation. If friction arises, facilitators are on hand to guide participants in identifying the components of successful collaborations and address any barriers to effectively implementing them.

Participants have positively responded to developing their own lessons collaboratively.

Having several people on [a] team was very beneficial to combin[ing] our experiences and resources together. --2013 participant

I loved the way our groups translated the research and data we were shown into really great lesson plans! Some were amazing! --2014 participant

I really, really, really enjoyed the time to work with my colleagues. In education, we often do not have the availability to work with our peers (so silly), and it was nice to have that time. --2015 participant

As we have described, not only time is necessary to develop teams that are successfully collaborative. Rather, facilitators must also ensure that participants have very clear goals and objectives to work towards and provide a safe environment in which all ideas and questions are welcomed. Facilitators must also ensure an atmosphere of mutual respect, which can be built through the appreciation of all team members' contributions.

EARTH PD Core Feature 5: Expectation of Accomplishment

The *expectation of accomplishment* addresses and supports all of the core characteristics of a successful PD: 1) increased content knowledge, 2) active learning experiences, 3) program and curriculum coherence, 4) sufficient duration and 5) collective participation in activities.

Often, traditional PD provides teachers with materials, step-by-step instructions, assessments, and sometimes even actual scripts to follow closely (e.g., Barrett-Tatum & Dooley, 2015). In his book "Teaching as a Subversive Activity," Neil Postman describes this type of professional development as an effort to "teacher proof" the curriculum (Postman, 2009). The intention is to provide a curriculum or activity that is not impacted by the variety of educators presenting the lesson. Somehow, the ideal has become removing the educator from the interaction so that learning opportunities can be presented uniformly to all students, which presumes that if this occurred, all students would then be treated equally and advance through the curriculum in a uniform way. As any educator will attest, this type of PD and educational ideal, though well-intentioned, does not work. All students do not learn in the same manner, and the variety of educators and teaching styles is important.

In the EARTH PD model, instead of trying to "teacher proof" the curriculum, the ideal is to support educators in enhancing their own expertise and increasing their pedagogical repertoire so that they can use the skills they have learned to engage with students long after participating in this PD. From the first interaction, during the application process, EARTH facilitators treat participants as professionals whose experience and skills are vital to the success of the PD. Over and over again, this model underscores the assumption that each participant is capable of excellence in learning and teaching. To this end, all expenses related to attending the PD (travel, lodging, and meals) are paid. In addition, participants receive a $500 stipend. This underscores that the participants are professionals who deserve to be compensated for their time and effort.

Scientists and facilitators provide science content that is often beyond what the participants have experienced previously. This requires a commitment from each participant to stretch outside his or her comfort zone to learn the information provided, thus signaling to the participants that this PD is not a spectator presentation but requires their full engagement and learning skills.

Participants are asked to develop and present a cohesive lesson plan in one and a half days. This activity challenges participants to expand their beliefs about how a lesson can be developed and what they are capable of accomplishing in a short period of time. Teachers not only synthesize and translate the science they have learned but also incorporate and develop new pedagogical strategies to successfully implement this new content.

Past participants are on-hand to provide support, encouragement, and reassurance. Often, there is a point in the PD at which some of the first-time participants feel overwhelmed or unsure if they can actually learn and synthesize the content to meet the goals of the PD. The past participants mentor their peers and help them to believe in their abilities to not only complete the task at hand but to excel and create something they take pride in. These interactions are important learning opportunities for all involved. They aid educators in remembering what it is like to learn something difficult, the excitement and satisfaction of completing a difficult task, and how we as educators can assist our students when they feel overwhelmed and lose confidence in their ability to learn. These experiences reinforce their compassion and professionalism as educators. As some teachers noted:

These discussions are so successful and meaningful because each person is treated as a professional and their expert opinions are important and add value to the overall discussion and learning of each person. 2015 participant

I think I have never been so inspired or treated so well. All of the participants were so great, and every speaker was informative and inspiring. --2015 participant

Overall, this was AMAZING, as usual! I am always impressed with the data directly from the scientists! I love it because it is so important as an educator to stay ahead of the curve to be effective for my students. MBARI and C-DEBI truly make teaching a "profession" as they treat us SO well – resources, materials, meals, lodging, [and] stipend [are] very much appreciated! I love the comradery and the friends that I make with fellow educators to be able to brainstorm and come up with amazing teaching tools for our students! Thank you! --2015 participant

The veterans welcomed and embraced the newbies like nothing I've seen before. I am leaving with so many new friendships that truly feel... like old friendships! Great job creating the teacher participant group. –2015 participant

Viewing and treating participants as intelligent professionals whose contributions are fundamental to the success of the group's learning and teaching experience is a central aspect of this PD. So often, educators report that they feel disenfranchised and without a voice (Richmond, 2014; Scholastic, Inc., 2016). The PD works to empower science educators by not only providing them with new skills, tools, and inquiry strategies but also by enabling them to perform as the professionals they are.

LIMITATIONS OF THE EARTH PD

Perhaps the greatest limitation of this PD model is the difficulty of scaling it up. It is inherently a small group process, with individuals specifically selected for their willingness to fully participate and stretch themselves to learn new science content and skills. The applicants selected exhibit an eagerness to grow and become more accomplished educators. The comradery developed during this PD is a byproduct of everyone's willingness to grapple with new ideas and sustained relationships that are built on mutual respect and a shared commitment to complete the big data lessons/activities.

However, even though the number of overall participants is small, the impact of this PD is not trivial. Recall that over half of those who responded to the post-2015-EARTH survey continue to use the big data lessons that they and their peers developed. Sixty-eight percent indicated that they use additional EARTH lessons posted on the website, and 28% of respondents have been motivated to take leadership roles in sharing the science content and skills learned in the EARTH PD with school and district colleagues. As one participant wrote, "This workshop has truly been a life changing experience for me! I learned from some awesome teachers, have been inspired to do even more marine biology in my classes, and to take more classes so that I have more background to 'breathe life into the content'" (2014 participant).

MBARI is now supporting 'satellite' EARTH workshops in which participants attempt to organize and coordinate their own one-day mini-EARTH workshops. They are responsible for finding and contacting researchers, organizing the venue, advertising the workshop, selecting participants, and then running the workshop. Each workshop will have one to two researchers and time for the participants to work on lessons that are then shared at the end of the day. MBARI provides some financial support to cover the workshop costs and a small stipend for participants. This has proven to be an excellent way to further disseminate big data tools and inquiry strategies at a minimal cost.

FUTURE RESEARCH DIRECTIONS

To meet the needs of the future, education must be as dynamic as the society in which it is embedded. In a recent McKinsey Report, approximately 190,000 workers with deep analytical experience will be needed in the USA, and 1.5 million managers will be required to become data-literate (Manyika et al., 2011). In order to prepare students for this reality, teachers must become knowledgeable and experienced in using big datasets in science contexts. As educators, we know that everyone cannot learn in the same manner and that instruction must be crafted to suit the needs of the learner. Though this PD model may not be applicable to all science educator audiences, the characteristics described and insights gained may be incorporated into other PD models to improve outcomes. Determination of the long-term impacts of this type of PD model is essential, ascertaining differences in how repeat and one-time participants use and share lessons learned in the classroom, and an examination of participant relationships (including those with participating scientists) that are maintained after the end of the PD are all intriguing directions for future research.

CONCLUSION

The characteristics of this PD model allow participants to assume primary control over their own professional growth and learning. This model recognizes the necessity and power of community and relationships in supporting teachers who take on this responsibility and persist in their efforts. It acknowledges that participants, scientists, and PD facilitators both teach and learn from one another. Rather than a one-way street, in which participants are to simply learn from the experts, this is a two-way street. Mutual respect is the foundational characteristic embodied in the EARTH PD model.

The EARTH PD model promotes awareness of local and online big data resources and offers time to practice using these resources; provides access to researchers, their stories, and their data; supports collaboration across grades and disciplines; and encourages educators to take leadership roles in sharing and disseminating their own learning. It affords opportunities to practice inquiry strategies such as self-directed learning, small-group collaborations, and active field experiences and to embed those strategies in relevant lesson plans. In addition, it is one of the few PDs that rewards educators for their willingness to struggle with new concepts and skills to develop themselves professionally. Participants develop deeper connections with their school colleagues, researchers, and other educators around the US who share common interests and experiences.

Though more work can always be done in science educator professional development, the EARTH PD offers an excellent model in which educators gain important skills that allow them to expose students to big datasets while engaged in inquiry explorations.

REFERENCES

Ball, D. L. (1996). Teacher learning and the mathematics reforms: What we think we know and what we need to learn. *Phi Delta Kappan*, *77*(7), 500–508.

Banilower, E., & Shimkus, E. (2004). *Professional development observation study*. Chapel Hill, NC: Horizon Research.

Barrett-Tatum, J., & Dooley, C. M. (2015). From idealized to actualized instruction: Two elementary teachers enactment of the ELA Common Core State Standards in instructional scripts. *Literacy Research and Instruction*, *54*(4), 257–284. doi:10.1080/19388071.2015.1062941

Chen, C. P., & Zhang, C. Y. (2014). Data-intensive applications, challenges, techniques and technologies: A survey on Big Data. *Information Sciences*, *275*, 314–347. doi:10.1016/j.ins.2014.01.015

Cukier, K., & Mayer-Schoenberger, V. (2013). Rise of big data. *Foreign Affairs*, *92*(3).

Darling-Hammond, L. (2000). Teacher quality and student achievement. *Education Policy Analysis Archives*, *8*, 1. doi:10.14507/epaa.v8n1.2000

Desimone, L. M. (2009). Improving impact studies of teachers professional development: Toward better conceptualizations and measures. *Educational Researcher, 38*(3), 181–199. doi:10.3102/0013189X08331140

Gardner, P. L. (1972). Structure-of-knowledge theory and science education. *Educational Philosophy and Theory, 4*(2), 25–46. doi:10.1111/j.1469-5812.1972.tb00053.x

Garet, M. S., Porter, A. C., Desimone, L., Birman, B. F., & Yoon, K. S. (2001). What makes professional development effective? Results from a national sample of teachers. *American Educational Research Journal, 38*(4), 915–945. doi:10.3102/00028312038004915

GitHub, Inc. (2016). *Awesome Public Datasets*. Retrieved June 15th, 2016 from https://github.com/caesar0301/awesome-public-datasets

Heller, J. I., Daehler, K. R., Wong, N., Shinohara, M., & Miratrix, L. W. (2012). Differential effects of three professional development models on teacher knowledge and student achievement in elementary science. *Journal of Research in Science Teaching, 49*(3), 333–362. doi:10.1002/tea.21004

Hill, H., Rowan, B., & Ball, D. (2005). Effects of teachers mathematical knowledge for teaching on student achievement. *American Educational Research Journal, 42*(2), 371–406. doi:10.3102/00028312042002371

Hill, H. C., Beisiegel, M., & Jacob, R. (2013). Professional development research consensus, crossroads, and challenges. *Educational Researcher, 42*(9), 476–487. doi:10.3102/0013189X13512674

Jagadish, H. V., Gehrke, J., Labrinidis, A., Papakonstantinou, Y., Patel, J. M., Ramakrishnan, R., & Shahabi, C. (2014). Big data and its technical challenges. *Communications of the ACM, 57*(7), 86–94. doi:10.1145/2611567

Joyce, B. R., & Showers, B. (2002). *Student achievement through staff development*. Ascd.

Kleickmann, T., Richter, D., Kunter, M., Elsner, J., Besser, M., Krauss, S., & Baumert, J. (2013). Teachers content knowledge and pedagogical content knowledge the role of structural differences in teacher education. *Journal of Teacher Education, 64*(1), 90–106. doi:10.1177/0022487112460398

Kunter, M., Klusmann, U., Baumert, J., Richter, D., Voss, T., & Hachfeld, A. (2013). Professional competence of teachers: Effects on instructional quality and student development. *Journal of Educational Psychology, 105*(3), 805–820. doi:10.1037/a0032583

Loucks-Horsley, S., Stiles, K. E., Mundry, S., Love, N., & Hewson, P. W. (2009). *Designing professional development for teachers of science and mathematics.* Corwin Press.

Makonin, S. (2013 -2016). *The Almanac of Minutely Power dataset.* Retrieved June 15th, 2016 from http://ampds.org/

Manyika, J., Chui, M., Brown, B., Bughin, J., Dobbs, R., Roxburgh, C., & Byers, A. H. (2011). Big data: The next frontier for innovation, competition, and productivity. *The McKinsey Global Institute, 5*(33), 222.

Newman, D., Finney, P. B., Bell, S., Turner, H., Jaciw, A. P., Zacamy, J. L., & Gould, L. F. (2012). Evaluation of the effectiveness of the Alabama Math, Science, and Technology Initiative (AMSTI). *Science, and Technology Initiative (AMSTI).*

Penuel, W. R., Fishman, B., Yamaguchi, R., & Gallagher, L. P. (2007). What makes professional development effective? Strategies that foster curriculum implementation. *American Educational Research Journal, 44*(4), 921–958. doi:10.3102/0002831207308221

Postman, N. (2009). *Teaching as a subversive activity.* Delta.

Richmand, E. (2014). New survey: Teachers say their voices aren't being heard. *Education Writers Association.* Retrieved June 16th 2015 from http://www.ewa.org/blog-educated-reporter/new-survey-teachers-say-their-voices-arent-being-heard

Saad, R., & BouJaoude, S. (2012). The relationship between teachers' knowledge and beliefs about science and inquiry and their classroom practices. *Eurasia Journal of Mathematics, Science & Technology Education, 8*(2), 113–128.

Scholastic, Inc. (2016). America's teachers on teaching in an era of change. *Primary Sources* (3rd ed.). Retrieved June 16th 2016 from http://www.scholastic.com/primarysources/

Smallen, D., & Leach, K. (1999). Making a NICER Transition to the Millennium: Five Keys to Successful Collaboration. *Educational Review, 34*(4).

Smith, T. M., Desimone, L. M., Zeidner, T., Dunn, A. C., Bhatt, M., & Rumyantseva, N. (2007). Inquiry-oriented instruction in science: Who teachers that way? *Educational Evaluation and Policy Analysis*, *9*(29), 169–199. doi:10.3102/0162373707306025

Turner, V., Gantz, J., Reinsel, D., & Minton, S. (2014). *The Digital Universe of Opportunities: Rich Data and the Increasing Value of the Internet of Things*. Retrieved June 15th, 2016 from http://www.emc.com/leadership/digital-universe/index.htm

KEY TERMS AND DEFINITIONS

Big Data: An evolving term that describes complex and extremely large datasets that combine information from a variety of sources. These datasets usually require that those attempting to collect and exploit them use non-traditional methods.

Coherence: The quality of consistency throughout a project. In this study, we use "coherence" to refer to the alignment of professional development implementation with the goals of professional development.

Inquiry Learning: A pedagogical method that poses questions, problems, or scenarios to assist participants in learning desired content through critical thinking and collaborative activities.

Pedagogical Content Knowledge (PCK): Knowledge regarding the strategies and methods needed for effective teaching practices.

Professional Development (PD): Ongoing learning for in-the-field educators to support their growth as they increase their expertise as educators.

Professional Learning Communities (PLC): Regularly scheduled meetings that allow educators to share expertise in an effort to advance their teaching skills and improve the academic performance of their students.

Subject Matter Knowledge (SMK): Deep understanding and knowledge of a selected domain of information, such as content knowledge regarding biology, reading, or anthropology.

Chapter 10
A Case Study of Infusing Collaborative STEM Inquiry Learning With Available Technology Into Undergraduate Student Learning:
Perceptions, Benefits, and Challenges

Katarin Alinta MacLeod
St. Francis Xavier University, Canada

ABSTRACT

This chapter describes the development, launch, and findings of an undergraduate course entitled "Teaching and Learning in the Physical Sciences". It was a second-year course that introduced students to content commonly taught within a Bachelor of Education program and connected this content to the literature found in the areas of Physics Educational Research and Chemistry Educational Research. It infused inquiry-based STEM learning allowing students to re-examine content and connect to context, while also examining the pedagogical significance and implications. This chapter provides a qualitatively based insight into students' experiences of the course content and students' voices of the transferability of content from this course to others citing lessons learned, implications for physical science students and recommendations for the future offering.

DOI: 10.4018/978-1-5225-2528-8.ch010

INTRODUCTION

Student engagement, learning and collaboration opportunities, as well as student achievement within a technology infused classrooms have been well documented within the research literature (Apple Education, 2010; MacLeod & Kraglund-Gauthier, 2015; Yerrick, 2010). However, little research exists on *how* educators, especially at the undergraduate level, infuse available technology and, connected to this, *how* to shift from a pedagogical standpoint from lecture and direct instruction within an undergraduate course to one that incorporates inquiry into Science, Technology, Engineering and Mathematics (STEM) content. This chapter begins by setting the context and describing the development and launch of a special topics undergraduate course, *PHYS 299/CHEM 299: TEACHING AND LEARNING IN THE PHYSICAL SCIENCES.* A discussion of how the content of the course connects to the research literature is given. Data collection, analysis, and discussion follows, explaining student perceptions and learning as well as classroom and course observations by the professor are given. From here, recommendation, next steps, and future possibilities are posed.

The purpose of PHYS 299/CHEM 299 was to introduce physical science students to teaching and learning principles as they pertain to 21st century learning within the sciences and rooted within the various connected literatures (physics education, chemistry education and science educational research). This course targeted students in both physics and chemistry who were not academically able to complete an honours program and who were interested in either teaching science as a possible career option or were considering a career in public health where teaching and facilitation skills would be an asset. Therefore, whenever possible, collaborative STEM inquiry lessons were taught to students with available technology, both digital and scientific devices, so that they could examine their own understanding of content and what it means to teach and learn in the physical sciences at both the secondary and post-secondary levels.

Setting the Context

Physics enrollments within the undergraduate program at the particular university where this research was conducted are not as strong today as they have been in the past. The author of this research is currently a professor in the Faculty of Education at this institution and also completed her undergraduate degree in the same Physics Department; therefore, she is familiar with the subject matter and context. In her current role, she prepares future science teachers, many who have commented that they had been inspired to become teachers from former teachers. Others have commented that they did not know what else to do with their degree so they decided to pursue an additional degree in Education because those who become science teachers usually are able to find full time employment within their chosen career. Most had little to no concept of what a 21st century teaching skill set should include.

It was with this in mind and with the author's own motivation to prepare the best science teacher possible, that she proposed a course, *Teaching and Learning in the Physical Sciences*, to the Dean of Education and the Dean of Science. According to the advertised course description:

An introduction to teaching and educational research with a focus on concepts and content from the physical sciences, this course is suitable for science students seeking opportunities to explore the art of teaching based on physical sciences educational research findings. Topics include principles and concepts of teaching and tutoring, moving content into context, STEM (Science, Technology, Engineering and Mathematics) education, and opportunities to connect with learners and educators in both the public-school system and at the undergraduate level. 3 credits, 36 hrs, no laboratory (MacLeod, 2015).

The course was open to undergraduate enrolment contingent on the prerequisites of first year calculus-based Physics or Chemistry as well as first year Calculus courses.

Content was organized under three main areas: (a) science education, whereby students gain a sense of teaching and learning including Bloom's Taxonomy, lesson planning and asking effective questions, classroom teaching strategies, assessment of/for learning and connection to evaluations, STEM, science, technology, society, and environment (STSE), and inquiry-based learning (IBL); (b) physics education research (PER), whereby students explore research that has been conducted at the post-secondary level and the impact of the research on today's undergraduate learning experiences and future research opportunities both in terms of the classroom setting, student laboratories, and tutorials; and (c) chemistry education research, which was similar to PER but with the focus on chemistry and an examination of all three areas using a chemistry lens.

Before approval was granted to advertise the course and launch it in September 2015, the instructional content and assessment plans were reviewed and approved by the chairs of the Physics Department and the Chemistry Department. As well, details of the course were circulated within both departments for professors to review so they were aware of the course content and how it could be linked to other physical science courses students may be taking concurrently from either department. No questions or concerns were brought to the author's attention, and the course was opened for enrollment in the summer of 2015.

At the beginning of classes in September 2015, 19 students had enrolled: 11 fourth year students, 5 third year students and 3 second year students. There was a combination of majors and advanced majors, most opted to take the course as a chemistry credit. Unfortunately, due to course scheduling, few physics students could fit the course into their schedules as it conflicted with required courses. This was noted by the chair as a point to keep in mind for the next iteration of the course.

Students were given a tentative schedule of how the course would unfold with supplementary materials placed on the course's Moodle® site. Supplementary materials included journal articles and class PowerPoint™ presentations to be reviewed prior to class (thus connecting with the notion of a flipped-classroom), handouts and images from in-class activities, pictures of board notes and student artifacts so that as the course progressed students could review the content as well as be reminded with visual queues of the learning which had taken place. In this way, auditory, visual and kinesthetic learners could relate to the content. The tentative schedule can be found in Table 1 with dates and highlights of content to be covered within that time frame.

Course content focused on the importance of science education and the growth and development of learners. Current curriculum issues, trends, and lesson delivery techniques were discussed and integrated into all components of the course via direct instruction, small group interaction, pair–share techniques, group brainstorming and problem solving, action research, co-operative learning and peer presentations, and inquiry-based learning (IBL) opportunities. The students engaged frequently in short IBL activities that connected the course content with the content that had been learned previously from either first-year courses or in their current physical science courses. IBL opportunities focused on outcomes previously covered and, therefore, assumed to be part of students' prior knowledge so that a deeper understanding of an inquiry-based learning task would occur. As the course progressed, it became evident that, at times, students were either relearning content during these lessons or learning the content for the first time, but the course design took these issues into account. IBL opportunities were always completed within a small group setting, and thus were collaborative in nature. Students had opportunities to create supportive networks, receive feedback from peers and the instructor and, through readings and in-class assignments showed growth and professional development using the reflective process. Further, they had multiple opportunities to change and challenge their current point of view as a student to that as a teacher or professor, examining content, activities, and delivery techniques from both their own perspective as an undergraduate student as well as from their professor's possible perspective. Connecting the literature from Science education to both PER and CER was a new process for the undergraduate students and the more they learned, the more questions they had.

Each of the IBLs had a STEM focus in that students had to draw from more than just their chemistry or physics backgrounds. Some of the IBLs required basic

Table 1. PHYS 299/CHEM 299 tentative topic schedule given to students as part of their course outline

Week	Dates	Topics
1	September 8–11	Introduction, survey to formulate co-construction of criteria for the course, course outline, education-based action research, data analysis, bias (from survey).
2	September 15–18	Bloom's Taxonomy, formulating questions, connecting Bloom to previous assessments and to current in-class queries.
3	September 22–25	Lesson planning and content development, laboratory activities connecting back to Bloom's Taxonomy, tutoring, introduction to STSE and STEM
4	September 29–2	Preparing to write or create a summative assessment, connecting to learning outcomes or goals and Micro teaching assignment (presentations)
5	October 6–9	Micro teaching assignment (presentations) and Review for midterm
6	October 13–16	Midterm and Formative Assessment connecting back to Bloom, formulation good questions, and learning outcomes.
7	October 20–23	Chemistry Education Research (CER) Chemistry to focus on stoichiometry, mechanisms before reactions, reduction-oxidation reactions, chemical bonding and thermodynamics
8	October 27–30	CER/ Physics Education Research (PER) Physics to focus on kinematics, dynamics, forces, light, static and current electricity, basic magnetism
9	November 3–6	PER continued
10	November 10	Reinventing assignment due, Connect CER, PER, STSE/STEM
11	November 17–20	Laboratories then and now
12	November 24—27	Issues-based education, scientific literacy and the nature of science, also what other countries are doing when it comes to education, culturally relevant content
13	December 1–4	Revisit/overflow and review for final

MacLeod, 2015.

laboratory equipment and others utilized online programs, mobile applications, or an interactive whiteboard. These lessons included but where not limited to the following:

- Capturing and measuring CO_2 given the formula $NaHCO_3$ + $HC_2H_3O_2$ gives…,
- Determining the ratio between a force and length of rope to be pulled in a pulley system,
- Extracting DNA from a banana,
- Determining the position and size of an image using a single lens,

- Ice, salt and ice-cream: Maximizing the potential and connecting science to food, and
- Save your egg: The classic egg drop experiment.

None of these IBLs were graded assignments, but instead were classroom activities used to highlight the content of the week and to provide scaffolding so that students could connect previous learning with pedagogy.

As part of their IBL activities, students were encouraged to use whatever form of digital technology they brought with them to class. These digital devices, including mobile phones, tablets, and laptops could be used for research, data collection and recording, or any other way that students felt was an appropriate way to carry out the assigned task. At times, this was simply using these devices to research part of the IBL problem; other times, these devices were integral to their activities in data collection, recording, or analysis. Infusing the available technology within the classroom setting was at times awkward in that students were not accustomed to using or being encouraged to use their own devices during class activities, and some of the technology brought into the classroom was completely new with a steep learning curve.

In terms of summative assessments, students were to complete all six tasks as per the course syllabus. A balance between typical undergraduate tests and other more collaborative assignments was set. Because the University Registrar had assessed PHYS 299/CHEM 299 to be a second year, 200-level course, university policy required that the course have an invigilated, time-restricted written midterm and final examination. The midterm was set at 25% of the course value and focused on the content that was taught from September until mid-October. It was comprised of a variety of questions spanning Bloom's Taxonomy and included a synthesis/evaluation question. No questions included the use of technology specifically; however, some questions focused on the IBL activities students completed during class time. The final examination was similar to the midterm examination, yet set at 35% of the course value and was cumulative in terms of content. It also required students to utilize a free, online formative assessment program to create an online quiz that was based on a previous task done during the course.

Of the remaining assessment activities comprising 40% of the course grade, two were done in pairs but could be done individually if pre-approved by the professor for a total of 25%. The remaining two assessment activities were individual assignments, and their tasks required students to observe middle school classrooms and to connect with middle school students in the capacity of a tutor/tutee. These observations and tutoring were completed at a local junior high school where the PHYS 299/CHEM 299 students had the opportunity to shadow a teacher and interact with secondary students between the ages of 11 and 14 for a half-day of school.

BACKGROUND

Integration of inquiry-based Science, Technology, Society, Environment (STSE) and Science, Technology, Engineering, and Mathematics (STEM) education within the K–12 education system has been well documented in the educational research literature (Aikenhead, 2006; Alsop, Bencze, & Pedretti, 2005; Bennett, Lubben, & Hogarth, 2007; Bloom, 2006; Bybee, 1993; 2006; Hofstein & Lunetta, 2004; MacLeod, 2012, 2013; Pedretti & Little 2008; Settler, 2003; Shipman, 2006). Studies have been done on the integration of different content areas to provide contextualization of the content for students within the K–12 system. Moving beyond contextualization into the notion of "action and agency" (Pedretti & Little, 2008) and—on a more macroscopic scale—the de-compartmentalization of science (MacLeod, 2012, 2013). Further, a number of undergraduate universities, including St. Francis Xavier University, offer science education methods courses to those enrolled in Bachelor of Education programs and these have also been well studied (Bencze & Hodson, 1999; Forbes & Davis, 2008; Franz, 2010; MacLeod, 2013; MacLeod & Kraglund-Gauthier, 2015). Again, when one examines the literature, moving students from their undergraduate experiences of indoctrination into the traditional sciences (Aikenhead, 2005) into a more holistic approach to understanding science and connecting science to a student's life (MacLeod, 2012, 2013; MacLeod & Kraglund-Gauthier, 2015) can be a daunting task. Finally, there is specific literature from the fields of Physics Education Research (PER) on how best to teach physics to undergraduates as discussed from the physicists' perspective (van Aalst, 2000; 2005; Aikenhead, 2003; Beichner, 2009; Etkina, 2005; Hawkes & O'Meara, 2005; Hillel, 2005; Holubova, 2005; Knight, 2004; Kortland, 2005; MacLeod, 2007, 2013; McDermott, 1998, 2001, 2006).

This research focuses on best practices within a university physics classroom so that students are able to understand and retain content. The literature surveys a vast array of specific physics content and, as undergraduate students moved through *Phys 299/Chem299*, students were given readings from this list as a way to connect the content of the course to best teaching practices as determined by researchers. This was also the case for the research literature as it pertained to CER. Here, detailed best practices to teach chemistry and specific chemistry topics to undergraduates from the chemist's perspective were given to the students as we worked our way through different chemistry topics (Aikenhead, 2006; Belt, Leisvik, Hyde, & Overton, 2005; Burcat, 2004, 2005; Ortiz, 2012). Interestingly, students did not realize that this type of research existed. Then, once we got into the literature, making connections with our own classroom practice and then extrapolating to other courses that they had taken, students began to make connections of when best practices had been used, or how teaching, lectures, or laboratories could be revised to include some of the suggestions from the literature.

For clarification purposes, pedagogical content knowledge, as defined by Shulman (1986), refers to "the ways of representing and formulating the subject that makes it comprehensible to others" (p. 9). And, from here, CIDER (2009) explained, "Scholarly learner-centered pedagogy represents the conscious creation of environments designed to foster learning through a focus on learner autonomy, social engagement, and cognitive processing, based on principles of teaching and learning developed through theoretical and empirical research." This fostering lead to student engagement, viewing the learning environment not simply as a place to learn but to extrapolate so that it can be seen as a place to become empowered and transformed (Freire, 1996; Kincheloe, 1998).

Yet, to date, little research has been done that specifically focuses on undergraduate student perceptions of a course such as what has been presented here, that is, a course that fuses the teaching and educational research with concepts and content from the physical sciences and physical sciences educational research, namely PER from the perspective of the physicist and CER from the perspective of the chemist. Connecting this research with undergraduates' current courses and the course "take aways", or rather, content that has direct transferability into other courses the student is required to complete or add to their professional skill set as they consider their possible career paths has not been well documented. This research is an attempt to fill this apparent void.

METHODOLOGY

To conduct this research, a constructivist paradigm was utilized as a "net that contains [the] researcher's epistemological, ontological and methodological premise" (Denzin & Lincoln, 2005, p. 24). It is understood that, by using this paradigm, the researcher adheres to a subjectivist or transactional epistemology, and a relativistic ontology is acknowledged where multiple realities exist and understandings are co-created. It can also be argued that the epistemology for this particular research is also partially interpretive since the "knower and known interact and shape one another" (Denzin & Lincoln, 2005, p. 22). The author employed an interpretive case study design (Novodvorsky, 2006). The specific phenomenon this case study examined and explored was the views, issues and challenges, and perception and attitudinal changes of the physical sciences undergraduate students voice as they reflected on *PHYS 299/CHEM 299.*

Merriam (1998) described a case study as "an empirical inquiry that investigates a contemporary phenomenon within its real-life context, especially when the boundaries between phenomenon and context are not clearly evident" (p. 27). This research is rooted in qualitative case study methodology and was focused "on discovery, insight, and understanding from the perspectives of those being studied and offers the greatest promise of making significant contributions to the knowledge base and practice of education" (Merriam, 1998, p. 1). Here, *those being studied* were the physical science undergraduate students who enrolled in *PHYS 299/CHEM 299.*

This research project can also be characterized as interpretive research (Merriam, 1998), since the physical science undergraduate students were learning about the content of their courses, how to be better students themselves, and considering their future career options. These are processes that take time and constantly evolve. Further, multiple realities voiced by the participants through the qualitative data helped to "understand and explain the meaning of the social phenomena with as little disruption of the natural setting as possible" (Merriam, 1998, p. 5). The students were asked to voice the complex meaning of teaching and learning within the physical sciences where there is an emphasis on IBL and STEM. "It is assumed that meaning is embedded in people's experiences and that this meaning is mediated through the investigator's own perceptions" (Merriam, 1998, p. 6). If we accept Merriam's (1998) suggestion, then the experiences and opportunities the students had to make connections between and among their various science classes and to relate their learning to their out-of-class experience will serve to bring forth meaning and understanding of and to the research questions. Namely, how do physical science students come to understand technology infused STSE, STEM and IBL? And, how do they connect this learning to other courses and their future career aspirations?

Methods

Data collection, once consent was given, consisted of an individual interview with both closed- and open-ended questions that focused on the participants' understanding of some of the terminology used during the course and how the content of the course connected to their current course of study as well as their future career plans. The interview questions were designed to allow for discussion and clarification. For accuracy, all participant interviews were recorded and transcribed verbatim. As well, the author, who was also the professor for this course, pulled from the course notes what her own views of the course were as it was taught including issues within the class, problems with content and technology, conceptual understanding difficulties, as well as logistical difficulties with Internet-based technologies (ICTs).

Using a constant comparative/grounded theory methodology developed by Glaser (1956), the transcripts were read and re-read for themes. As discussed by Glaser, this method has been specifically designed to "aid analysts with these abilities in generating a theory which is integrated, consistent, plausible, close to the data, and in a form which is clear enough to be readily operationalized for testing in quantitative research" (pp. 437–438). This sense of the creative possibility of theory generation is a strength in that it is "concerned with generating and plausibly suggesting many properties and hypotheses about a general phenomenon (Glaser, 1956, p. 438). Some of these properties may be considered causes, some are considered conditions, consequences, dimensions, types, or process, etc. However, all of these, when considered together, will result in an integrated theory.

The Participants

The participants were those students enrolled in *PHYS 299/CHEM 299: Teaching and Learning in the Physical Sciences* in the fall of 2015, and the author as the course instructor. There were 19 students enrolled in the course (N=19), of those 7 volunteered to participate (P=7).

Research Process

Phase 1 of this research began in early February 2016 after research ethics was granted by the university's research ethics board. An email invitation was sent to those who had been enrolled in the course asking them to participate in an individual interview of approximately 45 minutes. This invitation was sent in early February 2016, 2 months after the course had concluded. Once students began to respond, Phase 2: February 19 – March 30, 2016, of this research project began. In Phase 2, individual 45-minute interviews were held with those who responded. Interviews took place in the professor's office at a time that was convenient for both the participant and the researcher and were then transcribed. Phase 3: March – April, 2016, included reviewing the instructional class notes to find connection between the author's experiences at the time and the students' perceptions as discussed within the interviews and as recorded from classroom activities.

DATA AND FINDINGS

In order to make sense of the data collected during the study, the author used a constant comparative method for data analysis. The data and findings are separated into three categories: the data and findings as voiced and seen from the students' perspectives, the data and findings as voiced from the professor's perspective, and recommendations and future research stemming from the two data sets.

Students' Perspectives

Using *students* as the unit of analysis, the structure of the course and technologies incorporated into teaching and learning methods helped students engage in STEM inquiry. Students were asked to recall, remember and explain specific terms from the course, which included STSE, STEM and IBL. In interviews, students were able to identify and explain both STSE and STEM; but when asked about IBL, they had a much more difficult time describing a generic IBL activity. Yet, when asked about examples from class, they were able to recall many with great accuracy in terms of what they were asked to do, what knowledge bases they needed to draw from, and the connections between the task and society and global awareness. For example, they described IBL activities such as capturing carbon dioxide in a balloon from the reaction of vinegar and baking soda, understanding momentum and impulse by dropping an egg within a home-made crumple zone, extracting DNA from bananas, and making homemade ice-cream by lowing the temperature through the use of ice and salt. After they described some of these activities, they were asked to identify the common factors. It was at this point that they began to understand how IBL sparks interest in science: students get to ask the questions and find their own answers, and through the process of inquiry, remember the content. The teacher's role is to guide the students who are in the leading roles.

After exploring their understanding of STSE, STEM and IBL, and their experiences within the course, students were asked whether they had had similar experiences within their other physical science classes, that is, if they could draw connections between these terms and the content of their other undergraduate courses. Of the students interviewed, all commented that they appreciate when professors help them see the connection between what they were learning in class and the larger, global picture of where this content is applicable; however, each also commented that this does not happen in many courses. Students were unsure if this was due to the course content, the professor assuming that the students should know these connections already, or time constraints. They did acknowledge that if STSE/STEM and IBL are to be incorporated into courses, then the professor would need to spend the time planning these activities and clearly communicate the connections.

Students also described the benefits and challenges of incorporating technology into classrooms, raising similar points as those often discussed in the literature (i.e., engagement, application to everyday life, distractions, time, money, learned helplessness). Interestingly, all students noted their hesitancy in using their devices during class. The author was told repeatedly that in almost all courses, students were not permitted to use their phones and that their professors did not want to see them using phones during class time. In contrast, students were encouraged and expected to use their phones or other mobile devices for PHYS 299/CHEM 299 course content and activities. The difference was that in this course, the technology had an educational purpose, and by the end of the term, they knew how their device could be used for tasks other than texting and status updates, often finding new features on their device to assist with the lessons, and when possible transferring this new set of skills to other classroom settings, if permitted.

Students were asked specifically about the technology used within the course and if the technology was beneficial to their learning. The technology infused within the course included any device that the student brought with them to be used at any time for referencing and an interactive whiteboard that was used to provide greater interaction than a standard electronic slide presentation. Formative assessments were often done via a Kahoot quiz. Other low-technology tools and science-based equipment for IBL activities included K'nex™ light benches, house hold materials for DNA extraction, specialized containers for preparing ice cream, egg drop experiment equipment, collectors of carbon dioxide from baking soda and vinegar, to name a few. Students commented that, in most cases, the technology added to their classroom experiences and deepened their learning. They also commented that sometimes they still need to write things to have it "sink into my brain" and that by having IBL and technology used within the class adds excitement. As one student said, "you never really know what is coming next so you need to pay attention and you don't want to miss a class because you might miss something really cool!"

When examining the data gathered during the student interviews, a clearer picture of the type of student taking this special topics course emerged. All the students who were interviewed were completing a Bachelor of Science, one was completing a major in Mathematics (in third year), one an honours in Biology with a minor in Chemistry (in second year), one a Human Kinetics degree with Biology and Chemistry (in third year), two were completing a major in Chemistry (one in fourth year, one in third year) and two were completing an advanced major in Chemistry (both in fourth year). Within the class as a whole, only two students took the course as a Physics credit and the other 17 as a Chemistry credit.

Students were asked about their future plans both during the course as well as during the interviews. From the class data, students commented that they were interested in (a) entering the teaching profession, (b) becoming a health care professional, or (c) staying in the area of the physical sciences (i.e., Masters of Science in Chemistry or Chemical Engineering). Compared to the data from the interviews, four of the seven indicated that they wanted to enter an Education degree program after graduation, one planned on doing Chemical engineering, another one graduate work in Chemistry, and one planned on going into a health care profession.

Each commented that they saw connections between the content of the course and their chosen career goals. For instance, they saw being able to explain, describe, and ask meaningful questions part of many professions. Further, each of the career areas involves working with the public; the students felt this course helped prepare them for that task in a way that other courses did not since they were required to speak and teach in front of their peers, work with a fellow student in the role of a tutor, and visit a middle school and act as a teacher assistant/classroom observer. Typically, students who are interested in pursuing a career in education have had some experiences with teaching others, and so the participants were asked about their background experiences. All interviewed commented that they had tutored while in high school as well as in university and many had also been hired to do laboratory demonstrating in different university departments.

In terms of content that students felt they could move forward from this course into future courses or careers, students commented that they now felt much more comfortable giving presentations and teaching small groups. They understood how to take a test and what the professor is looking for when asking certain questions, and they felt more confident asking questions or asking for clarification. Students also commented that the course content and activities gave them a clear picture of the teaching profession and how it has changed from when they were in secondary school. Students commented that they were very happy with the opportunity to work with peers on different projects and within the daily classroom activities instead of the entire class content being driven by an electronic slide presentation. They felt that they had "learned during class instead of memorizing at home." Students also commented on the overall nature of the course and the freedom within certain assignments; this was something that they had not experienced since secondary school and some found it to be very stressful because the course did not follow the "typical physical science course routine" or have a "cookbook style," rather, the professor "required us to think all the time!"

Professor's Perspective

Designing and delivering PHYS 299/CHEM 299 course content was a rewarding experience as it was an opportunity to combine literature that connects teaching and learning from the Science educator, Physicist, and Chemist perspectives. In retrospect, the first half of the course seemed quite content-heavy and theoretical, while the second half of the course felt much more applied. Hands-on activities and IBL processes presented students with opportunities to apply their knowledge of content and pedagogy to new situations. They were willing to try new things, take on new challenges, and expand their thinking about *doing* science. They worked well together and there was little time spent "off task," notably, less texting was observed. In fact, students often showed up early for class and had to be reminded that class was over so that they would not be late for their next lecture. At times, the classroom was very quiet when students were trying to understand new theoretical content, and at other times, the noise level increased when students were put into groups and given a challenge.

For one particular challenge, building a pulley system with K'nex™ to determine the connection between force, length of string, and number of active pulleys, the students were engaged, chatting, on task, making progress in connecting theory to practice. Suddenly, the classroom door was opened by another professor who complained we were "too loud in a place where learning was supposed to be happening". This baffled the students, and some commented on how much they were learning. The incident became a "teachable" moment, and we discussed the perception and reality of science teaching and learning in post-secondary spaces compared to how science is taught and learned in school. The questions then arose in terms of what should learning "look like" and whether is there a hierarchy of learning or teaching strategies that should be used. They questioned why university class delivery methods gravitate towards a lecture method when research suggests using a variety of teaching strategies. Students left that day discussing their education, connecting it to literature, and wondering how the system could be changed.

Students embraced, sometimes hesitatingly, all challenges both planned and unplanned as is sometimes the case when using technology. Students were and are deeply attached to their chosen device and the various apps and programs. This can be seen at any time outside of classrooms, yet when asked to use their device in class, students were in unfamiliar territory. When asked about their hesitation, students responded that they were never *asked* to use their devices in class, in fact they were told to put them away or leave them at home as they were a distraction.

For the professor for this course, incorporating digital technologies into teaching and learning was sometimes very frustrating because of limiting infrastructures. For example, limited wifi bandwidth meant not all students could stay logged on at the same time. As well, in order to use the interactive whiteboard device, the class was re-located from a room with a black chalkboard to a room with a whiteboard. Further, the lack of technological resources did, at times, result in a more "old-fashion" type of lesson instead of having the newest gadgets. However, with patience and an alternate plan ready at a moment's notice, most of the available technology could be infused as was planned.

Students completed course assignments with a great deal of thought and detail; some were even of publishable quality. In their responses, students connected content from this course to their other courses, describing how the other courses could be made better through using findings within the literature and how they, as individuals, could be better students. On multiple occasions, students commented that no teacher or professor had ever explained to them how to study for and how to take a test. Because we had spent a class on summative assessment, they now knew more about taking tests. During the interviews for this research, all commented that they felt that their marks had improved because they "knew what they were doing instead of just guessing." Students commented repeatedly on how the content of this course has helped and improved their standing in other courses because they now understood how to be better students.

FUTURE RESEARCH DIRECTIONS

When considering the next steps of this research, two are obvious. The first is to reconnect with the students who participated in this study to see what impact this course and the content covered had on their educational outlook and career choices. It would be interesting to explore what the students found to be of value from this course as they moved forward in both their post-secondary studies and as they enter their careers. It would also be interesting to learn if and how the content of the course changed their understanding of teaching and learning, and what advice they would give to the course professor, future students, and university administration concerning either a course of this nature or an infusion of the course ideas into other university science courses.

The second future research opportunity is to connect and share the findings from this research with Faculty of Science professors and to see what conversations and research possibilities arise. As professors, we assume students who have completed secondary school know "the basics," for example, how to study, how to ask questions, how to write a test, how to connect theory with practice, and to somehow apply theory and practice to everyday life. Some students come to university with these skillsets, some do not. In addition to "the basics", there is a host of research literature from all three fields (Education Research literature, Physics Education literature and Chemistry Education literature) that advocates to faculty the benefits of IBL opportunities, technology in the classroom, and the power of peer instruction/discussion. As faculty, we have the opportunity to infuse components of the course described within this chapter, and these specific components into any undergraduate science course. By examining the pre/post inclusion of previously well-tested pedagogical strategies, this, in turn, could provide new research opportunities in the areas of both physics and chemistry undergraduate teaching as well as within the required laboratories sessions.

CONCLUSION

This chapter described the development and launch of an undergraduate course entitled *Teaching and Learning in the Physical Sciences.* The course introduced students to ideas commonly discussed within a Bachelor of Education program and connected this content with the Science education research literature, PER, and CER. It infused IBL–STEM into its content so that students could experience and reexamine their understanding of various concepts previously learned. Students' and the professor's perceptions were given and findings presented.

When considering general conclusions from this research and the "take aways" that any professor or teacher at any level of education could consider prior to walking into the classroom to teach a lesson, there are many. Students are in our classrooms because they want to learn or they need to learn the content so they can achieve an outcome or obtain their chosen career; however, sometimes they do not come with the skillsets needed to learn efficiently. These skillsets include understanding Bloom's Taxonomy, knowing how to ask effective questions, understanding higher order thinking skills, and knowing how to take and prepare for a test. Taking time to review these skills could seem as a waste, yet from the data presented here, students found this to be extremely useful and commented that their overall grades in other courses improved by understanding these concepts.

In addition to these ideas is the notion of relevancy. When students understand *why* they need to learn and understand a concept or topic and how it is connected to the real world, they are much more engaged and better able to understand and remember. As a professor or a teacher, if we cannot answer the question "Why are we learning this?" then why are we teaching it? Students do not always see or are able to make the connections we assume they should be able to make; therefore, it is our professional responsibility to be explicit and state the relevancy and real world application of the content we deliver in classrooms.

Connected to these concepts of relevancy and professional responsibilities are the notions of technology and of infusing technology into our pedagogy. The typical student will come to class with some sort of electronic device, typically a smart phone. The students of the 21st century are digital natives since they have not known life without technology and the Internet. However, it is a fallacy to believe that just because students have a device that they know how to maximize its use or its educational potential. Students are on a continuum of digital expertise, similar to faculty, and therefore need to be taught when and how technology can be used for educational purposes either in the classroom or when doing research. As a professor or teacher, when faced with an electronic "distraction", there are two options: to ban the item, which typically makes it more of a distraction because it is akin to a forbidden fruit, or to embrace it. By embracing the technology that students have with them, having very specific tasks during class time that requires their appropriate use, the distraction becomes a tool to augment learning.

As scientists conducting research, we read and analyze what others have done in our field, compare this to our own findings, and explore how these ideas can move our own understanding and research forward. We test, we tweak, and then we tell others what we have found. This same process can be applied to our classrooms and our teaching. Our classrooms can be our educational research laboratories in which we seek to improve our own practice, where we embrace our professional responsibilities as not just researchers, but as educators too. It can become a place where we think of teaching as an active learning process, where passive lecture is reserved to quietly relay concepts, but where learning to apply those concepts is sometimes noisy with engaged students exuberantly exploring science together.

An understanding of effective pedagogical processes that address 21st century student learning and content delivery techniques such as inquiry processes that engage students and create opportunities to develop critical thinking and questioning skills can serve to invigorate our own work and strengthen scientific processes beyond our classrooms and degree programs. By engaging our students in the scientific process, we can perhaps influence the *way* society thinks of science and its connection to society and the environment. Our own students can become teachers of science, influencing future generations of students to develop a love of science and want to learn to research science too.

REFERENCES

Aikenhead, G. S. (2003). Chemistry and physics instruction: Integration, ideologies, and choices. *Chemical Education: Research and Practices*, *4*(2), 115–130.

Aikenhead, G. S. (2005). Research into STS Science Education. *Educación en la Química*, *16*, 384–397.

Aikenhead, G. S. (2006). *Science education for everyday life: Evidence-based practice*. New York, NY: Teachers College Press.

Alsop, S., Bencze, L., & Pedretti, E. (Eds.). (2005). *Analysing exemplary science teaching* (1st ed.). New York, NY: Open University Press.

Apple Education. (2010). *Getting started: Apple technology for diverse learners.* Available from http://www.apple.com/education/resources/

Beichner, R. J. (2009). An introduction to physics education research. In C. Henderson & K. A. Harper (Eds.), Getting started in PER. College Park, MD: American Association of Physics Teachers. Available from http://www.percentral.org/document /ServeFile.cfm?ID=8806

Belt, S. T., Leisvik, M. J., Hyde, A. J., & Overton, T. L. (2005). Using a context-based approach to undergraduate chemistry teaching: A case study for introductory physical chemistry. *Chemistry Education Research and Practice*, *6*(3), 166–179. doi:10.1039/B5RP90007G

Bencze, L., & Hodson, D. (1999). Changing practice by changing practice: Towards more authentic science and science curriculum development. *Journal of Research in Science Teaching*, *36*(5), 521–539. doi:10.1002/(SICI)1098-2736(199905)36:5<521::AID-TEA2>3.0.CO;2-6

Bennett, J., Lubben, F., & Hogarth, S. (2007). Bringing science to life: A synthesis of the research evidence on the effects of context-based and STSE approaches to science teaching. *Science Education*, *91*(3), 347–370. doi:10.1002/sce.20186

Bloom, J. W. (2006). *Creating a community of young scientists* (2nd ed.). New York, NY: Trifolium Books.

Bucat, R. (2004). Pedagogical content knowledge as a way forward: Applied research in chemistry education. *Chemistry Education Research and Practice*, *5*(3), 215–228. doi:10.1039/B4RP90025A

Bucat, R. (2005). Implications of chemistry education research for teaching practice: Pedagogical content knowledge as a way forward. *Chemistry Education International*, *6*(1), 1–2.

Bybee, R. W. (1993). *Reforming science education*. New York, NY: Teachers College Press.

Bybee, R. W. (2006). Scientific inquiry and science teaching. In L. B. Flick & N. G. Lederman (Eds.), Scientific inquiry and nature of science: Implications for teaching, learning, and teacher education (Vol. 25, pp. 1–15). Dordrecht, The Netherlands: Springer.

Center for Instructional Development and Educational Research (CIDER). (2009). *Pedagogy*. Retrieved May 13, 2014, from http://www.cider.vt.edu/pedagogy.html

Denzin, N., & Lincoln, Y. (2005). Introduction: The discipline and practice of qualitative research. In *The sage handbook of qualitative research* (3rd ed.; pp. 1–33). Thousand Oaks, CA: Sage.

Etkina, E. (2005). Physics teacher preparation: Dreams and reality. *Journal of Physics Teacher Education Online, 3*(2), 3–9.

Forbes, C. T., & Davis, E. A. (2008). Exploring preservice elementary teachers critique and adaption of science curriculum materials in respect to socioscientific issues. *Science and Education, 17*(8-9), 829–854. doi:10.1007/s11191-007-9080-z

Franz, J. (2010). Improving K-12 science education. *American Physical Society, 19*(7). Retrieved from: http://www.aps.org/publications/apsnews/201007/backpage.cfm

Freire, P. (1996). Pedagogy of the oppressed (New rev. ed.; M. B. Ramos, Trans.). New York, NY: Penguin Books.

Glaser, B. (1956). The constant comparative method of qualitative analysis. *Social Problems, 12*(4), 436–445. doi:10.2307/798843

Haralambos, M., & Holborn, M. (1991). *Sociology: Themes and Perspectives*. New York, NY: Collins Educational.

Hawkes, B., & O'Meara, J. (2005). *Physics education in Canada*. Retrieved from http://www.cap.ca/pic/Archives/61.2(2005)/editorial.html

Hillel, J. (2005). *Physics education research: A comprehensive study*. Toronto, Canada: University of Toronto.

Hofstein, A., & Lunetta, V. N. (2004). The laboratory in science education: Foundations for the twenty-first century. *Science education, 88*(1), 28–54. doi:10.1002/sce.10106

Holubova, R. (2005). Environmental physics: Motivation in physics teaching and learning. *Journal of Physics Teacher Education Online, 3*(1), 17–20.

Kincheloe, J. (1998). *Critical pedagogy primer* (2nd ed.). New York, NY: Peter Lang.

Knight, R. (2004). *Five easy lessons: Strategies for successful physics teaching* (1st ed.). Toronto, Canada: Addison Wesley.

Kortland, J. (2005). Physics in personal, social and scientific contexts: A retrospective view on the Dutch physics curriculum development project PLON. In P. Nentwig & D. Waddington (Eds.), *Making it relevant: Context based learning of science* (pp. 67–89). Munster, Germany: Waxmann.

MacLeod, K. A. (2007). A practical example to understand momentum in 1D: The watergun experiment. *Physics Education, 42*(5), 492–495. doi:10.1088/0031-9120/42/5/008

MacLeod, K. A. (2012). *Integrating science, technology, society and environment (STSE) into physics teacher education: Pre-service teachers' perceptions and challenges* (PhD Dissertation). University of Toronto, Toronto, Canada.

MacLeod, K. A. (2013). Physics education and STSE: Perspectives from the literature. *European Journal of Physics Education*, *4*(4), 1–12. doi:10.20308/ejpe.21926

MacLeod, K.A. (2015). *Teaching and Learning in the Physical Sciences* [Syllabus]. Antigonish, NS: Physics Department and Chemistry Department, St. Francis Xavier University.

MacLeod, K. A., & Kraglund-Gauthier, W. L. (2015). A case study of infusing technology into pre-service secondary science teacher learning: Conceptions and attitudes while navigating changing digital landscapes. *Journal of European Education*, *5*(2), 1–15. doi:10.18656/jee.24828

McDermott, L. C. (1998). Research in physics education. *APS News*, *7*(1), 8.

McDermott, L. C. (2001). Oersted Medal Lecture 2001. Physics education research: The key to student learning. *American Journal of Physics*, *69*(11), 1127–1137. doi:10.1119/1.1389280

McDermott, L. C. (2006). Preparing K–12 teachers in physics: Insights from history, experience, and research. *American Journal of Physics*, *74*(9), 748–762. doi:10.1119/1.2209243

Merriam, S. (1998). *Qualitative research and case study applications in education*. San Francisco, CA: Jossey-Bass.

Novodvorsky, I. (2006). Shifts in beliefs and thinking of beginning physics teacher. *Journal of Physics Teacher Education On-line*, *3*(3), 11–17.

Ortiz Nieves, E. L., Barreto, R., & Medina, Z. (2012). JCE classroom activity #111: Redox reactions in three representations. *Journal of Chemical Education*, *89*(5), 643–645. doi:10.1021/ed100694m

Pedretti, E., & Little, C. (2008). *From engagement to empowerment*. Toronto, Canada: Pearson Education.

Seatter, C. S. (2003). Constructivist science teaching: Intellectual and strategic teaching acts. *Interchange*, *34*(1), 63–87. doi:10.1023/A:1024506417791

Shipman, H. L. (2006). Inquiry learning in college classrooms: For the times, they are, a changing. In L. B. Flick & N. G. Lederman (Eds.), Scientific inquiry and nature of science: Implications for teaching, learning, and teacher education (Vol. 25, pp. 357–389). Dordrecht, The Netherlands: Springer.

van Aalst, J. (2000). An introduction to physics education research. *Canadian Journal of Physics*, *78*(1), 57–71. doi:10.1139/p00-005

van Aalst, J. (2005). Critical details and physics teaching. *Canadian Journal of Science. Mathematics and Technology Education*, *5*(3), 417–422. doi:10.1080/14926150509556670

Yerrick, R. (2010). *How notebook computers, digital media, and probeware can transform science learning in the classroom.* Available from http://www.apple.com/education/ resources/

KEY TERMS AND DEFINITIONS

21st Century Teaching: The skills of the teachers of today that are to be explicitly modeled to the students we teach. These include information and technology literacy, communication, critical thinking and problem solving, creativity and invention, global awareness, collaboration, and self-direction.

Chemistry Education Research: Educational research that is typically facilitated by chemistry faculty with students who are enrolled in undergraduate or graduate level chemistry courses. The aim of CER is to improve chemistry teaching at the university level.

Flipped Classroom: A new term for an old strategy, this is where the teacher has developed an activity, video, simulation, or a reading that students are required to complete prior to class, then during the class the teacher or professor pulls content from the pre-class assignment to move the class forward into an exercise or activity. The content from the pre-class assignment is not taught during class, but used for application or synthesis purposes. This puts responsibility onto to student to come to class prepared; else they will not understand the in-class component.

IBL-STEM: Inquiry Based Learning – Science Technology Engineering and Mathematics. This is the use of a teaching strategy known as inquiry based learning whereby students are given a task to complete with little to no guidance by the teacher. The task will infuse the use of science content, available technology and usually requires the student(s) to problem solve connecting the science to applications and using engineering principles as well as various calculations (mathematics).

Physical Sciences: Physics, chemistry, astronomy and earth sciences make up the physical sciences.

Physics Education Research: Educational research that is typically facilitated by physics faculty with students who are enrolled in undergraduate or graduate level physics courses. The aim of PER is to improve physics teaching at the university level.

STSE Education: Science, Technology, Society and Environment Education. This is the study of how science content is connected to technology, society and the environment as well as the impact of the science on each of these areas. It is a section found in most Canadian curriculum documents at the elementary and secondary level. Its goal is to help students see the relevancy of the science to their everyday world.

Chapter 11
The Use of Tablet Technology to Support Inquiry Science for Students Incarcerated in Juvenile Justice Settings

Michael Krezmien
University of Massachusetts – Amherst, USA

Christina Bosch
University of Massachusetts – Amherst, USA

Wardell Powell
University of Massachusetts – Amherst, USA

Tracey Hall
Center for Applied Special Technology, USA

Martina Nieswandt
University of Massachusetts – Amherst, USA

ABSTRACT

This chapter describes the challenges in implementing science instruction in juvenile corrections settings and present a tablet-based model for meeting the complex challenges. Project RAISE is a Project-based Inquiry Science (PBIS) curriculum designed in the Universal Design for Learning framework. It is developed in a tablet platform, and is designed to meet the unique needs of incarcerated learners. The chapter describes the juvenile justice educational setting, the characteristics of the classrooms, the learners, and the teachers. It provides an overview of one iBook that has been co-designed and tested with incarcerated learners.

DOI: 10.4018/978-1-5225-2528-8.ch011

INTRODUCTION

There is a dearth of research examining science instruction in juvenile corrections settings. However, the limited evidence of education in these settings reveals substandard instructional practices and low percentage of highly qualified teachers (Krezmien, 2008). In juvenile justice education, even the potential for high-quality science instruction is limited by a general overreliance on independent worksheet completion, the consequential lack of content-specific skill instruction, and the delimited ability of educators working outside their subject area to support student science learning. Furthermore, most juvenile corrections settings are operated from a safety and security perspective that does not allow laboratory opportunities or the use of equipment or materials for classroom experiments. Finally, because the students are confined in a secure placement because of infractions of the law, students cannot leave the facility to do field based experiments that would be particularly powerful in Biological Science classes. In this chapter, we illustrate how tablet-based technology can be utilized to circumvent some of these issues and develop content knowledge, inquiry skills, and increased interest and engagement in science among a group of incarcerated youth.

The work described in this chapter is part of a National Science Foundation (NSF) funded research project (DRL-1418152) titled *Reclaiming Access to Inquiry-based Science Education (RAISE) for Incarcerated Students: An Investigation of Project-Based Inquiry Science within a Universal Design for Learning (UDL) Framework in Juvenile Corrections Settings*. Project RAISE is a design and development project that focuses on developing, implementing and researching a Project-Based Inquiry Science (PBIS) curriculum within a Universal Design for Learning (UDL) framework. We worked collaboratively with the administrators, teachers, and students in a juvenile justice agency in all stages of the curriculum development, an approach that has been identified as a strategy for creating knowledge that is more relevant to classroom practice than the knowledge generated by research institutes (Enthoven & de Bruijn, 2010). The ongoing feedback from the critical stakeholders in our juvenile justice settings was used to revise all piloted components of the curriculum and thus informed the development of all instructional components.

The Project RAISE curriculum is delivered digitally through a tablet platform that replaces textbooks, workbooks, worksheets, and paper assessments, while at the same time integrating text-based content with multiple media (e.g., pictures, videos, animations, and interactive diagrams). The curriculum is tailored to the full range of learning needs in a diverse population of learners, provides real-time feedback to students and teachers, and enables students to participate in virtual inquiry experiences unavailable in typical juvenile corrections settings. Furthermore, our curriculum engages incarcerated youth in scientific practices that support scientific thinking, which is critical to preparing students with the 21st century skills that enhance their college and career readiness as outlined by the Committee on Defining Deeper Learning and 21st Century Skills (NRC, 2012). Educators have a duty to prepare students incarcerated in juvenile justice settings to become scientific thinkers. The scientific mindset aligns with the knowledge and skills necessary to integrate thoughtfully back into society and to become active, critical and engaged members of a rapidly transforming world.

In this chapter, we will show science education researchers and policy makers and other professionals how to utilize technology while forming partnerships with stakeholders with the ultimate aim of developing student inquiry skills and an engagement in science, even in a setting as constrained as juvenile corrections. Through a technology-based curriculum, we engaged incarcerated students in interactive instructional experiences that delivered complex science content, all within a UDL framework that addresses the numerous academic and behavioral barriers typical of marginalized incarcerated learners. Furthermore, we outline how we employed a unique co-development model to develop the curriculum in ways that enlisted students as authentic collaborators and sources of expertise, thus giving creative control and voice to this under examined population of learners.

The information we provide in this chapter can support policy makers to reform the educational system in juvenile justice settings to better support and engage difficult to serve learners. Science education researchers can use the model we describe to engage in further research to better understand relationships between digital technology, scientific literacy, and incarcerated students. For educators, engaging incarcerated students in science education that promotes inquiry skills and scientific thinking is critical to preparing these youth with skills necessary for the 21st Century workforce. This is paramount in preparing them to integrate back into society and to contribute to the development of society in general and the enhancement of their community in particular.

To overcome the fundamental problems confronting educators in juvenile justice settings, our NSF grant team developed a project-based inquiry science (PBIS) curriculum based on the principles of universal design for learning (UDL) and utilizing a mobile technology platform. This chapter will provide educators, science teachers, academicians, researchers, school administrations, and technology developers with an understanding of the theoretical and practical aspects of our curriculum. We will include examples of our curriculum and will discuss the anticipated impact that the curriculum will have on incarcerated learners.

This chapter will:

1. Describe the population of learners in juvenile corrections settings and the learning environments that characterize correctional facilities.

2. Discuss how the curriculum increased access to and mastery of science content, concepts, and inquiry skills critical for careers in the 21st Century STEM workforce.
3. Describe the technology-based UDL supports embedded within the science curriculum.
4. Discuss the prototyping process of using student feedback to modify and enhance the final curriculum.

BACKGROUND

Student Demographics

Incarcerated learners are disproportionately low-income students of color with limited educational opportunities (Gregory, Skiba, & Noguera, 2010; Krezmien, Leone, & Achilles, 2006). Many of these students also have disabilities that impact their education (Bullock & McArthur, 1994; Krezmien, Mulcahy, Travers, Wilson, & Wells, 2013; Quinn, Rutherford, Leone, Osher, & Poirier, 2005). Incarcerated learners do not typically possess a basic understanding of science and scientific concepts (Anderman, Greisinger, & Westerfield, 1998) and they usually lack the inquiry skills necessary to support scientific thinking (Brigham, Scruggs, & Mastropieri, 2011; Therrien, Taylor, Watt, & Kaldenberg, 2013). These factors frequently prevent incarcerated students from passing high-stakes tests in science and obtaining a high school diploma. Consequently, students in juvenile corrections facilities are unable to pursue a post-secondary degree in the STEM fields and are ill-equipped to participate in the 21st Century workforce. The work described in this chapter was conducted in partnership with the juvenile corrections education provider in one state. The youth in the system were from impoverished urban communities. Students ranged from 14 to 18 years old, in the 9th through 12th grades, and almost all were students were students of color. Very few of the youth had passed the State's Biology examination required to earn a high school diploma.

Learner Characteristics

Most incarcerated learners have a history of trauma and face a host of academic barriers to learning science content. Recorded rates of disability classification amongst students in juvenile corrections are about three to four times that reported in public schools, with the rate of youth with emotional and behavioral disorders being about six times higher (Gagnon & Barber, 2010). Substantial percentages of the students had emotional/behavioral disorders (EBD) and learning disabilities (LD), with high numbers of students identified as having ADHD (Quinn et al., 2005; Gagnon, Barber, Van Loan, & Leone, 2009; Krezmien, Mulcahy, & Leone, 2008). High school students with ED and LD have been found to display academic deficits of comparable magnitude, placing them far below grade level in reading and mathematics (Lane, Carter, Pierson & Glaeser, 2006). Among this group, students with reading disabilities and dyslexia are typical (Krezmien et al., 2013; Krezmien et al., 2008). These learners demonstrate limited phonological awareness (facility discerning and manipulating sounds in oral language and relating them to written language), poor decoding skills, (converting letters into sounds) and deficient orthographic skills (spelling and conventions of written language). These foundational reading competencies are critical for reading fluency, since without accuracy and automaticity, reading is a laborious and slow process, a cognitive load that impedes comprehension of text. Of course, for some students with disabilities, "reading to learn" or engaging in other typical instructional activities across the content areas may be challenging for other reasons. Depending on how tasks are delivered and structured, students with disabilities may encounter barriers to learning that stem from executive functioning issues related to prioritizing and organizing information, or challenges related to attentional management or memory processing. The high rates of special education and associated behavioral health needs create numerous issues for incarcerated learners and their teachers. They often require intensive behavioral and therapeutic supports consistent with those in residential treatment schools.

Many adjudicated students have also not gone to school for several years, and have experienced substantial numbers of disciplinary problems which resulted in removal from class, suspensions from school, or expulsion (Gagnon, Murphy, Steinberg, Gaddis, & Crockett, 2013). Incarcerated youth in both general and special education are often reading several years below their grade level or more than one standard deviation below the normative sample mean as measured by standardized reading assessments (Krezmien et al., 2008; Krezmien et al., 2013; Houchins, Jolivette, Krezmien & Baltodano, 2008; Baltodano, Harris & Rutherford, 2005). They also are substantially behind in mathematics (Mulcahy, Krezmien & Travers, 2015; Gagnon & Barber, 2010; Maccini, Gagnon, Mulcahy, & Leone, 2006). Research indicates that about half of this population of learners can be expected to have co-occurring mental health and behavioral disorders, as well as substance use disorders (Mulcahy & Krezmien, 2009; Gagnon & Barber, 2010). These issues create numerous challenges in science classes. Specifically, students are typically unable to read the complex science texts, lack the necessary vocabulary to comprehend the science content, and lack the basic math skills necessary to participate meaningfully in science classes, especially the with the new Next Generation Science Standards (NGSS).

STEM Learning Characteristics

The previously described learning deficits are associated with students who are ill prepared to participate in most science curricula, and cannot typically succeed in a science class that requires the conceptual understanding and procedural skills and abilities required for individuals to address relevant personal, social, and global issues (Bybee, 2010). According to Bybee (2010), STEM literacy involves the integration of STEM disciplines as four interrelated and complementary components. These include:

1. Acquiring scientific, technological, engineering, and mathematical understandings and using them to identify issues, acquire new knowledge, and apply the knowledge to STEM-related issues.
2. Understanding the characteristic features of STEM disciplines as forms of human endeavors that include the processes of inquiry, design, and analysis.
3. Recognizing how STEM disciplines shape our material, intellectual, and cultural world.
4. Engaging in STEM-related issues with the ideas of science, technology, engineering, and mathematics as concerned, affective, and constructive citizens.

Achieving these standards is difficult even for knowledgeable and skilled learners. They are often extremely difficult for disadvantaged learners and learners with disabilities -- the students who predominate the juvenile justice settings. These students frequently have limited prior knowledge, are reluctant to pose questions, are less likely to have a plan for solving problems, struggle to implement teacher recommendations, have difficulties with inductive and deductive reasoning, and seldom transfer knowledge to other contexts (Dalton, Morocco, Tivnan, & Mead, 1997). Additionally, incarcerated learners often have fundamental misconceptions about scientific phenomenon, which leads to further struggles during the inquiry process (Jacobson & Archodidou, 2000). Our PBIS-UDL Biology model was designed to address these learning barriers intervention and to provide incarcerated learners with the supports necessary to fully participate in science learning.

Characteristics of Juvenile Justice Facilities

Juvenile corrections education is complex. The types of facilities vary across regions, districts, and states. They differ by purpose. Some facilities are detention programs that hold students awaiting adjudication. Detention facilities are generally short-term placements, although some youth have been in detention settings for as long as a year or more. Other facilities are commitment facilities, which hold youth who have been adjudicated to a locked facility for a determinate or indeterminate amount of time. Some facilities are large, housing several hundred youth. Other facilities are small, serving as few as one student as a time. Some facilities are local, some are regional, and some are statewide. Some facilities run like adult prisons, while other run like residential treatment programs. Finally, some are publically operated, while others are privately operated by providers with and without experience in juvenile justice education.

All students are mandated to participate in a full day educational programs in compliance with state and federal requirements. The education programs vary. Some educational programs have extensive vocational and technical programs, while other have no vocational opportunities. Some programs focus on a high school diploma credit based schools approach, while many focus on alternative education certificate like GED or HiSET. Most facilities place youth into educational programs for criminogenic purposes, so classrooms are typically comprised of students of varying ages, grades, abilities, skills, and course needs. For instance, science class might consist of students from 14 to 18 years of age, from 9th grade to 12th grade, and in classes ranging from 8th grade mathematics to trigonometry. The students may have the reading skills of a 4th or 5th grade student, while others read on grade level. Finally, most classrooms will have a substantial number of students with disabilities.

Researchers have documented concerns about appropriate provision of basic educational materials and practices in secure care for over thirty years. Educational services tend to be limited by issues of physical space, uninterrupted and sufficient instructional time, scheduling conflicts, as well as behavior-related interruptions and the associated punitive responses, which can include the isolation and segregation of youth (Gagnon & Barber, 2010; Mulcahy, Krezmien, Leone, Houchins & Baltodano, 2008). Students in these facilities leave and arrive at different times, and can be very mobile across sites within the system (Leone, Krezmien, Mason & Meisel, 2005). Instruction in these facilities is often worksheet-based (Coffey & Gemignani, 1994, as cited in Leone & Weinberg, 2012; Tannis, 2014). Furthermore, the lack of rigorous applied instructional intervention studies in juvenile corrections facilities and/or with students with EBD limits the quality of educational programming even in systems that attempt to prioritize it (Mulcahy et al., 2015; Krezmien & Mulcahy, 2008). Inadequately completed, lapsed, or missing school records that might have guided the provision of Free and Appropriate Public Education (IDEA, 2004) in correctional facilities is also a common barrier to adequate instruction (Krezmien, et al., 2013; Leone & Weinberg, 2012). Special education in juvenile corrections can potentially provide academic instruction and supports as well as access to behavioral and mental health services, but identification of special needs in these settings can be impeded by student histories of truancy and the aforementioned paperwork problems (Krezmien et al., 2008). Teachers in these facilities are often isolated and have limited access to training and professional development specific to their needs (Gagnon & Barber, 2014; Tannis, 2014). Because facilities are designed with safety and behavior modification as a priority over education, controlling behavior is often a prerequisite to learning that limits the materials teachers can employ in their classes and the research that can be effectively carried out in these facilities (Mulcahy et al., 2008; Houchins, et al., 2008).

Characteristics of Science Classroom Environments

Most science classrooms in juvenile corrections schools are small and lack the materials and equipment typical of a high school science class. Science classes are typically devoid of any equipment that could be used as a weapon, like microscopes, chemistry glassware, burners, and chemical supplies, petri dishes, specimen containers and biological specimens. The classrooms in most facilities are multi-purpose and not specifically dedicated to a single subject, and most are small, with limited space to move around and engage in differentiated activities. Most classrooms have a computer for the teacher only, though many have interactive whiteboards. The partnering classroom for Project RAISE was typical of juvenile corrections settings.

The classrooms in the facilities are generally small. The rooms usually have a small white board, an LCD projector, a few desks for students, an additional chair for the custodial/security staff, and a chair and desk for the teacher. Science class lasted for 45 minutes and usually included 4-5 students. The teacher typically used a textbook and worksheets, and the students were instructed to read the textbook and answer the questions in the worksheets. The teacher responded to student questions and helped them to complete the worksheets as needed. During class, it was a common occurrence for custody staff to call students out of the classroom to attend to various issues.

Characteristics of Teachers

There is limited research on the characteristics and qualities of teachers in juvenile corrections settings. Teachers in juvenile corrections settings are often required to teach in subjects outside of their content area (Gagnon et al., 2009). Teachers are sometimes second career teachers. Sometimes teachers are not licensed teachers, and many are on waivers as they pursue a teaching license. Facilities compete for a generally limited pool of certified science teachers, who are in high demand and can get preferred teaching jobs in public schools. The partner science teacher for Project RAISE was a licensed veteran teacher certified to teach high school mathematics. The teacher was not licensed to teach science. The teacher had 28 years of teaching experience in juvenile detention facilities, and had taught mathematics and science for the last eight years.

Technology in Facilities

Juvenile justice educators typically face three major challenges with respect to technology. First, most juvenile corrections facilities lack the necessary fiscal and structural resources necessary to employ technology in the classroom. For instance, educators typically lack an instructional budget to purchase computers and interactive whiteboards. Additionally, facilities often lack the technological infrastructure to install technology or the technical support to maintain the technology. Second, juvenile justice educators that do have access to technology typically lack the intensive training, support, and ongoing professional development necessary to successfully employ the technology to promote engagement and learning. So, interactive whiteboards are often only used as a typical whiteboard. Interactive software is not adequately utilized because the teachers lack resources to support student use. Third, internet access is typically not available for student use. Because juvenile corrections settings are heavily focused on safety and security, security measures often lock Internet access. In facilities that allow instructional use of the Internet, security breaches typically result in system-wide shutdowns of the Internet on both a temporary and permanent basis. As a result, science teachers in juvenile justice schools typically lack access to the conventional online resources available in public schools or cannot consistently plan lessons around the utilization of online resources even when the facility has internet access. Consequently, teachers often revert to the use of worksheets and independent textbook reading as the primary means of science instruction.

THE PROJECT RAISE MODEL

We designed Project RAISE to address the challenges with providing rigorous science education in juvenile corrections settings. Project RAISE addressed the four major challenges of effective science instruction in juvenile corrections settings: (1) the science learning profile of incarcerated students, (2) the dearth of effective inquiry-based science instruction, (3) the limitations of the classroom and facilities environments, and (4) the inconsistency with technology hardware and software consistency and support. The Project RAISE model involved a Project Based Inquiry Science (PBIS) instructional design based upon the Universal Design for Learning (UDL) framework and delivered through a standalone tablet-based platform. We will describe the components of the model in the following sections, and will provide initial findings about student interest and engagement in science using the model. We will discuss PBIS, UDL, and Technology separately, but we will include examples from one of our tablet-based PBIS-UDL lessons to show how these frameworks function in concert to respond to student needs. The lesson addresses the 10% rule, a critical concept in ecology that is also part of the Next Generation Science Standards (NGSS Lead States, 2013). In the 10% Rule iBook, students learn how energy is transferred in an ecosystem, and that in any ecosystem, about 10% of the energy is transferred from one trophic level to the next. Students are required to understand the mathematical model in the context of an oceanic ecosystem, as well as to calculate multistep math problems involving the transfer of energy. The RAISE 10% Rule iBook addresses parts of two NGSS standards:

- HS-LS2-6 asked students to demonstrate understanding regarding how to evaluate the claims, evidence, and reasoning that the complex interactions in ecosystems maintain relatively consistent numbers and types of organisms in stable conditions, but changing conditions may result in a new ecosystem.
- MS-LS2-1 asked students to demonstrate understanding regarding how to analyze and interpret data and to provide evidence for the effects of resource availability on organisms and populations of organisms in an ecosystem.

The 10% Rule lesson incorporates Project Based Inquiry Science and Universal Design for Learning delivered in a tablet-based e-book.

Student Learning Goals

Despite the numerous challenges around science education in juvenile corrections, we adopted the NGSS learning goals for Project RAISE. We did not modify or adapt the learning goals of the NGSS in any way. Rather, we employed the use of PBIS, UDL, and technology to modify the content, delivery, and supports in a manner that would allow incarcerated students to meet the standards set by the NGSS. Incarcerated students can meet rigorous learning goals if the instruction is designed in a manner that supports all learners and engages students in authentic and meaningful ways.

Technology and Usability

Project RAISE adopted tablet-based technology that allowed us to overcome the numerous individual and structural barriers to learning in juvenile corrections settings. Internet access is inconsistent in juvenile corrections settings, so we have developed a curriculum that functions offline, without the use of any online supports.

We have adopted the Apple iPad because it provides a single device per student at a relatively low cost, does not require the use of a keyboard, allows for split screens so that students to use multiple tools simultaneously, and accepts a variety of inputs including: a touch screen interface, drag and drop tasks, typed notes and responses, handwritten responses using a finger or a stylus, audio recording, and visual recording through pictures and video. The incarcerated learners involved in our project have responded positively to the use of the iPad tablets. They reported (1) the touch screen is easy to use, (2) the interactive tools are engaging and useful, (3) the opportunity to write, type, speak, or video are helpful, (4) and the ability to access information via video, images, multimedia, and text make information more engaging and accessible.

Project-Based Inquiry Science

The 10% Rule iBook utilizes the principles of Project Based Inquiry Science by giving students opportunities to engage in content organized by active investigations to answer driving questions (Marx, Blumenfeld, Krajcik, & Soloway, 1997) and an extended process of responding to questions, problems, or challenges (Buck Institute for Education, 2013). The process requires students to think critically as they generate questions, generate hypotheses, revise hypotheses, collect data, evaluate findings, communicate, and collaborate with peers. Rather than being the source of instruction, the teacher serves as a facilitator (Barron & Darling-Hammond, 2008) by structuring tasks that build knowledge, promoting utilization of social skills, maximizing engagement of all students, and conducting ongoing assessment of student performance. Consequently, students who engage in Project Based Learning acquire a deep understanding of concepts, develop a lifelong interest in learning, learn to use technology, acquire skills vital to careers and employment, and become increasingly motivated to solve problems of the 21-century (Buck Institute for Education, 2013).

While project-based learning has the potential to enhance students' use of critical thinking skills as they apply knowledge and skills to real-world problems, there are formidable barriers that must be overcome when exposing incarcerated youth to this type of learning. First, the secure nature of the facilities prevents students from experiencing the natural world and conducting field-based science activities. Second, within the classrooms, security protocols restrict the use of any living organisms, glassware, and microscopes within the classroom, and even prohibit the use of virtual simulations that require the internet. Consequently, students cannot participate in any typical scientific hands-on laboratory experiences. Third, many project-based learning protocols require teachers to be skillful with technology and to understand its use as a tool to enhance learning and thinking (Blumenfeld, Fishman, Krajcik,

Marx & Soloway, 2000). To overcome these barriers, we developed a final Universal Design for Learning Project-Based Inquiry Science (UDL-PBIS) curriculum that includes virtual learning environments, virtual laboratories, and digital scaffolds and supports that promote scientific learning.

The 10% Rule iBook is built within a PBIS framework in multiple ways. First, the 10% Rule iBook is part of an overarching four-week unit project focused on the human impact on the environment. In that project, students will create Identity Maps—visual representations of their own identity and their place in the ecological world. Over the course of the unit, students revise the Identity Maps by incorporating information learned during the lessons into their Map. The Identity Maps are completed using various apps on the iBooks that allow students to create visual art projects (e.g., self-portraits with symbolic references), multimedia projects (e.g., audio-video presentations), or audio projects such as music or spoken word. The Identity Map unit project addresses multiple aspects of PBIS, including (1) sustained inquiry, (2) student voice and choice, (3) opportunities for reflection on the content and on their individual projects, (4) continuous opportunities to critique and revise their identity molecule projects, and (5) a publicly presented product that allows students to share and explain their project to facility personnel, managerial personnel, members of the community, and parents (Buck Institute for Education, 2013). Additionally, the project is authentic and specific to the unique perspectives of the incarcerated learner, allowing each student to explore the conceptual and content knowledge of the unit within the context of their own experiences and perspectives (Buck Institute for Education, 2013).

Finally, the 10% Rule iBook incorporates aspects of PBIS into the discrete components of the lesson. For instance, the 10% Rule is introduced to the students through a short video that shows a single ocean ecosystem and that describes the energy relationships between algae, krill, mackerel, seal, and a great white shark in an aquatic ecosystem. The video is accompanied by a challenging driving question (Buck Institute for Education, 2013), "Where is most of the energy in an energy pyramid?" The students are engaged in a discussion and independent reporting activity about the video using one of the Visible Thinking routines from Harvard's Project Zero, known as "See, Think, Wonder." This routine encourages students to explore the concept of energy and energy transfer in a single ecosystem by listing observations (in this case, from the video), using their observations as justifications for thoughts, and extending these insights into questions. The students then move on to an exploration of a model of the energy pyramid, with embedded video-based supports that teach students how the energy pyramid model explains the transfer of energy they observed in the video. Then the students engage with an interactive energy pyramid tool (See Figure 1) that allows the students to explore the 10% rule

Figure 1. Interactive energy pyramid
Source: RAISE 10% Rule iBook

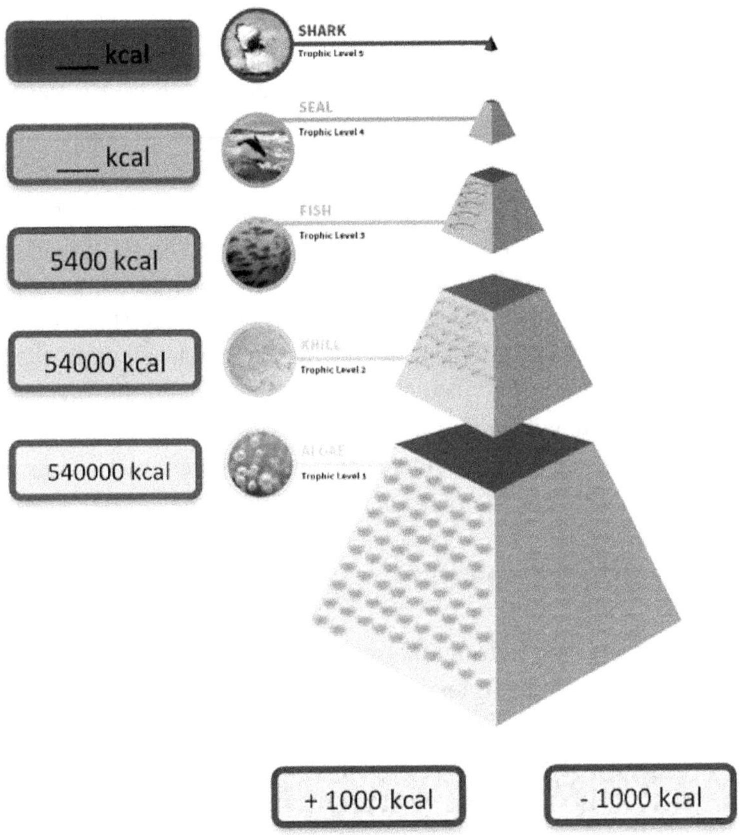

through manipulation of the organisms and energy in the ocean ecosystem from the video.

The energy pyramid tool in Figure 1 allows students to add or subtract energy to trophic level 1 of the pyramid by clicking on the buttons at the bottom. As they add energy, images of the representative organisms (algae, krill, mackerel, seal, shark) appear in the respective trophic level, and the energy needed at that trophic level appears in the text boxes next to the images in each respective level. The energy pyramid tool is accompanied by a series of multiple choice and open-response questions. The multiple choice questions provide instantaneous feedback to students. For instance, if a student chose the incorrect choice "C", the tool will provide them with feedback specific to the errors in choice "C", will give them some feedback to help them rethink the answer, and give them another opportunity to respond to the question. The students can answer using by a whiteboard tool to do longhand computations, using a calculator, and/or using the interactive pyramid tool, all within the lesson's interface. After the students complete the questions, the students are asked to revisit the answer they developed for the driving question, "Where is most of the energy in an energy pyramid?" The question appears in the iBook and their original response is shown in the integrated whiteboard tool. Students discuss what they learned through the interactive tool, and make changes to their response to the driving question or add to it by reflecting on what they have learned. The interactive aspects of the activities, the utilization of video representations of ecosystems that the students cannot visit, and the responsive multiple-choice questions which provide instantaneous feedback allow the students to participate in a rich and meaningful experience despite the restrictions of the environment. The students then have the opportunity to watch part of a video about human impact on the environment, and to incorporate what they have learned in this lesson into their identity molecule. These PBIS activities allow the students to develop a personal connection to the content, which can typically feel abstract or esoteric.

Universal Design for Learning

Because the students face substantial and complex issues that impede learning, we designed all Project RAISE lessons within the Universal Design for Learning framework (UDL). UDL is an educational design framework for the development of curriculum and instruction that expects and accounts for learner variability (Meyer, Rose, & Gordon, 2014). UDL views human variability as a natural phenomenon and considers differences to be the norm. The significance of this view lies in its implications for the initial design of learning experiences: curricula can be developed with flexible options and supports that anticipate variations in how learners will perceive, act on, and internalize information in the environment. UDL has three guiding principles, corresponding to three neural networks, that promote options in learning: (1) multiple means of representation (corresponding to the recognition network, or the "what" of learning); (2) multiple means of student action and expression (corresponding to the strategic network, or the "how" of learning); and (3) multiple means of student engagement (corresponding to the affective network, or the "why" of learning) (Rose & Meyer, 2002). UDL further posits that the use of digital technology can help facilitate the development of flexible curricula by creating dynamic learning opportunities for students and teachers (Meyer & Rose, 2005). The 10% Rule lesson has numerous examples that show how options for perception (representation), action and expression, and engagement were established or enhanced through the integration of the three guiding principles into the curricular design.

UDL Principle I: Multiple Means of Representation

There are multiple ways that Project RAISE has incorporated multiple means of representation and thus options for comprehension into the lesson. First, we aligned the textbook text with the NGSS, and eliminated any text that was not directly related to the standards. This resulted in a reduction of about 80% of the text, thereby reducing the number of words and sentences. Additionally, we re-authored the text so that non-essential (not directly linked to science knowledge or concepts) complex multisyllabic words were eliminated and sentences were written as short declarative sentences. Finally, we limited the amount of text per page of the e-book to reduce the perceptual load, and added audio and audiovisual versions of the text as alternative representations of the same information. For example, our final ocean ecosystem video is accompanied by a standards-aligned oral narration, and has the option for subtitles. The changes to the text and the associated media dramatically increased accessibility for students with below grade level reading skills, and substantially decreased the literacy barriers for students with learning disabilities and dyslexia. Instead of dedicating substantial time and effort trying to decode and understand complex text, students spend time engaged in a multimedia experience that delivers the content in multiple ways and in a straightforward design.

Figure 2 also shows multiple ways that multiple means of representation are integrated into a single interactive tool. Second, each level is differentiated by text-based titles, color, shading patterns, space between each level, and images of the organisms in each level drawn from the original video. Additionally, the images of the organisms are interactive: students can touch an image to replay the section of the video that is related just to that organism in the video. Finally, when a student taps on a specific level, the level is highlighted as shown separately in the right hand pyramid in Figure 2. These features guide information processing through multiple means, and also allow students to experience multiple representations of the same concepts in a single interactive tool. As with the multiple representations of the text, the broadened representations of the energy pyramid provide more entry-points into the content. The options to customize the display of information help individualize the content to the learner, and thus decrease the amount of time and effort that might be spent trying to understand a single fixed model, while increasing the amount of time on learning.

UDL Principle II: Multiple Means of Expression

The 10% Rule iBook also provides multiple means of expression, which allows students to convey their knowledge, ideas, and understanding in multiple ways. For example, the 10% Rule iBook has multiple types of questions and multiple types

Figure 2. UDL representations of the energy pyramid
Source: RAISE 10% Rule iBook

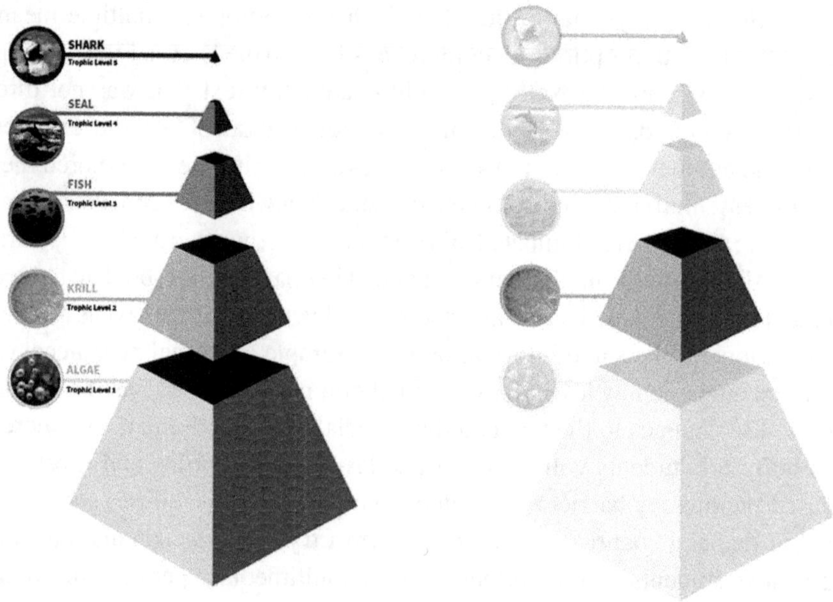

of response systems. Figure 3 displays two different interactive types of questions. The right side shows a drag and drop tool that allows students to demonstrate their knowledge about the energy chain by dragging images into the chain itself. The images are the same images used in the video and in the energy pyramid diagram, and the chain is color coded to match the pyramid, to maximize transfer and generalization throughout the iBook. Additionally, students can use the "check answer" feature to

Figure 3. Drag and drop tool and multiple choice tool
Source: RAISE 10% Rule iBook

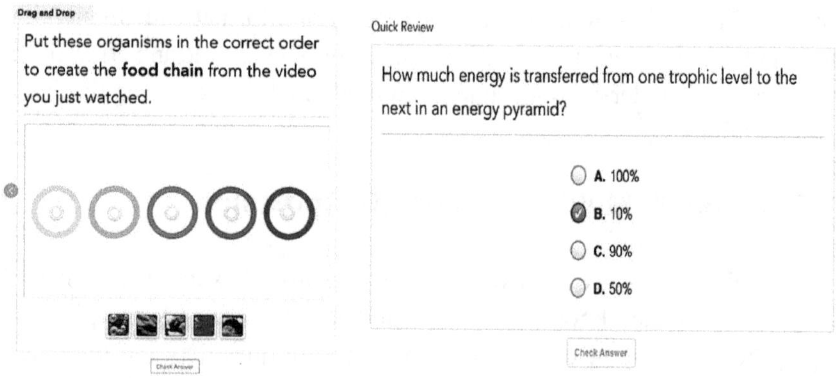

self-monitor their understanding. The 10% iBook also utilizes an app-based tool that allows students to answer open-ended questions by typing, writing with their fingers or a stylus, or drawing. The tool also includes images and figures from the iBook, such as the energy pyramid in Figure 2, which allows the students to make notes or annotations next to or over the figures that they encounter in the iBook. Additionally, students can use the tool to record audio or video of themselves explaining answers to questions. These various methods of response provide students with multiple ways to communicate their knowledge. Immediate feedback supports their expressive strategy development. Data storage systems, built-in to the iBook, prevent work from being lost or not turned in, a behavior typical of students with LD or ADHD. These options for action and expression also allow students to convey their thoughts, knowledge, and ideas in more individualized ways. These tools increase the amount of time and effort students can dedicate to higher-order thinking tasks like answering questions, and allows the teachers to have multiple data sources by which to evaluate student achievement.

Additionally, the 10% Rule iBook includes tools and supports that help students to answer conceptual and mathematical questions even if they have limited or no computational skills, which is typical of many of the students in juvenile corrections settings. The supports for expression and executive functioning are clear in the math questions related to the 10% rule. First, the iBook has an embedded video that provides students with on-demand support for how to calculate 10% of a number, and the video contextualizes this instruction by using organisms from the lesson. Second, the iBook contains a calculator app that students can use if they struggle with hand and/or mental calculations. Third, Figure 4 shows how the energy in the 10% model can be represented in multiple ways that can help a student to better express their understanding of the 10% rule. Figure 4 shows the kilocalories available at each level of the energy pyramid in a graphic representation. The graphic representation of kilocalories as boxes (like math manipulative blocks) helps students to understand the magnitude of the energy at each level. Furthermore, the arrangement of the energy boxes in the pyramid orientation with the same color coding and images of the representative organisms from each trophic level maintains a pattern of representation that can serve as a scaffold for students to understand where most of the energy is found in an energy pyramid, the driving question of the lesson. The students can use their understanding of the graphic representation to explain their understanding of the 10% rule, and can use the figure to answer multiple choice and open-ended questions about the 10% rule.

Figure 5. shows an example of one of the whiteboard features used to answer the driving question, "Where is most of the energy in an energy pyramid." The student whose work is shown in Figure 5. used the typing tool to write, revisit, and revise his response to the driving question. In comparison to Figure 1, Figure 5 shows

Figure 4. Graphic Representation of kilocalories in an energy pyramid
Source: RAISE 10% Rule iBook

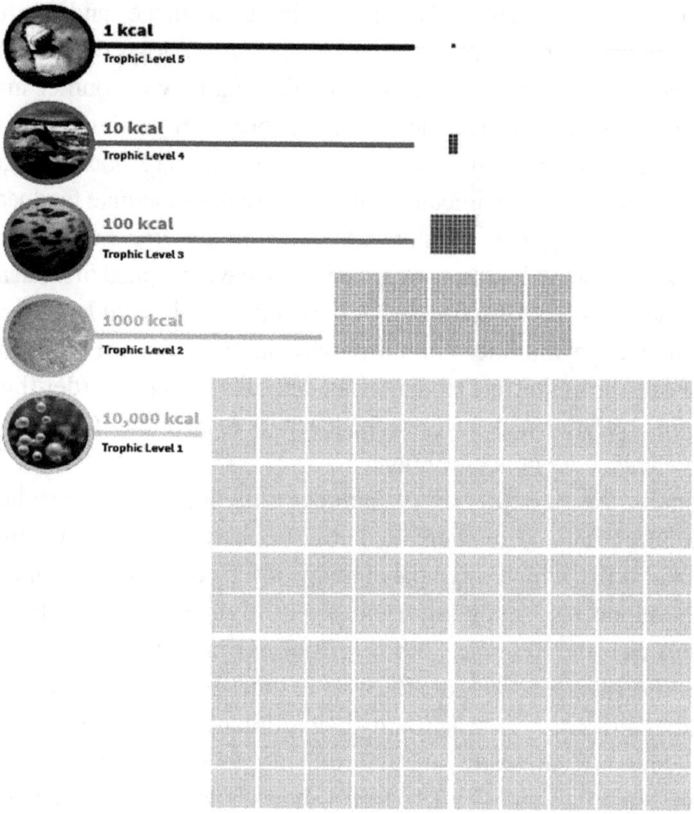

how the student's understanding of the energy pyramid developed over the course of the lesson, and also shows that the student was able refine his understanding by reading some of the material available in the 10% Rule iBook. Figure 5 also shows how the tool allows the student to reply in multiple ways, through typing, writing, drawing tools, or recording tools (located in the bottom bar).

Finally, Figure 1 shows the interactive energy pyramid that allows students to explore how the 10% rule functions. However, this interactive model can also be used as a calculation tool, allowing students to change the numbers of representative organisms in the first trophic level, observe the changes in the representative organisms in higher levels, and use the energy totals in the number boxes at the top to determine how much energy is at each level. By adding or subtracting algae in the model, students can determine how much energy is used at each subsequent trophic level. So, if a student struggles with computation and even struggles with

Figure 5. Whiteboard tool with an example of a student response to the driving question
Source: RAISE 10% Rule iBook

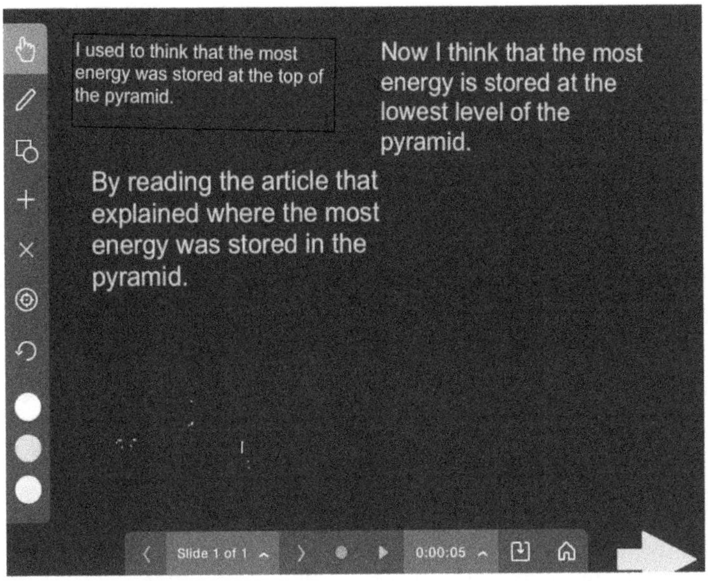

the use of a calculator, the student can use the interactive energy pyramid tool to answer math questions.

UDL Principle III: Multiple Means of Engagement

The entire Project RAISE curriculum is designed with attention to providing multiple means of engagement throughout all activities. The UDL framework considers self-regulation, effort and persistence, and interest as broad components of engagement. Project RAISE lessons are designed to address these aspects of engagement. For example, Project RAISE iBooks recruit interest by utilize various sorts of videos that show students examples of living organisms in authentic environments as well as tutorial videos that have been aligned with the lesson content. The videos are presented in ways that activate or supply background knowledge, which in turn optimizes the relevance of and minimizes aversive responses to the instructional materials. The iBooks support self-regulation and persistence by allowing students to access videos at any time. This option supports their self-assessment and develops their coping skills and learning strategies because it encourages students to question their understanding and utilize resources in the environment to support their learning. This option also helps sustain their efforts to answer questions, because it facilitates students' drawing on evidence from the lesson. Also, the iBook pages have been designed with the other two UDL principles in mind, and as such they reduce the cognitive load normally required during the reading of textually and visually complex science textbooks. This makes the relevant text more accessible and subsequently more likely to be engaged with by incarcerated learners. By making complex content more accessible, we are recruiting student interest and thus increase the likelihood that learners will feel motivated to participate in difficult assessments that they typically skip in a textbook and worksheet driven model of instruction.

Project RAISE also incorporates numerous interactive features, like the interactive energy pyramid (See Figure 1), which increase the choices students have over how they acquire and engage with concepts. All of these features are incorporated within a single student driven Identity Map project that optimizes the relevance, value and authenticity of the learning experience as students relate the content to their individual lives and experiences.

Curricular Co-Design and the Project RAISE Prototyping Process

In curriculum design, the UDL framework helps educators and instructional designers consider and develop as many options for learning as appropriate, given the instructional goal, to insure the engagement of the widest range of learners possible. In digital design (e.g., the development of web sites and mobile applications), User Experience (UX) techniques emphasize iterative cycles of prototype development and user testing to insure the best product experience possible. There are emerging efforts to enlist older students as co-creators of curricula (Bovill, Cook-Sather, & Felten, 2011), but Project RAISE has been unique in that it combined UDL and UX approaches into a co-design model, and engaged juvenile corrections students and teachers as authentic collaborators throughout the curricular development stages.

In addition to bridging the gap between pedagogical practices and methods of technology development, the RAISE co-design model rests on established theories in the learning sciences. It represents the fruition of Donald Campbell's (1969) position that, to solve complex problems, scholars should bridge their disciplinary knowledge with others in related but distinct fields – rather than just pursue continued, siloed expertise solely in their discipline. Expanding on Campbell's idea, Gerard Fischer endorsed a theory of distributed intelligence, which contends that collaboration with other people and technology extends the power of the unaided human mind (Fischer, 2006). More recently, scholars at the intersection of education and technology have described the potential of "knowledge communities," or "socially-oriented" learning environments, enabled by contemporary advances in technology, in which "the production and aggregation of content emerges from the collective contributions of all members of the community, rather than from a single authoritative source" (Tissenbaum, Lui, & Slotta, 2012, p. 328). By including marginalized students and teachers as experts alongside traditional education "experts" in our curriculum development process, we are drawing on a more comprehensive pool human intelligence to more comprehensively address the multifaceted challenges associated with science education in the juvenile justice system. In conjunction with what can be described as the "artificial" intelligence available through mobile technologies, the RAISE team is then able to innovate and creatively re-define what constitutes a UDL-PBIS digital curriculum, as well as what is possible in juvenile justice education.

We piloted the co-design model by testing the 10% Rule iBook prototype with students in two juvenile corrections facilities. A RAISE team researcher with over a decade of science teaching experience introduced the lesson by projecting the cover and first page of the iBook for the whole class to see. The researcher engaged the students in some discussion about what they might already know about the topic, and then stated the objectives of the lesson. After discussing the driving question, the students proceeded through the lesson individually, with the researcher and classroom teacher providing individual support to students as needed. A second researcher observed and recorded behaviors and comments. Finally, we posed questions in a focus-group format at the conclusion of the lesson to collect feedback from both students and the teacher. Figure 6 displays the records from one 20-minute observation. The Student behavior columns show that the students were engaged in task related behaviors for the entire 20-minute session. Student comments were directly related to the instructional content, and demonstrate some capacity to understand the potential ethical issues associated with the film, which is critical to the ability to relate to the content in an authentic and meaningful manner.

RECOMMENDATIONS

Researchers evaluated student responses by collecting both feedback as well as artifacts of student navigation and engagement captured by the technology in the iBooks. Students generally commented positively on the 10% rule iBook, but they also provided substantial informative and critical feedback. Students appreciated the videos and the interactive and responsive quizzes. However, they found that some of the video narrations were difficult to understand. As a response, the team revised the narration that accompanied the videos. The students responded positively to the look and feel of the interactive pyramid, and they used the embedded instructional videos associated with the calculation of the 10% rule. The students were critical of the use of a separate whiteboard apps. Specifically, they reported problems with opening and closing the iBook in order to access the whiteboard tools. Based on student

Figure 6. Example of direct observation of prototype session
Source: RAISE 10% Rule iBook

Time Interval	Instructional Activity and Observer Notes	Student Behavior		Student Discussion / Responses	
		Student 1	Student 2	Student 1	Student 2
0-2 min	Introduction of Lesson and 10% rule by instructor	1.Talk to Teacher	1.iPad Use 2.Talk to Teacher		In response to instructor question. Student says: " It's gonna lower it. Cuz 10% is really a lot. Cuz then the balance and the fish and stuff wouldn't be equal."
2-4 min	Film projected on screen	1.Watch Film	1.Watch Film	Watched a diver swimming with a whale shark. Says, "That's dope"	
4-6 min	Film projected on screen	Watch Film	Watch Film		
6-8 min	Film projected on screen	1.Watch Film	1.iPad		
8-10 min	Instructor shows film and leads discussion.	1.Watch Film	1.Watch Film	In Response to Instructor Question about Morality of Killing Tope PredatorsSays, "Yeah. Take out another species of animals when they already took out another one [sharks]."	Student watches scene of underwater hunting. Student Says, "That's crazy"
10-12 min	Insturctor leads discussion and supports iBook use	1.Talk to Teacher	1.Talk to Teacher		
12-14 min	Insturctor leads discussion and supports iBook use	1.Talk to Teacher	1.Talk to Teacher		
14-16 min	Insturctor leads discussion and supports iBook use	1.Talk to Teacher	1.iPad Use		
16-18 min	Instructor shows film and leads discussion.	1.Watch Film	1.Watch Film	Student Says, "They're not seeing things both ways. After a while they're not gonna get no type of money or food cuz there's not gonna be none. But if there aren't any sharks you can swim in the water. But you don't wanna swim in all that plankton."	Student Says, ""This shit's nasty" S1: "Do they do this in the US?"
18-20 min	Instructor shows film and leads discussion.	1.Watch Film	1.Watch Film		

suggestions, the development team redesigned all of the iBooks and accompanying apps to function via split screen, so that the iBook can always be open and used as a resource in associated activities. However, the students also indicated that the tools were difficult to control using fingers and suggested styli. As a result, the development team procured styli as an option for all tablet activities. The RAISE researchers and instructional designers also analyzed evidence of student use of features in the iBook (e.g., which videos were played? How often were calculator and/or white board tools used?) and student responses to questions to evaluate not just what students said about the iBook, but what they did when it was in front of them during class. We recommend capturing both kinds of information as our data resonates with established social-science understandings of human behavior: people often say one thing, and do another. We took into account student's articulated reactions to the curriculum in tandem with the work they produced within the curriculum, compared this with the insights of teachers and science education experts, and used this multi-layered feedback to redesign lessons and re-test the revisions. We suggest that this iterative and inclusive approach, when bounded by deadlines and time tables, yields better quality learning experiences and produces improved learner outcomes in science.

FUTURE RESEARCH DIRECTIONS

The process for the development of the 10% Rule lesson, described in this chapter, was repeatedly applied to each of the lessons in this four-week curriculum to develop a final version for piloting in juvenile corrections classrooms. Following the analysis of results, we anticipate expanding our co-development method to create semester and year-long curricula, with the intention of capturing more long-term outcomes. In the age of ephemeral digital media, it is especially important for educators to track what students learn and how; otherwise, future technology development efforts might miss opportunities for innovatively supporting the widest possible range of learners. One interesting example might lie in considering how dosage compares to quality of experience: how might short and long-term student learning outcomes compare between, say, a multi-level complex game or simulation, or a set of lessons in an e-book, to a single short engaging video clip, or a high-definition three-dimensional image? It will also be interesting to try to disentangle the elements that constitute engagement: are students interacting with educational mobile technologies because they are novel, or because their curiosity in the academic content is peaked? Does distinguishing these kinds of engagement even matter? And are digital platforms presenting students with opportunities to engage with content in unique and dynamic ways, or are they merely reproducing conventional didactic instruction in a shiny new package? As the possibilities for learning expand with the development of mobile technologies, research along these lines will need to use nuanced methods to better approximate the outcomes and implications of what technology enables marginalized learners to do.

CONCLUSION

Mobile tablet-based technology has dramatically enhanced the ability to provide engaging and enriching science instruction and associated experiences for incarcerated learners. The developmental phase of Project RAISE has led to innovations in the ability to provide meaningful project based inquiry science opportunities to our most vulnerable and disengaged learners. The tablet technology fully supports a plethora of UDL features that can decrease or eliminate barriers to learning for a population of learners characterized by disability, gaps in factual and conceptual knowledge, and major deficiencies in reading and mathematics skills. The Project RAISE PBIS-UDL curriculum built on the tablet platform has also provided us with the ability to provide engaging and meaningful virtual experiences despite the facility strictures that prevent laboratory opportunities, the use of equipment or materials for classroom experiments, and limit access to typical Internet based resources available in public schools. Project RAISE utilizes tablet-based technology to develop content knowledge, inquiry skills, and increased interest and engagement in science among a group of incarcerated youth. The PBIS-UDL Biology curriculum will provide incarcerated learners with the supports necessary to fully participate in science learning. It has the capacity to promote increased interest and engagement in science and other STEM fields and the potential for postsecondary STEM education or STEM careers.

REFERENCES

Anderman, E. M., Griesinger, T., & Westerfield, G. (1998). Motivation and cheating during early adolescence. *Journal of Educational Psychology*, *90*(1), 84–93. doi:10.1037/0022-0663.90.1.84

Baltodano, H. M., Harris, P. J., & Rutherford, R. B. (2005). Academic achievement in juvenile corrections: Examining the impact of age, ethnicity and disability. *Education & Treatment of Children*, *28*(4), 361–379.

Barron, B., & Darling-Hammond, L. (2008). *Teaching for Meaningful Learning: A Review of Research on Inquiry-Based and Cooperative Learning. Book Excerpt.* George Lucas Educational Foundation.

Blumenfeld, P., Fishman, B. J., Krajcik, J., Marx, R. W., & Soloway, E. (2000). Creating usable innovations in systemic reform: Scaling up technology-embedded project-based science in urban schools. *Educational Psychologist*, *35*(3), 149–164. doi:10.1207/S15326985EP3503_2

Bovill, C.-S., Cook-Sather, A., & Felten, P. (2011). Students as co-creators of teaching approaches, course design, and curricula: Implications for academic developers. *The International Journal for Academic Development*, *16*(2), 133–145. doi:10.1080/1360144X.2011.568690

Brigham, F. J., Scruggs, T. E., & Mastropieri, M. A. (2011). Science education and students with learning disabilities. *Learning Disabilities Research & Practice*, *26*(4), 223–232. doi:10.1111/j.1540-5826.2011.00343.x

Buck Institute for Education. (2013). *Project-based learning for the 21st century.* Retrieved from http://www.bie.org

Bullock, L. M., & McArthur, P. (1994). Correctional special education: Disability prevalence estimates and teacher preparation programs. *Education & Treatment of Children*, *17*(3), 347–355.

Bybee, R. W. (2010). Advancing STEM Education: A 2020 Vision. *Technology and Engineering Teacher*, *70*(1), 30–35.

Campbell, D. T. (1969). Ethnocentrism of disciplines and the fish-scale model of omniscience. *Interdisciplinary Relationships in the Social Sciences, 328*, 348.

Dalton, B., Morocco, C. C., Tivnan, T., & Mead, P. L. R. (1997). Supported Inquiry Science Teaching for Conceptual Change in Urban and Suburban Science Classrooms. *Journal of Learning Disabilities*, *30*(6), 670–684. doi:10.1177/002221949703000611 PMID:9364905

David, J. (2008). What research says about project-based learning. *Educational Leadership*, *65*, 80–82.

Enthoven, M., & de Bruijn, E. (2010). Beyond locality: The creation of public practice-based knowledge through practitioner research in professional learning communities and communities of practice. A review of three books on practitioner research and professional communities. *Educational Action Research*, *8*(2), 289–298. doi:10.1080/09650791003741822

Fischer, G. (2006, May). Distributed intelligence: extending the power of the unaided, individual human mind. In *Proceedings of the working conference on Advanced visual interfaces* (pp. 7-14). ACM. doi:10.1145/1133265.1133268

Gagnon, J. C., & Barber, B. R. (2010). Characteristics of and services provided to youth in secure care facilities. *Behavioral Disorders*, *36*(1), 7–19.

Gagnon, J. C., & Barber, B. R. (2014). Instructional practice guide for teaching reading and mathematics in juvenile correctional schools. *Journal of Correctional Education*, *65*(3), 5–23.

Gagnon, J. C., Barber, B. R., Van Loan, C., & Leone, P. E. (2009). Juvenile correctional schools: Characteristics and approaches to curriculum. *Education & Treatment of Children*, *32*(4), 673–696. doi:10.1353/etc.0.0068

Gregory, A., Skiba, R. J., & Noguera, P. A. (2010). The achievement gap and the discipline gap two sides of the same coin? *Educational Researcher*, *39*(1), 59–68. doi:10.3102/0013189X09357621

Houchins, D. E., Jolivette, K., Krezmien, M. P., & Baltodano, H. A. (2008). A Multi-State study examining the impact of explicit reading instruction with incarcerated students. *Journal of Correctional Education*, *59*(1), 65–85.

Jacobson, M. J., & Archodidou, A. (2000). The design of hypermedia tools for learning: Fostering conceptual change and transfer of complex scientific knowledge. *Journal of the Learning Sciences*, *9*(2), 145–199. doi:10.1207/s15327809jls0902_2

Krezmien, M. P. (2008, January). *National perspectives of youth with disabilities and their interaction with the juvenile system: A statement of the issues and need.* Presentation at the 1st conference of the Ohio Juvenile Justice Project, Columbus, OH.

Krezmien, M. P., Leone, P. E., & Achilles, G. M. (2006). Suspension, race, and disability: Analysis of statewide practices and reporting. *Journal of Emotional and Behavioral Disorders*, *14*(4), 217–226. doi:10.1177/10634266060140040501

Krezmien, M. P., & Mulcahy, C. A. (2008). Literacy and Delinquency: Current status of reading interventions with detained and incarcerated youth. *Reading & Writing Quarterly*, *24*(2), 219–238. doi:10.1080/10573560701808601

Krezmien, M. P., Mulcahy, C. A., & Leone, P. E. (2008). Detained and Committed Youth: Examining Differences in Achievement, Mental Health Needs, and Special Education Status. *Education & Treatment of Children*, *31*(4), 1–20.

Krezmien, M. P., Mulcahy, C. A., Travers, J., Wilson, M., & Wells, C. (2013). Reading performance of incarcerated youth: Understanding and responding to a unique population of readers. *Journal of Special Education Leadership*, *26*(2), 71–81.

Lane, K. L., Carter, E. W., Pierson, M. R., & Glaeser, B. C. (2006). Academic, social, and behavioral characteristics of high school students with emotional disturbances or learning disabilities. *Journal of Emotional and Behavioral Disorders*, *14*(2), 108–117. doi:10.1177/10634266060140020101

Leone, P. E., Krezmien, M., Mason, L., & Meisel, S. M. (2005). Organizing and delivering empirically based literacy instruction to incarcerated youth. *Exceptionality*, *13*(2), 89–102. doi:10.1207/s15327035ex1302_3

Leone, P. E., & Weinberg, L. A. (2012). *Addressing the unmet educational needs of children and youth in the juvenile justice and child welfare systems, 2012 Edition*. Georgetown Public Policy Institute, Center for Juvenile Justice Reform.

Maccini, P., Gagnon, J. C., Mulcahy, C. A., & Leone, P. E. (2006). Math instruction for committed youth within juvenile correctional schools. *Journal of Correctional Education, 57*(3), 210-229. Retrieved from http://www.jstor.org/stable/23282753

Marx, R. W., Blumenfeld, P. C., Krajcik, J. S., & Soloway, E. (1997). Enacting project-based science. *The Elementary School Journal*, *97*(4), 341–358. doi:10.1086/461870

Meyer, A., & Rose, D. H. (2005). The future is in the margins: The role of technology and disability in educational reform. In D. H. Rose, A. Meyer, & C. Hitchcock (Eds.), *The universally designed classroom: Accessible curriculum and digital technologies* (pp. 13–35). Cambridge, MA: Harvard Education Press.

Meyer, A., Rose, D. H., & Gordon, D. (2014). Universal design for learning. *Theory into Practice*.

Mulcahy, C. A., & Krezmien, M. P. (2009). Effects of a contextualized instructional package on the mathematics performance of secondary students with EBD. *Behavioral Disorders*, *34*(3), 136–150.

Mulcahy, C. A., Krezmien, M. P., Leone, P. E., Houchins, D. E., & Baltodano, H. (2008). Lessons learned: Barriers and solutions for conducting reading investigations in juvenile corrections settings. *Reading & Writing Quarterly*, *24*(2), 239–252. doi:10.1080/10573560701808635

Mulcahy, C. A., Krezmien, P. K., & Travers, J. (2015). Improving mathematics performance among secondary students with EBD: A methodological review. *Remedial and Special Education*, *37*(2), 113–128. doi:10.1177/0741932515579275

National Research Council. (2012). *Education for life and work: Developing transferable knowledge and skills in the 21st century*. Washington, DC: The National Academies Press. doi:10.7226/13398

NGSS Lead States. (2013). Next Generation Science Standards: For States by States. Author.

Quinn, M. M., Rutherford, R. B., Leone, P. E., Osher, D. M., & Poirier, J. M. (2005). Youth with disabilities in juvenile corrections: A national survey. *Exceptional Children*, *71*(3), 339–345. doi:10.1177/001440290507100308

Rose, D. H., & Meyer, A. (2002). *Teaching every student in the digital age: Universal design for learning*. Alexandria, VA: Association for Supervision and Curriculum Development.

Tannis, L. 2014. Educating incarcerated youth: Exploring the impact of relationships, expectations, resources and accountability. Palgrave Macmillan UK. doi:10.1057/9781137451026

Therrien, W. J., Taylor, J. C., Watt, S., & Kaldenberg, E. R. (2013). Science instruction for students with emotional and behavioral disorders. *Remedial and Special Education*, *35*(1), 15–27. doi:10.1177/0741932513503557

Tissenbaum, M., Lui, M., & Slotta, J. D. (2012). Co-Designing Collaborative Smart Classroom Curriculum for Secondary School Science. *Journal of Universal Computer Science*, *18*(3), 327–352.

KEY TERMS AND DEFINITIONS

Co-Design: A process based in generative design research methods that seeks to collect the explicit as well as tacit knowledge of users, as well as entrust users as colleagues and collaborators in the development of new experiences and technologies.

Human Variability: The normally occurring vast range of differences between and within individuals that result from the varying conditions surrounding human development.

Project-Based Learning: A method of instruction in which students learn content by working for an extended period of time to investigate and respond to complex, real-world issues.

Prototyping: The creation and testing, with users, of preliminary models that elicit insights about improvements and lead to iterative development of a final product.

Scientific Inquiry: The process by which scientists study and explain phenomena; the activities students undertake, such as formulating questions, designing investigations, conducting observations, analyzing data, drawing conclusions, and communicating results.

Special Education: Instruction designed to meet the unique needs of a student with a disability, at no cost to the child's primary caregiver(s).

Universal Design for Learning (UDL): A framework for designing learning environments and materials that addresses human variability by providing multiple options for perception, action and expression, and engagement.

Compilation of References

Abraham, M. R. (1998). The learning cycle approach as a strategy for instruction in science. In K. Tobin & B. Fraser (Eds.), *International Handbook of Science Education* (pp. 513–524). Kluwer. doi:10.1007/978-94-011-4940-2_30

Achieve, Inc. (2013). *Next Generation Science Standards*. Washington, DC: Achieve, Inc. Retrieved from http://www.nextgenscience.org/

Achieve, Inc. (2013). *Next generation science standards*. Washington, DC: Author.

Adejumo, G., Duimering, R. P., & Zhong, Z. (2008). A balance theory approach to group problem solving. *Social Networks*, *30*(1), 83–99. doi:10.1016/j.socnet.2007.09.001

Aikenhead, G. S. (2003). Chemistry and physics instruction: Integration, ideologies, and choices. *Chemical Education: Research and Practices*, *4*(2), 115–130.

Aikenhead, G. S. (2005). Research into STS Science Education. *Educación en la Química*, *16*, 384–397.

Aikenhead, G. S. (2006). *Science education for everyday life: Evidence-based practice*. New York, NY: Teachers College Press.

Akaygun, S., & Jones, L. L. (2013). Animation or Simulation: Investigating the Importance of Interactivity for Learning Solubility Equilibria. In J. P. Suits & M. J. Sanger (Eds.). *Pedagogic Role of Animations and Simulations in Chemistry Courses. ACS Symposium Series*, (*vol. 1142*, pp, 127-159). Washington, DC: American Chemical Society. doi:10.1021/bk-2013-1142.ch006

Almog, H. A. (2011). *Learning about complex systems through exploring models and taking part in participatory simulation in the domain of ecology* (Unpublished master's thesis). Faculty of Education, University of Haifa.

Alsop, S., Bencze, L., & Pedretti, E. (Eds.). (2005). *Analysing exemplary science teaching* (1st ed.). New York, NY: Open University Press.

Anderman, E. M., Griesinger, T., & Westerfield, G. (1998). Motivation and cheating during early adolescence. *Journal of Educational Psychology*, *90*(1), 84–93. doi:10.1037/0022-0663.90.1.84

Anderson, L. W., & Krathwohl, D. R. (2001). *A taxonomy for learning, teaching, and assessing: A revision of Bloom's taxonomy of educational objectives*. New York: Addison Wesley Longman.

Anderson, R. (2002). Reforming science teaching: What research says about inquiry. *Journal of Science Teacher Education, 13*(1), 1–12. doi:10.1023/A:1015171124982

Annetta, L. A., Cook, M., & Schultz, M. (2007). Video Games: A Vehicle for Problem-Based Learning. *e-Journal of Instructional Science and Technology, 10*(1).

Annetta, L. A., & Cheng, M. (2008). Why educational video games? In L. A. Annetta (Ed.), *Serious educational games: From theory to practice* (pp. 1–11). Rotterdam, The Netherlands: Sense Publishers.

Annetta, L. A., Frazier, W. M., Folta, E., Holmes, S., Lamb, R., & Cheng, M.-T. (2013). Science teacher efficacy and extrinsic factors toward professional development using video games in a design-based research model: The next generation of STEM learning. *Journal of Science Education and Technology*, *22*(1), 47–61. doi:10.1007/s10956-012-9375-y

Apple Education. (2010). *Getting started: Apple technology for diverse learners*. Available from http://www.apple.com/education/resources/

Ares, N., Stroup, W. M., & Schademan, A. R. (2009). The power of mediating artifacts in group-level development of mathematical discourses. *Cognition and Instruction*, *27*(1), 1–24. doi:10.1080/07370000802584497

ARIS. (2016). *Mobile learning experiences*. Retrieved July 15, 2016, from http://arisgames.org/

Arnold, J. C., Kremer, K., & Mayer, J. (2014). Understanding students experiments - What kind of support do they need in inquiry tasks? *International Journal of Science Education*, *36*(16), 2719–2749. doi:10.1080/09500693.2014.930209

Aronson, E., & Patnoe, S. (2011). Cooperation in the classroom: The jigsaw method. London: Pinter & Martin, Ltd.

Aronson, E., & Patnoe, S. (2011). *Cooperation in the classroom: The jigsaw method* (3rd ed.). London, UK: Pinter & Martin Ltd.

Arvaja, M., Hakkinen, P., Rasku-Puttonen, H., & Etelapelto, A. (2002). Social processes and knowledge building during small group interaction in a school science project. *Scandinavian Journal of Educational Research*, *46*(2), 161–179. doi:10.1080/00313830220142182

Asbell-Clarke, J., Edwards, T., Larsen, J., Rowe, E., Sylvan, E., & Hewitt, J. (2012). Martian Boneyards: Scientific inquiry in an MMO game. *International Journal of Game-Based Learning*, *2*(1), 52–76. doi:10.4018/ijgbl.2012010104

Ashe, C. A., & Yaron, D. J. (2013). Designing analogy-based simulations to teach abstractions. In J. P. Suits, & M. J. Sanger. (Eds.), *Pedagogic Roles of Animations and Simulations in Chemistry Courses. ACS Symposium Series*, (*vol. 1142*, pp, 367-388). Washington, DC: American Chemical Society. doi:10.1021/bk-2013-1142.ch015

Avouris, N., Dimitracopoulou, A., & Komis, V. (2003). On analysis of collaborative problem solving: An object-oriented approach. *Computers in Human Behavior*, *19*(2), 147–167. doi:10.1016/S0747-5632(02)00056-0

Ayuso, E., & Banet, E. (2002). Alternativas a la enseñanza de la Genética en educación secundaria. *Ensenañza de las Ciencias*, *20*(1), 133–157.

Ball, D. L. (1996). Teacher learning and the mathematics reforms: What we think we know and what we need to learn. *Phi Delta Kappan*, *77*(7), 500–508.

Baltodano, H. M., Harris, P. J., & Rutherford, R. B. (2005). Academic achievement in juvenile corrections: Examining the impact of age, ethnicity and disability. *Education & Treatment of Children*, *28*(4), 361–379.

Banet, E., & Ayuso, E. (2000). Teaching genetics at secondary school: A strategy for teaching about the location of inheritance information. *Science Education*, *84*(3), 313–351. doi:10.1002/(SICI)1098-237X(200005)84:3<313::AID-SCE2>3.0.CO;2-N

Bang, E., & Luft, J. A. (2013). Secondary Science Teachers Use of Technology in the Classroom during Their First 5 Years. *Journal of Digital Learning in Teacher Education*, *29*(4), 118–126. doi:10.1080/21532974.2013.10784715

Banilower, E., & Shimkus, E. (2004). *Professional development observation study*. Chapel Hill, NC: Horizon Research.

Bank, T. W. (2012). Turn down the heat: why a 4 C warmer world must be avoided. The World Bank, 74455.

Barab, S., Sadler, T., Heiselt, C., Hickey, D., & Zuiker, S. (2007). Relating narrative, inquiry, and inscriptions: Supporting consequential play. *Journal of Science Education and Technology*, *16*(1), 59–82. doi:10.1007/s10956-006-9033-3

Barab, S., Scott, B., Siyahhan, S., Goldstone, R., Ingram-Goble, A., Zuiker, S., & Warren, S. (2009). Transformational play as a curricular scaffold: Using videogames to support science education. *Journal of Science Education and Technology*, *18*(4), 305–320. doi:10.1007/s10956-009-9171-5

Barak, M. (2013). Making the unseen seen: Integrating 3D molecular visualizations in elementary, high school, and higher education. In J. P. Suits, & M. J. Sanger (Eds.), *Pedagogic Roles of Animations and Simulations in Chemistry Courses. ACS Symposium Series*, (*vol. 1142*, pp. 273-291). Washington, DC: American Chemical Society. doi:10.1021/bk-2013-1142.ch011

Barnea, N., & Dori, Y. J. (1999). High school chemistry students performance and gender differences in a computerized molecular modeling learning environment. *Journal of Science Education and Technology*, *8*(4), 257–271. doi:10.1023/A:1009436509753

Barrett-Tatum, J., & Dooley, C. M. (2015). From idealized to actualized instruction: Two elementary teachers enactment of the ELA Common Core State Standards in instructional scripts. *Literacy Research and Instruction*, *54*(4), 257–284. doi:10.1080/19388071.2015.1062941

Barron, B., & Darling-Hammond, L. (2008). *Teaching for Meaningful Learning: A Review of Research on Inquiry-Based and Cooperative Learning. Book Excerpt.* George Lucas Educational Foundation.

Bar-Yam, Y. (2003). *Dynamics of complex systems.* New York: Perseus Publishing.

Beichner, R. J. (2009). An introduction to physics education research. In C. Henderson & K. A. Harper (Eds.), Getting started in PER. College Park, MD: American Association of Physics Teachers. Available from http://www.percentral.org/document /ServeFile.cfm?ID=8806

Bell, R. L., Smetana, L., & Binns, I. (2005). Simplifying inquiry instruction. *Science Teacher (Normal, Ill.), 72*(7), 30–33.

Belt, S. T., Leisvik, M. J., Hyde, A. J., & Overton, T. L. (2005). Using a context-based approach to undergraduate chemistry teaching: A case study for introductory physical chemistry. *Chemistry Education Research and Practice, 6*(3), 166–179. doi:10.1039/B5RP90007G

Bencze, L., & Hodson, D. (1999). Changing practice by changing practice: Towards more authentic science and science curriculum development. *Journal of Research in Science Teaching, 36*(5), 521–539. doi:10.1002/(SICI)1098-2736(199905)36:5<521::AID-TEA2>3.0.CO;2-6

Bennani, F., & Novak, K. (2016). Project Outbreak: Take biology out of the classroom and into the world. In H. Crompton & J. Traxler (Eds.), *Mobile learning and STEM: Case studies in practice* (pp. 125–138). New York, NY: Routledge.

Bennett, J., Lubben, F., & Hogarth, S. (2007). Bringing science to life: A synthesis of the research evidence on the effects of context-based and STSE approaches to science teaching. *Science Education, 91*(3), 347–370. doi:10.1002/sce.20186

Bloom, B. S. (1956). *Taxonomy of educational objectives. Handbook I: The cognitive domain.* New York: David McKay.

Bloom, J. W. (2006). *Creating a community of young scientists* (2nd ed.). New York, NY: Trifolium Books.

Blumenfeld, P., Fishman, B. J., Krajcik, J., Marx, R. W., & Soloway, E. (2000). Creating usable innovations in systemic reform: Scaling up technology-embedded project-based science in urban schools. *Educational Psychologist, 35*(3), 149–164. doi:10.1207/S15326985EP3503_2

Blumer, H. (1969). The methodological position of symbolic interactionism. In Symbolic interactionism: Perspective and Method (pp. 1-60). Englewood Cliffs, NJ: Prentice-Hall.

Boo, H. K. (1998). Students understandings of chemical bonds and the energetics of chemical reactions. *Journal of Research in Science Teaching, 35*(5), 569–581. doi:10.1002/(SICI)1098-2736(199805)35:5<569::AID-TEA6>3.0.CO;2-N

Borovoy, R., McDonald, M., Martin, F., & Resnick, M. (1996). Things that blink: Computationally augmented name tags. *IBM Systems Journal, 35*(3.4), 488-495.

Borovoy, R., Martin, F., Vemuri, S., Resnick, M., Silverman, B., & Hancock, C. (1998). Meme tags and community mirrors: moving from conferences to collaboration. In *Proceedings of the 1998 ACM conference on Computer supported cooperative work* (pp. 159-168). ACM. doi:10.1145/289444.289490

Bou Jaoude, S., & Barakat, H. (2003). Students' problem solving strategies on stoichiometry and their relationships to conceptual understandings and learning approaches. *Electronic Journal of Science Education, 7*(3), 1–42.

Bovill, C.-S., Cook-Sather, A., & Felten, P. (2011). Students as co-creators of teaching approaches, course design, and curricula: Implications for academic developers. *The International Journal for Academic Development, 16*(2), 133–145. doi:10.1080/1360144X.2011.568690

Bower, M., & Sturman, D. (2015). What are the educational affordances of wearable technologies? *Computers & Education, 88*, 343–353. doi:10.1016/j.compedu.2015.07.013

Boykoff, M. T., & Boykoff, J. M. (2004). Balance as bias: Global warming and the US prestige press. *Global Environmental Change, 14*(2), 125–136. doi:10.1016/j.gloenvcha.2003.10.001

Bransford, J., Brown, A. L., Cocking, R. R., & National Research Council (U.S.). (1999). *How people learn: Brain, mind, experience, and school.* Washington, DC: National Academy Press.

Bransford, J. D., Brown, A. L., & Cocking, R. R. (Eds.). (2000). *How people learn: Brain, mind, experience, and school.* Washington, DC: National Academy.

Bressler, D. M. (2014). Better than business-as-usual: Improving the scientific practices during discourse and writing by playing a collaborative mystery game. *Double Helix: A Journal of Critical Thinking and Writing, 2*, 1-13.

Bressler, D. M., & Bodzin, A. M. (2014, April). *Continuing to flow: Student experience during a scaled-up INPLACE mobile game.* Paper presented at the National Association for Research in Science Teaching (NARST), Pittsburgh, PA.

Bressler, D. M. (2015). School Scene Investigators: A collaborative AR game for middle school science inquiry. In C. Holden, S. Dikkers, J. Martin, & B. Litts (Eds.), *Mobile Media Learning: Innovation and Inspiration* (pp. 87–100). Pittsburgh, PA: ETC Press.

Bressler, D. M., & Bodzin, A. M. (2013). A mixed methods assessment of students flow experience during a mobile augmented reality science game. *Journal of Computer Assisted Learning, 29*(6), 505–517. doi:10.1111/jcal.12008

Bressler, D., & Bodzin, A. (2016). Investigating flow experience and scientific practices during a mobile Serious Educational Game. *Journal of Science Education and Technology, 25*(5), 795–805. doi:10.1007/s10956-016-9639-z

Brigham, F. J., Scruggs, T. E., & Mastropieri, M. A. (2011). Science education and students with learning disabilities. *Learning Disabilities Research & Practice, 26*(4), 223–232. doi:10.1111/j.1540-5826.2011.00343.x

Brown, A. L. (1992). Design experiments: Theoretical and methodological challenges in creating complex interventions in classroom settings. *Journal of the Learning Sciences*, *2*(2), 141–178. doi:10.1207/s15327809jls0202_2

Brulle, R. J. (2014). Institutionalizing delay: Foundation funding and the creation of US climate change counter-movement organizations. *Climatic Change*, *122*(4), 681–694. doi:10.1007/s10584-013-1018-7

Bucat, R. (2004). Pedagogical content knowledge as a way forward: Applied research in chemistry education. *Chemistry Education Research and Practice*, *5*(3), 215–228. doi:10.1039/B4RP90025A

Bucat, R. (2005). Implications of chemistry education research for teaching practice: Pedagogical content knowledge as a way forward. *Chemistry Education International*, *6*(1), 1–2.

Buck Institute for Education. (2013). *Project-based learning for the 21st century*. Retrieved from http://www.bie.org

Bullock, L. M., & McArthur, P. (1994). Correctional special education: Disability prevalence estimates and teacher preparation programs. *Education & Treatment of Children*, *17*(3), 347–355.

Burke, K., Greenbowe, T. J., & Windschitl, M. (1998). Developing and using conceptual computer animations for chemistry instruction. *Journal of Chemical Education*, *75*(12), 1658–1661. doi:10.1021/ed075p1658

Burns, J., Okey, J., & Wise, K. (1985). Development of an integrated process skill test: TIPS II. *Journal of Research in Science Teaching*, *22*(2), 169–177. doi:10.1002/tea.3660220208

Bybee, R. W. (2006). Scientific inquiry and science teaching. In L. B. Flick & N. G. Lederman (Eds.), Scientific inquiry and nature of science: Implications for teaching, learning, and teacher education (Vol. 25, pp. 1–15). Dordrecht, The Netherlands: Springer.

Bybee, R. W. (1993). *Reforming science education*. New York, NY: Teachers College Press.

Bybee, R. W. (2006). Scientific Inquiry And Science Teaching. In L. B. Flick & N. G. Lederman (Eds.), *Scientific Inquiry and Nature of Science: Implications for Teaching, Learning, and Teacher Education* (pp. 1–14). Dordrecht: Springer Netherlands.

Bybee, R. W. (2010). Advancing STEM Education: A 2020 Vision. *Technology and Engineering Teacher*, *70*(1), 30–35.

Campbell, D. T. (1969). Ethnocentrism of disciplines and the fish-scale model of omniscience. *Interdisciplinary Relationships in the Social Sciences, 328*, 348.

Campbell, T., Oh, P. S., & Neilson, D. (2013). Reification of five types of modeling pedagogies with model based inquiry (MBI) modules for high school science classrooms. In M. S. Khine & I. M. Saleh (Eds.), *Approaches and strategies in next generation science learning* (pp. 106–126). Hershey, PA: IGI Global. doi:10.4018/978-1-4666-2809-0.ch006

Cartier, J., Rudolph, J., & Stewart, J. (2001). *The nature and structure of scientific models.* National Center for Improving Student Learning and Achievement in Mathematics and Science.

CCS Insight. (2015, September 1). *Wearables Market to Be Worth $25 Billion by 2019.* Retrieved July 21, 2016, from http://www.ccsinsight.com/press/company-news/2332-wearables-market-to-be-worth-25-billion-by-2019-reveals-ccs-insight

Center for Instructional Development and Educational Research (CIDER). (2009). *Pedagogy.* Retrieved May 13, 2014, from http://www.cider.vt.edu/pedagogy.html

Chang, C.-S., Chen, T.-S., & Hsu, W.-H. (2011). The study on integrating WebQuest with mobile learning for environmental education. *Computers & Education, 57*(1), 1228–1239. doi:10.1016/j.compedu.2010.12.005

Chang, K.-E., Chen, Y.-L., Lin, H.-Y., & Sung, Y.-T. (2008). Effects of learning support in simulation-based physics learning. *Computers & Education, 51*(4), 1486–1498. doi:10.1016/j.compedu.2008.01.007

Charistos, N. D., Tsipis, C. A., & Sigalas, M. P. (2005). 3D molecular symmetry Shockwave: A web application for interactive visualization and three dimensional perception of molecular symmetry. *Journal of Chemical Education, 82*(11), 1741–1742. doi:10.1021/ed082p1741.2

Chauncey, A., & Azevedo, R. (2010). Emotions and motivation on performance during multimedia learning: How do I feel and why do I care? In V. Aleven, J. Kay, & J. Mostow (Eds.), *Intelligent Tutoring Systems* (Vol. 6094, pp. 369–378). Springer Berlin Heidelberg. doi:10.1007/978-3-642-13388-6_41

Chen, C. P., & Zhang, C. Y. (2014). Data-intensive applications, challenges, techniques and technologies: A survey on Big Data. *Information Sciences, 275*, 314–347. doi:10.1016/j.ins.2014.01.015

Chen, D., & Stroup, W. (1993). General system theory: Toward a conceptual framework for science and technology education for all. *Journal of Science Education and Technology, 2*(3), 447–459. doi:10.1007/BF00694427

Cheng, K., & Tsai, C. (2013). Affordances of augmented reality in science learning: Suggestions for future research. *Journal of Science Education and Technology, 22*(4), 449–462. doi:10.1007/s10956-012-9405-9

Chi, M. T., Feltovich, P. J., & Glaser, R. (1981). Categorization and representation of physics problems by experts and novices. *Cognitive Science, 5*(2), 121–152. doi:10.1207/s15516709cog0502_2

Ching, C. C., & Schaefer, S. (2014). Identities in motion, identities at rest: Engaging bodies and minds in fitness gaming research and design. In V. R. Lee (Ed.), Learning technologies and the body: Integration and implementation in formal and informal learning environments (pp. 201–219). Routledge.

Chinn, C. A., & Malhotra, B. A. (2002). Epistemologically authentic inquiry in schools: A theoretical framework for evaluating inquiry tasks. *Science Education, 86*(2), 175–218. doi:10.1002/sce.10001

Chiu, M. M., & Khoo, L. (2003). Rudeness and status effects during group problem solving: Do they bias evaluations and reduce the likelihood of correct solutions? *Journal of Educational Psychology, 95*(3), 506–523. doi:10.1037/0022-0663.95.3.506

Choe, E. K., Lee, N. B., Lee, B., Pratt, W., & Kientz, J. A. (2014). Understanding Quantified-selfers' Practices in Collecting and Exploring Personal Data. In *Proceedings of the 32nd Annual ACM Conference on Human Factors in Computing Systems* (pp. 1143–1152). New York, NY: ACM. doi:10.1145/2556288.2557372

Christensen, C. M., Horn, M. B., & Johnson, C. W. (2008). *Disrupting class: How disruptive innovation will change the way the world learns*. New York, NY: McGraw-Hill.

Clarke, T., Ayres, P., & Sweller, J. (2005). The impact of sequencing and prior knowledge on learning mathematics through spreadsheet applications. *Educational Technology Research and Development, 53*(3), 15–24. doi:10.1007/BF02504794

Clement, J. (2000). Analysis of Clinical Interviews: Foundations and Model Viability. In A. E. Kelly & R. Lesh (Eds.), *Handbook of Research Design in Mathematics and Science Education* (pp. 547–589). Mahwah, NJ: Lawrence Erlbaum Associates.

Clough, E. E., & Wood-Robinson, C. (1985). Children's understanding of inheritance. *Journal of Biological Education, 19*(4), 305–310.

Colella, V. (2000). Participatory simulations: Building collaborative understanding through immersive dynamic modeling. *Journal of the Learning Sciences, 9*(4), 471–500. doi:10.1207/S15327809JLS0904_4

Cole, R. S., & Todd, J. B. (2003). Effects of web-based multimedia homework with immediate rich feedback on student learning in general chemistry. *Journal of Chemical Education, 80*(11), 1338–1343. doi:10.1021/ed080p1338

Cook-Deegan, R., & Heaney, C. (2010). Patents in genomics and human genetics. *Annual Review of Genomics and Human Genetics, 22*(1), 383–425. doi:10.1146/annurev-genom-082509-141811 PMID:20590431

Cooke, N. J., Kiekel, P. A., Salas, E., Stout, R., Bowers, C., & Cannon-Bowers, J. (2003). Measuring team knowledge: A window to the cognitive underpinnings of team performance. *Group Dynamics, 7*(3), 179–219. doi:10.1037/1089-2699.7.3.179

Corbin, J., & Strauss, A. C. (2015). *Basics of Qualitative Research: Techniques and Procedures for Developing Grounded Theory* (4th ed.). Thousand Oaks, CA: Sage Publications.

Costa-Font, M., Gil, J. M., & Traill, W. B. (2008). Consumer acceptance, valuation of and attitudes towards genetically modified food: Review and implications for food policy. *Food Policy, 33*(2), 99–111. doi:10.1016/j.foodpol.2007.07.002

Crawford, B. A. (2007). Learning to teach science as inquiry in the rough and tumble of practice. *Journal of Research in Science Teaching, 44*(4), 613–642. doi:10.1002/tea.20157

Crawford, C. (1984). *The art of computer game design*. Berkeley, CA: Osborne/McGraw-Hill.

Creswell, J. W., & Plano Clark, V. L. (2011). *Designing and Conducting Mixed Methods Research*. SAGE.

Crompton, H. (2013). Mobile learning: New approach, new theory. In Z. Berge & L. Muilenburg (Eds.), *Handbook of Mobile Learning* (pp. 47–57). New York, NY: Routledge.

Csikszentmihalyi, M. (1990). *Flow: The psychology of optimal experience*. New York, NY: Harper and Row.

Csikszentmihalyi, M. (1996). *Creativity: Flow and the psychology of discovery and invention*. New York: Harper Collins.

Cukier, K., & Mayer-Schoenberger, V. (2013). Rise of big data. *Foreign Affairs*, *92*(3).

Dalton, B., Morocco, C. C., Tivnan, T., & Mead, P. L. R. (1997). Supported Inquiry Science Teaching for Conceptual Change in Urban and Suburban Science Classrooms. *Journal of Learning Disabilities*, *30*(6), 670–684. doi:10.1177/002221949703000611 PMID:9364905

Darling-Hammond, L. (2000). Teacher quality and student achievement. *Education Policy Analysis Archives*, *8*, 1. doi:10.14507/epaa.v8n1.2000

David, J. (2008). What research says about project-based learning. *Educational Leadership*, *65*, 80–82.

De Boer, G. E., Quellmalz, E. S., Davenport, J. L., Timms, M. J., Herrmann-Abell, C. F., Buckley, B. C., & Flanagan, J. C. et al. (2014). Comparing three online testing modalities: Using static, active, and interactive online testing modalities to access middle school students understanding of fundamental ideas and use of inquiry skills related to ecosystems. *Journal of Research in Science Teaching*, *51*(4), 523–554. doi:10.1002/tea.21145

de Jong, T. (Ed.). (2014). *Preliminary inquiry classroom scenarios and guidelines. D1.3. Go-Lab Project (Global Online Science Labs for Inquiry Learning at School)*. Retrieved from http://www.go-lab-project.eu/sites/default/files/files/deliverable/file/Go-Lab%20D1.3.pdf

de Jong, T. (2006). Computer simulations – Technological advances in inquiry learning. *Science*, *312*(5773), 532–533. doi:10.1126/science.1127750 PMID:16645080

de Jong, T., Sotiriou, S., & Gillet, D. (2014). Innovations in STEM education: The Go-Lab federation of online labs. *Smart Learning Environments*, *1*(1), 1–16. doi:10.1186/s40561-014-0003-6

de Jong, T., & van Joolingen, W. R. (1998). Scientific discovery learning with computer simulations of conceptual domains. *Review of Educational Research*, *68*(2), 179–201. doi:10.3102/00346543068002179

de Jong, T., van Joolingen, W. R., Giemza, A., Girault, I., Hoppe, U., Kindermann, J., & van der Zanden, M. et al. (2010). Learning by creating and exchanging objects: The SCY experience. *British Journal of Educational Technology*, *41*(6), 909–921. doi:10.1111/j.1467-8535.2010.01121.x de Jong, T., Weinberger, A., van Joolingen, W. R., Ludvigsen, S., Ney, M., Girault, I., & Zacharia, Z. et al. (2012). Designing complex and open learning environments based on scenarios. *Educational Technology Research and Development*, *60*, 883–901. doi:10.1007/s11423-012-9258-1

Dede, C. (2009). Immersive interfaces for engagement and learning. *Science*, *323*(5910), 66–69. doi:10.1126/science.1167311 PMID:19119219

Denzin, N., & Lincoln, Y. (2005). Introduction: The discipline and practice of qualitative research. In *The sage handbook of qualitative research* (3rd ed.; pp. 1–33). Thousand Oaks, CA: Sage.

Desimone, L. M. (2009). Improving impact studies of teachers professional development: Toward better conceptualizations and measures. *Educational Researcher*, *38*(3), 181–199. doi:10.3102/0013189X08331140

Diaz, K. M., Krupka, D. J., Chang, M. J., Ma, Y., Goldsmith, J., Schwartz, J. E., & Davidson, K. W. (2015). Fitbit: An Accurate and Reliable Device for Wireless Physical Activity Tracking. *Circulation*, *131*(Suppl 1), AMP11–AMP11. PMID:25795203

Dieterle, E. (2009). Neomillennial learning styles and River City. *Children, Youth and Environments*, *19*(1), 245–278.

Dillenbourg, P. (1999). What do you mean by "collaborative learning"? In P. Dillenbourg (Ed.), *Collaborative learning: Cognitive and computational approaches* (pp. 1–16). Amsterdam, The Netherlands: Pergamon, Elsevier Science.

Dillenbourg, P. (Ed.). (1999). *Collaborative learning: Cognitive and computational approaches*. Amsterdam: Pergamon, Elsevier Science.

Dillenbourg, P., & Traum, D. (2006). Sharing solutions: Persistence and grounding in multi-modal collaborative problem solving. *Journal of the Learning Sciences*, *15*(1), 121–151. doi:10.1207/s15327809jls1501_9

Donahue, D. M. (1993). Serving Students, Science, or Society? The Secondary School Physics Curriculum in the United States, 193065. *History of Education Quarterly*, *33*(3), 321–352. doi:10.2307/368196

Donnelly, D., OReilly, J., & McGarr, O. (2013). Enhancing the Student Experiment Experience: Visible Scientific Inquiry through a Virtual Chemistry Laboratory. *Research in Science Education*, *43*(4), 1571–1592. doi:10.1007/s11165-012-9322-1

Duncan, R. G., & Reiser, B. J. (2007). Reasoning across ontologically distinct levels: Students understandings of molecular genetics. *Journal of Research in Science Teaching*, *44*(7), 938–959. doi:10.1002/tea.20186

Duncan, R. G., Rogat, A. D., & Yarden, A. (2009). A learning progression for deepening students understandings of modern genetics across the 5th–10th grades. *Journal of Research in Science Teaching*, *46*(6), 655–674. doi:10.1002/tea.20312

Dunleavy, M., & Dede, C. (2013). Augmented reality teaching and learning. In J. M. Spector, M. D. Merrill, J. Elen, & M. J. Bishop (Eds.), *Handbook of research on educational communications and technology* (4th ed.; Vol. 2). New York, NY: Springer.

Dunleavy, M., Dede, C., & Mitchell, R. (2009). Affordances and limitations of immersive participatory augmented reality simulations for teaching and learning. *Journal of Science Education and Technology, 18*(1), 7–22. doi:10.1007/s10956-008-9119-1

Edelson, D. C., Gordon, D. N., & Pea, R. D. (1999). Addressing the challenge of inquiry-based learning. *Journal of the Learning Sciences, 8*, 392–450. doi:10.1080/10508406.1999.9672075

Elby, A. (2000). What students learning of representations tells us about constructivism. *The Journal of Mathematical Behavior, 19*(4), 481–502. doi:10.1016/S0732-3123(01)00054-2

Engel Clough, E., & Wood-Robinson, C. (1985). Childrens understanding of inheritance. *Journal of Biological Education, 19*(4), 304–310. doi:10.1080/00219266.1985.9654757

Enthoven, M., & de Bruijn, E. (2010). Beyond locality: The creation of public practice-based knowledge through practitioner research in professional learning communities and communities of practice. A review of three books on practitioner research and professional communities. *Educational Action Research, 8*(2), 289–298. doi:10.1080/09650791003741822

Etkina, E. (2005). Physics teacher preparation: Dreams and reality. *Journal of Physics Teacher Education Online, 3*(2), 3–9.

Etkina, E., Karelina, A., Ruibal-Villasenor, M., Rosengrant, D., Jordan, R., & Hmelo-Silver, C. E. (2010). Design and reflection help students develop scientific abilities: Learning in introductory physics laboratories. *Journal of the Learning Sciences, 19*(1), 54–98. doi:10.1080/10508400903452876

Facer, K., Joiner, R., Stanton, D., Reid, J., Hull, R., & Kirk, D. (2004). Savannah: Mobile gaming and learning? *Journal of Computer Assisted Learning, 20*(6), 399–409. doi:10.1111/j.1365-2729.2004.00105.x

Fall, R., Webb, N., & Chudowsky, N. (1997). *Group discussion and large-scale language arts assessment: Effects on students' comprehension.* CSE Technical Report 445. Los Angeles, CA: CRESST.

Falvo, D. A. (2008). Animations and simulations for teaching and learning molecular chemistry. *International Journal of Technology in Teaching and Learning, 4*(1), 68–77.

Falvo, D. A., & Suits, J. P. (2009). Gender and spatial ability and the use of specific labels and diagrammatic arrows in a micro-level chemistry animation. *Journal of Educational Computing Research, 41*(1), 83–102. doi:10.2190/EC.41.1.d

Falvo, D. A., Urban, M. J., & Suits, J. P. (2011). Exploring the impact of and perceptions about interactive, self-explaining environments in molecular-level animations. *Center for Educational Policy Studies Journal, 1*(4), 45–61.

Fawcett, L. M., & Garton, A. F. (2005). The effects of peer collaboration on childrens problem solving ability. *The British Journal of Educational Psychology, 75*(2), 157–169. doi:10.1348/000709904X23411 PMID:16033660

Feldman, A., Cooke, M. L., & Ellsworth, M. s. (2010). The Classroom Sandbox: A physical model for scientific inquiry. *Science Teacher (Normal, Ill.), 77*(9), 58–62.

Feynman, R., Leighton, R., & Sands, M. (1964). Atoms in Motion. In *The Feynman Lectures on Physics, Volume I.* Retrieved from the Feynman Lectures Website: http://www.feynmanlectures.caltech.edu/I_01.html

Finley, F. N., Stewart, J., & Yarroch, W. L. (1982). Teachers perceptions of important and difficult science content. *Science Education, 66*(4), 551–538. doi:10.1002/sce.3730660404

Fiore, S. M., & Schooler, J. W. (2002). How did you get here from there? Verbal overshadowing of spatial mental models. *Applied Cognitive Psychology, 16*(8), 897–910. doi:10.1002/acp.921

Fiore, S., Rosen, M., Smith-Jentsch, K., Salas, E., Letsky, M., & Warner, N. (2010). Toward an understanding of macrocognition in teams: Predicting process in complex collaborative contexts. *The Journal of the Human Factors and Ergonomics Society, 53*(2), 203–224. doi:10.1177/0018720810369807

Fiore, S., & Schooler, J. W. (2004). Process mapping and shared cognition: Teamwork and the development of shared problem models. In E. Salas & S. M. Fiore (Eds.), *Team cognition: Understanding the factors that drive process and performance* (pp. 133–152). Washington, DC: American Psychological Association. doi:10.1037/10690-007

Fischer, G. (2006, May). Distributed intelligence: extending the power of the unaided, individual human mind. In *Proceedings of the working conference on Advanced visual interfaces* (pp. 7-14). ACM. doi:10.1145/1133265.1133268

Fishman, B., Best, S., Marx, R., & Tal, R. (2001). *Design research on professional development in a systemic reform context.* Paper presented at AERA 2001, Seattle, WA.

Foltz, P. W., & Martin, M. J. (2008). Automated communication analysis of teams. In E. Salas, G. F. Goodwin, & S. Burke (Eds.), *Team effectiveness in complex organizations and systems: Cross-disciplinary perspectives and approaches* (pp. 411–431). New York: Routledge.

Forbes, C. T., & Davis, E. A. (2008). Exploring preservice elementary teachers critique and adaption of science curriculum materials in respect to socioscientific issues. *Science and Education, 17*(8-9), 829–854. doi:10.1007/s11191-007-9080-z

Franz, J. (2010). Improving K-12 science education. *American Physical Society, 19*(7). Retrieved from: http://www.aps.org/publications/apsnews/201007/backpage.cfm

Freire, P. (1996). Pedagogy of the oppressed (New rev. ed.; M. B. Ramos, Trans.). New York, NY: Penguin Books.

Furtak, E. M. (2006). The problem with answers: An exploration of guided scientific inquiry teaching. *Science Education, 90*(3), 453–467. doi:10.1002/sce.20130

Gabel, D. L., Sherwood, R. D., & Enochs, L. (1984). Problem-solving skills of high school chemistry students. *Journal of Research in Science Teaching, 21*(2), 221–233. doi:10.1002/tea.3660210212

Gabel, D. S., & Bunce, D. M. (1994). Research on Problem Solving: Chemistry. In D. L. Gabel (Ed.), *Handbook of Research on Science Teaching and Learning. A Project of the National Science Teachers Association*. New York: Macmillan Publishing Company.

Gagnon, J. C., & Barber, B. R. (2010). Characteristics of and services provided to youth in secure care facilities. *Behavioral Disorders, 36*(1), 7–19.

Gagnon, J. C., & Barber, B. R. (2014). Instructional practice guide for teaching reading and mathematics in juvenile correctional schools. *Journal of Correctional Education, 65*(3), 5–23.

Gagnon, J. C., Barber, B. R., Van Loan, C., & Leone, P. E. (2009). Juvenile correctional schools: Characteristics and approaches to curriculum. *Education & Treatment of Children, 32*(4), 673–696. doi:10.1353/etc.0.0068

Gardner, P. L. (1972). Structure-of-knowledge theory and science education. *Educational Philosophy and Theory, 4*(2), 25–46. doi:10.1111/j.1469-5812.1972.tb00053.x

Garet, M. S., Porter, A. C., Desimone, L., Birman, B. F., & Yoon, K. S. (2001). What makes professional development effective? Results from a national sample of teachers. *American Educational Research Journal, 38*(4), 915–945. doi:10.3102/00028312038004915

Gatto, J. T. (2009). *Weapons for mass instruction: A schoolteacher's journey through the dark world of compulsory schooling*. Gabriola Island, BC, Canada: New Society Publishers.

Gee, J. P. (2003). *What video games have to teach us about learning and literacy*. New York: Palgrave Macmillan.

Gee, J. P. (2010). *New digital media and learning as an emerging area and "worked examples" as one way forward.* Cambridge, MA: The MIT Press.

Gehrke, N. J., Knapp, M. S., & Sirotnik, K. A. (1992). In Search of the School Curriculum. *Review of Research in Education, 18*, 51–110. doi:10.2307/1167297

Gerard, L. F., Spitulnik, M., & Linn, M. C. (2010). Teacher use of evidence to customize inquiry science instruction. *Journal of Research in Science Teaching, 47*(9), 1037–1063. doi:10.1002/tea.20367

Gerjets, P. H., & Hesse, F. W. (2004). When are powerful learning environments effective: The role of learner activities and of students' conceptions of educational technology. *International Journal of Educational Research, 41*(6), 445–465. doi:10.1016/j.ijer.2005.08.011

Germann, P. J., Aram, A., & Burke, G. (1996). Identifying patterns and relationships among the responses of seventh-grade students to the science process skill of designing experiments. *Journal of Research in Science Teaching*, *33*(1), 79–99. doi:10.1002/(SICI)1098-2736(199601)33:1<79::AID-TEA5>3.0.CO;2-M

GitHub, Inc. (2016). *Awesome Public Datasets*. Retrieved June 15th, 2016 from https://github.com/caesar0301/awesome-public-datasets

Glaser, B. (1956). The constant comparative method of qualitative analysis. *Social Problems*, *12*(4), 436–445. doi:10.2307/798843

Glaser, R., Schauble, L., Raghavan, K., & Zeitz, C. (1992). Scientific reasoning across different domains. In E. De Corte, M. Linn, H. Mandl, & L. Verschaffel (Eds.), *Computer-based learning environments and problem solving* (pp. 345–373). Berlin: Springer-Verlag. doi:10.1007/978-3-642-77228-3_16

Gopnick, A., Meltzoff, A. N., & Kuhl, P. K. (1999). *The scientist in the crib: What early learning tells us about the mind.* New York: HarperCollins Publishers.

Govaerts, S., Cao, Y., Vozniuk, A., Holzer, A., Zutin, D. G., Ruiz, E. S. C., . . . Gillet, D. (2013). Towards an online lab portal for inquiry-based STEM learning at school. In J.-F. Wang & R. Lau (Eds.), *Advances in web-based learning – ICWL 2013, 12th International Conference, Kenting, Taiwan, October 6-9, 2013, Proceedings* (pp. 244-253). Berlin: Springer. doi:10.1007/978-3-642-41175-5_25

Goyal, N. (2012). *One size does not fit all: A student's assessment of school* [Kindle version]. Retrieved from Amazon.com

Graesser, A. C. (2016). Conversations with AutoTutor help students learn. *International Journal of Artificial Intelligence in Education*, *26*(1), 124–132. doi:10.1007/s40593-015-0086-4

Graesser, A. C., Jeon, M., & Dufty, D. (2008). Agent technologies designed to facilitate interactive knowledge construction. *Discourse Processes*, *45*(4), 298–322. doi:10.1080/01638530802145395

Gregory, A., Skiba, R. J., & Noguera, P. A. (2010). The achievement gap and the discipline gap two sides of the same coin? *Educational Researcher*, *39*(1), 59–68. doi:10.3102/0013189X09357621

Griffin, P., Care, E., & McGaw, B. (2012). The changing role of education and schools. In P. Griffin, B. McGaw, & E. Care (Eds.), *Assessment and teaching 21st century skills* (pp. 1–15). Heidelberg, Germany: Springer. doi:10.1007/978-94-007-2324-5_1

Groves, F. H., & Pugh, A. F. (1999). Elementary pre-service teacher perceptions of the greenhouse effect. *Journal of Science Education and Technology*, *8*(1), 75–81. doi:10.1023/A:1009433705790

Gulacar, O., Eilks, I., & Bowman, C. R. (2014). Differences in General Cognitive Abilities and Domain-Specific Skills of Higher- and Lower-Achieving Students in Stoichiometry. *Journal of Chemical Education*, *91*(7), 961–968. doi:10.1021/ed400894b

Guo, Y., Wagh, A., Brady, C., Author, S. T., Horn, M. S., & Wilensky, U. (2016, June). Frogs to Think with: Improving Students' Computational Thinking and Understanding of Evolution in A Code-First Learning Environment. In *Proceedings of the 15th International Conference on Interaction Design and Children* (pp. 246-254). ACM. doi:10.1145/2930674.2930724

Gupta, T. (2012). *Guided-inquiry based laboratory instruction: Investigation of critical thinking skills, problem solving skills, and implementing student roles in chemistry* (Unpublished doctoral dissertation). Iowa State University, Ames, IA.

Gupta, T., Burke, K. A., Mehta, A., & Greenbowe, T. J. (2015). Impact of Guided-Inquiry-Based Instruction with a Writing and Reflection Emphasis on Chemistry Students Critical Thinking Abilities. *Journal of Chemical Education*, *92*(1), 32–38. doi:10.1021/ed500059r

Haga, S. B. (2006). Teaching resources for genetics. *Nature Reviews. Genetics*, *7*(3), 223–229. doi:10.1038/nrg1803 PMID:16462854

Halfhide, T., Wilkie, A., & Ergas, S. (2016). *Indigenous Algal Growth on Municipal Sludge Centrate and Measuring Lipid Productivity Using Fluorospectroscopy and Gravimetric Analyses.* Paper presented at the Algal Biomass, Biofuels and Bioproducts (ABBB), San Diego, CA.

Hall, R., Ma, J. Y., & Nemirovsky, R. (2014). Rescaling bodies in/as representational instruments in GPS drawing. In V. R. Lee (Ed.), Learning technologies and the body: Integration and implementation in formal and informal learning environments (pp. 112–131). Routledge.

Hallar, A. G., McCubbin, I. B., & Wright, J. M. (2011). CHANGE: A place-based curriculum for understanding climate change at Storm Peak Laboratory, Colorado. *Bulletin of the American Meteorological Society*, *92*(7), 909–918. doi:10.1175/2011BAMS3026.1

Hanson, R.M. (2010). Jmol-a paradigm shift in crystallographic visualization. *Journal of Applied Crystallography, 43*(5-2), 1250–1260.

Haralambos, M., & Holborn, M. (1991). *Sociology: Themes and Perspectives.* New York, NY: Collins Educational.

Harrison, A. G., & Treagust, D. F. (2000). A typology of school science models. *International Journal of Science Education*, *22*(9), 1011–1026. doi:10.1080/095006900416884

Hasler, B. S., Kersten, B., & Sweller, J. (2007). Learner control, cognitive load and instructional animation. *Applied Cognitive Psychology*, *21*(6), 713–729. doi:10.1002/acp.1345

Hattie, J. A. C. (2009). *Visible learning: A synthesis of 800+ meta-analyses on achievement.* London: Routledge.

Hawkes, B., & O'Meara, J. (2005). *Physics education in Canada.* Retrieved from http://www.cap.ca/pic/Archives/61.2(2005)/editorial.html

Hegarty, M., Kriz, S., & Cate, C. (2003). The roles of mental animations and external animations in understanding mechanical systems. *Cognition and Instruction*, *21*(4), 325–360. doi:10.1207/s1532690xci2104_1

Heller, J. I., Daehler, K. R., Wong, N., Shinohara, M., & Miratrix, L. W. (2012). Differential effects of three professional development models on teacher knowledge and student achievement in elementary science. *Journal of Research in Science Teaching*, *49*(3), 333–362. doi:10.1002/tea.21004

Herman, B. C., Feldman, A., & Vernaza-Hernandez, V. (2015). Article. *International Journal of Science and Mathematics Education*, 1–21.

Herraez, A. (2006). Biomolecules in the computer: Jmol to the rescue. *Biochemistry and Molecular Biology Education*, *34*(4), 255–261. doi:10.1002/bmb.2006.494034042644 PMID:21638687

Herron, J. D. (1975). Piaget for chemists. *Journal of Chemical Education*, *52*(3), 146–150. doi:10.1021/ed052p146

Hessesn, D. O., Ågren, G. I., Anderson, T. R., Elser, J. R., & Ruiler, P. C. (2004). Carbon sequestration in ecosystems: The role of stoichiometry. *Ecology*, *97*(11), 1179–1192. doi:10.1890/02-0251

Hickey, D. T., Kindfield, A. C. H., Horwitz, P., & Christie, M. A. T. (2003). Integrating curriculum, instruction, assessment, and evaluation in a technology-supported genetics learning environment. *American Educational Research Journal*, *40*(2), 495–538. doi:10.3102/00028312040002495

Higgins, D., & Higgins, H. (2001). Intermedia. *Leonardo, 34*(1), 49-54. Retrieved from Project MUSE database website: https://muse.jhu.edu/article/19618

Hillel, J. (2005). *Physics education research: A comprehensive study*. Toronto, Canada: University of Toronto.

Hill, H. C., Beisiegel, M., & Jacob, R. (2013). Professional development research consensus, crossroads, and challenges. *Educational Researcher*, *42*(9), 476–487. doi:10.3102/0013189X13512674

Hill, H., Rowan, B., & Ball, D. (2005). Effects of teachers mathematical knowledge for teaching on student achievement. *American Educational Research Journal*, *42*(2), 371–406. doi:10.3102/00028312042002371

Hofstein, A., & Lunetta, V. N. (2003). The laboratory in science education: Foundations for the twenty-first century. *Science Education*, *88*(1), 28–54. doi:10.1002/sce.10106

Hofstein, A., Navon, O., Kipnis, M., & Mamlok-Naaman, R. (2005). Developing students ability to ask more and better questions resulting from inquiry-type chemistry laboratories. *Journal of Research in Science Teaching*, *42*(7), 791–806. doi:10.1002/tea.20072

Holubova, R. (2005). Environmental physics: Motivation in physics teaching and learning. *Journal of Physics Teacher Education Online*, *3*(1), 17–20.

Houchins, D. E., Jolivette, K., Krezmien, M. P., & Baltodano, H. A. (2008). A Multi-State study examining the impact of explicit reading instruction with incarcerated students. *Journal of Correctional Education*, *59*(1), 65–85.

Hovardas, T. (2016). A learning progression should address regression: Insights from developing non-linear reasoning in ecology. *Journal of Research in Science Teaching*, *53*(10), 1447–1470. doi:10.1002/tea.21330

Howell, D. C. (2007). *Statistical Methods for Psychology*. SAGE.

Huddle, P. A., & Pillay, A. E. (1996). An in-depth study of misconceptions in stoichiometry and chemical equilibrium at a South African university. *Journal of Research in Science Teaching*, *33*(1), 65–77. doi:10.1002/(SICI)1098-2736(199601)33:1<65::AID-TEA4>3.0.CO;2-N

Hug, B., & McNeill, K. L. (2008). Use of First-hand and Second-hand Data in Science: Does data type influence classroom conversations? *International Journal of Science Education*, *30*(13), 1725–1751. doi:10.1080/09500690701506945

IPCC. (2013). The physical science basis. Contribution of working group I to the fifth assessment report of the Intergovernmental Panel on Climate Change.

Ito, M., Baumer, S., Bittanti, M., Boyd, D., Cody, R., Herr-Stephenson, B., & Tripp, L. (2010). *Hanging out, messing around, geeking out: Kids living and learning with new media*. Cambridge, MA: MIT Press.

Jackson, S., Eklund, R., & Martin, A. (2010). *The FLOW manual*. Queensland, Australia: Mind Garden, Inc.

Jacobson, M. J., & Archodidou, A. (2000). The design of hypermedia tools for learning: Fostering conceptual change and transfer of complex scientific knowledge. *Journal of the Learning Sciences*, *9*(2), 145–199. doi:10.1207/s15327809jls0902_2

Jagadish, H. V., Gehrke, J., Labrinidis, A., Papakonstantinou, Y., Patel, J. M., Ramakrishnan, R., & Shahabi, C. (2014). Big data and its technical challenges. *Communications of the ACM*, *57*(7), 86–94. doi:10.1145/2611567

Johnson, D. W., Johnson, R. T., & Holubec, E. J. (1993). *Circles of learning: Cooperation in the classroom* (4th ed.). Edina, MN: Interaction Book Company.

Johnstone, A. H., MacDonald, J. J., & Webb, G. (1977). Misconceptions in school thermodynamics. *Physics Education*, *12*(4), 248–251. doi:10.1088/0031-9120/12/4/011

Jollands, N., Ruth, M., Bernier, C., & Golubiewski, N. (2005) Climate's long-term impacts on New Zealand infrastructure—a Hamilton City case study. In Proceedings ecological economics in action. New Zealand Centre for Ecological Economics.

Jordan, R. C., Ruibal-Villasenor, M., Hmelo-Silver, C. E., & Etkina, E. (2011). Laboratory materials: Affordances or constraints? *Journal of Research in Science Teaching*, *48*(9), 1010–1025. doi:10.1002/tea.20418

Joyce, B. R., & Showers, B. (2002). *Student achievement through staff development*. Ascd.

Kaiser, D., & Wasserman, L. (2016). The Rockefeller Family Fund vs. Exxon. *The New York Review of Books*, *63*(19). Retrieved from http://www.nybooks.com/articles/2016/12/08/the-rockefeller-family-fund-vs-exxon/

Kali, Y., Linn, M. C., & Roseman, J. E. (2008). *Designing coherent science education: Implications for curriculum, instruction, and policy*. New York: Teachers College Press.

Kalyuga, S. (2007). Expertise reversal effect and its implications for learner-tailored instruction. *Educational Psychology Review*, *19*(4), 509–539. doi:10.1007/s10648-007-9054-3

Kalyuga, S., & Sweller, J. (2004). Measuring knowledge to optimize cognitive load factors during instruction. *Journal of Educational Psychology*, *96*(3), 558–568. doi:10.1037/0022-0663.96.3.558

Kamarainen, A. M., Metcalf, S., Grotzer, T., Browne, A., Mazzuca, D., Tutwiler, M. S., & Dede, C. (2013). EcoMOBILE: Integrating augmented reality and probeware with environmental education field trips. *Computers & Education*, *68*, 545–556. doi:10.1016/j.compedu.2013.02.018

Kanari, Z., & Millar, R. (2004). Reasoning from data: How students collect and interpret data in science investigations. *Journal of Research in Science Teaching*, *41*(7), 748–769. doi:10.1002/tea.20020

Kanter, R. M. (2001). *Evolve! Succeeding in the digital culture of tomorrow*. Cambridge, MA: Harvard Business School Press.

Kara, Y., & Yesilyurt, S. (2008). Comparing the impacts of tutorial and edutainment software programs on students achievements, misconceptions, and attitudes towards biology. *Journal of Science Education and Technology*, *17*(1), 32–41. doi:10.1007/s10956-007-9077-z

Kargbo, D. B., Hobbs, D., & Erickson, G. L. (1980). Childrens beliefs about inherited characteristics. *Journal of Biological Education*, *14*(2), 137–146. doi:10.1080/00219266.1980.10668980

Katz, J. (2006). *Magic in the air: Mobile communication and the transformation of social life*. New Brunswick, NJ: Transaction Publishers.

Kearney, M. (2004). Classroom use of multimedia-supported predict-observe-explain tasks in a social constructivist learning environment. *Research in Science Education*, *34*(4), 427–453. doi:10.1007/s11165-004-8795-y

Keeney-Kennicutt, W. L., & Merchant, Z. H. (2013).Using virtual worlds in the general chemistry classroom. In J. P. Suits, & M. J. Sanger (Eds.), *Pedagogic Roles of Animations and Simulations in Chemistry Courses. ACS Symposium Series*, (*vol. 1142*, pp, 181-204). Washington, DC: American Chemical Society. doi:10.1021/bk-2013-1142.ch008

Kesselheim, A. S., Cook-Deegan, R. M., Winickoff, D. E., & Mello, M. M. (2013). Gene patenting – The Supreme Court finally speaks. *The New England Journal of Medicine*, *369*(9), 869–875. doi:10.1056/NEJMhle1308199 PMID:23841703

Khandelwal, S., & Wickstrom, N. (2016). Gait Event Detection in Real-World Environment for Long-Term Applications: Incorporating Domain Knowledge into Time-Frequency Analysis. *IEEE Transactions on Neural Systems and Rehabilitation Engineering*. https://doi.org/10.1109/TNSRE.2016.2536278

Kincheloe, J. (1998). *Critical pedagogy primer* (2nd ed.). New York, NY: Peter Lang.

Kirriemuir, J., & McFarlane, A. (2004). *Report 8: Literature review in games and learning*. Bristol, UK: NESTA Futurelab.

Kirschner, P. A., Sweller, J., & Clark, R. E. (2006). Why minimal guidance during instruction does not work: An analysis of the failure of constructivist, discovery, problem-based, experiential, and inquiry-based teaching. *Educational Psychologist*, *41*(2), 75–86. doi:10.1207/s15326985ep4102_1

Klahr, D., & Nigam, M. (2004). The equivalence of learning paths in early science instruction: Effect of direct instruction and discovery learning. *Psychological Science*, *15*(10), 661–667. doi:10.1111/j.0956-7976.2004.00737.x PMID:15447636

Kleickmann, T., Richter, D., Kunter, M., Elsner, J., Besser, M., Krauss, S., & Baumert, J. (2013). Teachers content knowledge and pedagogical content knowledge the role of structural differences in teacher education. *Journal of Teacher Education*, *64*(1), 90–106. doi:10.1177/0022487112460398

Klopfer, E., Osterweil, S., & Salen, K. (2009). Moving learning games forward. Cambridge, MA: Education Arcade. Retrieved from http://education.mit.edu/papers/MovingLearningGamesForward_EdArcade.pdf

Klopfer, E., Perry, J., Squire, K., Jan, M.-F., & Steinkuehler, C. (2005). *Mystery at the Museum: A collaborative game for museum education*. Paper presented at the annual conference of Computer Supported Collaborative Learning, Taipei, Taiwan. doi:10.3115/1149293.1149334

Klopfer, E., Squire, K., & Jenkins, H. (2002). *Environmental detectives: PDAs as a window into a virtual simulated world*. Paper presented at the International Conference on Wireless Mobile Technologies in Education, Växjö, Sweden. doi:10.1109/WMTE.2002.1039227

Klopfer, E. (2008). *Augmented learning: Research and design of mobile educational games*. Cambridge, MA: The MIT Press. doi:10.7551/mitpress/9780262113151.001.0001

Klopfer, E., & Sheldon, J. (2010). Augmenting your own reality: Student authoring of science-based augmented reality games. *New Directions for Youth Development*, *2010*(128), 85–94. doi:10.1002/yd.378 PMID:21240956

Klopfer, E., Yoon, S., & Perry, J. (2005). Using palm technology in participatory simulations of complex systems: A new take on ubiquitous and accessible mobile computing. *Journal of Science Education and Technology*, *14*(3), 285–297. doi:10.1007/s10956-005-7194-0

Klopfer, E., Yoon, S., & Rivas, L. (2004). Comparative analysis of Palm and wearable computers for Participatory Simulations. *Journal of Computer Assisted Learning*, *20*(5), 347–359. doi:10.1111/j.1365-2729.2004.00094.x

Knight, R. (2004). *Five easy lessons: Strategies for successful physics teaching* (1st ed.). Toronto, Canada: Addison Wesley.

Kohut, A., Keeter, S., Doherty, C., & Dimock, M. (2009). Scientific achievements less prominent than a decade ago: Public praises science; scientists fault public, media. Washington, DC: The Pew Research Center for the People & the Press. Retrieved from http://www. people-press. org/files/legacy-pdf/528. pdf

Koksal, E. A., & Berberoglou, G. (2014). The effect of guided inquiry instruction on 6th grade Turkish students achievement, science process skills, and attitudes toward science. *International Journal of Science Education*, *36*(1), 66–78. doi:10.1080/09500693.2012.721942

Konold, C., & Miller, C. D. (2005). TinkerPlots. Emeryville, CA: Key Curriculum Press. Retrieved from http://scholar.google.com/scholar?cluster=5929212600541009408&hl=en&oi=scholarr

Konold, C., & Pollatsek, A. (2002). Data Analysis as the Search for Signals in Noisy Processes. *Journal for Research in Mathematics Education*, *33*(4), 259–289. doi:10.2307/749741

Kortland, J. (2005). Physics in personal, social and scientific contexts: A retrospective view on the Dutch physics curriculum development project PLON. In P. Nentwig & D. Waddington (Eds.), *Making it relevant: Context based learning of science* (pp. 67–89). Munster, Germany: Waxmann.

Koutromanos, G., Sofos, A., & Avraamidou, L. (2015). The use of augmented reality games in education: A review of the literature. *Educational Media International*, *52*(4), 253–271. doi:10.1080/09523987.2015.1125988

Kozma, R. B., & Russell, J. (1997). Multimedia and understanding: Expert and novice responses to different representations of chemical phenomena. *Journal of Research in Science Teaching*, *34*(9), 949–968. doi:10.1002/(SICI)1098-2736(199711)34:9<949::AID-TEA7>3.0.CO;2-U

Krahn, G. L., Hammond, L., & Turner, A. (2006). A cascade of disparities: Health and health care access for people with intellectual disabilities. *Mental Retardation and Developmental Disabilities Research Reviews*, *12*(1), 70–82. doi:10.1002/mrdd.20098 PMID:16435327

Krajcik, J., Blumenfeld, P. C., Marx, R. W., Bass, K. M., Fredricks, J., & Soloway, E. (1998). Inquiry in Project-Based Science Classrooms: Initial Attempts by Middle School Students. *Journal of the Learning Sciences*, *7*(3–4), 313–350. doi:10.1080/10508406.1998.9672057

Kremer, K., Specht, C., Urhahne, D., & Mayer, J. (2014). The relationship in biology between the nature of science and scientific inquiry. *Journal of Biological Education*, *48*(1), 1–8. doi:10.1080/00219266.2013.788541

Krezmien, M. P. (2008, January). *National perspectives of youth with disabilities and their interaction with the juvenile system: A statement of the issues and need.* Presentation at the 1st conference of the Ohio Juvenile Justice Project, Columbus, OH.

Krezmien, M. P., Leone, P. E., & Achilles, G. M. (2006). Suspension, race, and disability: Analysis of statewide practices and reporting. *Journal of Emotional and Behavioral Disorders*, *14*(4), 217–226. doi:10.1177/10634266060140040501

Krezmien, M. P., & Mulcahy, C. A. (2008). Literacy and Delinquency: Current status of reading interventions with detained and incarcerated youth. *Reading & Writing Quarterly, 24*(2), 219–238. doi:10.1080/10573560701808601

Krezmien, M. P., Mulcahy, C. A., & Leone, P. E. (2008). Detained and Committed Youth: Examining Differences in Achievement, Mental Health Needs, and Special Education Status. *Education & Treatment of Children, 31*(4), 1–20.

Krezmien, M. P., Mulcahy, C. A., Travers, J., Wilson, M., & Wells, C. (2013). Reading performance of incarcerated youth: Understanding and responding to a unique population of readers. *Journal of Special Education Leadership, 26*(2), 71–81.

Kuhn, D. (2007). Reasoning about multiple variables: Control of variables is not the only challenge. *Science Education, 91*(5), 710–726. doi:10.1002/sce.20214

Kunter, M., Klusmann, U., Baumert, J., Richter, D., Voss, T., & Hachfeld, A. (2013). Professional competence of teachers: Effects on instructional quality and student development. *Journal of Educational Psychology, 105*(3), 805–820. doi:10.1037/a0032583

Kuş, Z. (2015). Science and social studies teachers' beliefs and practices about teaching controversial issues: Certain comparisons. *JSSE-Journal of Social Science Education, 14*(3), 84–97.

Lancaster, K., Moore, E. B., Parson, R., & Perkins, K. (2013). Insights from using PhET's design principles for interactive chemistry simulations. In J. P. Suits, & M. J. Sanger (Eds.), *Pedagogic Roles of Animations and Simulations in Chemistry Courses. ACS Symposium Series*, (*vol. 1142*, pp, 97-126). Washington, DC: American Chemical Society. doi:10.1021/bk-2013-1142.ch005

Landriscina, F. (2013). *Simulation-Based Learning. In Simulation & Learning: A model-centered approach* (pp. 99–36). New York: Springer. doi:10.1007/978-1-4614-1954-9_5

Lane, K. L., Carter, E. W., Pierson, M. R., & Glaeser, B. C. (2006). Academic, social, and behavioral characteristics of high school students with emotional disturbances or learning disabilities. *Journal of Emotional and Behavioral Disorders, 14*(2), 108–117. doi:10.1177/10634266060140020101

Langbeheim, E., & Author, S. T. (2015). *Learning Affordances related to Participation-in and Observation-of Particle Simulations: Hints from Seemingly Off-task Talk.* Paper presented at the National Association for Research in Science Teaching (NARST) annual conference, Chicago, IL.

Langbeheim, E., & Author, S. T. (2016). ConfChem Conference on Interactive Visualizations for Chemistry Teaching and Learning: Learning by Being: Playing Particles in the MeParticle–WeMatter Simulation. *Journal of Chemical Education, 93*(6), 1145–1147. doi:10.1021/acs.jchemed.5b00443

Laughlin, J. S. (2001). Breaking out of the box: Teaching biology with web-based active learning modules. *The American Biology Teacher, 63*(2), 110–115. doi:10.1662/0002-7685(2001)063[0110:BOOTBT]2.0.CO;2

Laurillard, D. (2009). The pedagogical challenges to collaborative technologies. *International Journal of Computer-Supported Collaborative Learning*, *4*(1), 5–20. doi:10.1007/s11412-008-9056-2

Lave, J., & Wenger, E. (1991). *Situated learning: Legitimate peripheral participation*. New York, NY: Cambridge University Press. doi:10.1017/CBO9780511815355

Lawson, A. E., Abraham, M. R., & Renner, J. W. (1989). *A theory of instruction: Using the learning cycle to teach science concepts and thinking skills*. National Association for Research in Science Teaching.

Lee, V. R., & Drake, J. (2013a). Digital physical activity data collection and use by endurance runners and distance cyclists. *Technology, Knowledge and Learning, 18*(1–2), 39–63. https://doi.org/10.1007/s10758-013-9203-3

Lee, V. R., Drake, J. R., & Thayne, J. (2016). Appropriating Quantified Self Technologies to Support Elementary Statistical Teaching and Learning. *IEEE Transactions on Learning Technologies*. https://doi.org/10.1109/TLT.2016.2597142

Lee, V. R., Drake, J., Thayne, J., & Cain, R. (2015). Opportunistic Uses of the Traditional School Day Through Student Examination of Fitbit Activity Tracker Data. *Proceedings of IDC 2015*. Retrieved from http://works.bepress.com/victor_lee/31

Lee, H. S., Linn, M. C., Varma, K., & Liu, O. L. (2010). How do technology-enhanced inquiry science units impact classroom learning? *Journal of Research in Science Teaching*, *47*(1), 71–90. doi:10.1002/tea.20304

Lee, V. R., & Briggs, M. (2014). Lessons Learned from an Initial Effort to Bring a Quantified Self "Meetup" Experience to a New Demographic. In *Proceedings of the 2014 ACM International Joint Conference on Pervasive and Ubiquitous Computing: Adjunct Publication* (pp. 707–710). New York, NY: ACM. doi:10.1145/2638728.2641321

Lee, V. R., & Drake, J. (2013b). Quantified recess: Design of an activity for elementary students involving analysis of their own movement data. In *Proceedings of the 12th International Conference on Interaction Design and Children* (pp. 273–276). New York, NY: ACM. doi:10.1145/2485760.2485822

Lee, V. R., & DuMont, M. (2010). An Exploration into How Physical Activity Data-Recording Devices Could be Used in Computer-Supported Data Investigations. *International Journal of Computers for Mathematical Learning*, *15*(3), 167–189. doi:10.1007/s10758-010-9172-8

Lehrer, R., Kim, M.-J., Ayers, E., & Wilson, M. (2014). Toward establishing a learning progression to support the development of statistical reasoning. In *Learning over Time: Learning Trajectories in Mathematics Education*. Charlotte, NC: Information Age Publishers.

Leiserowitz, A. A. (2005). American risk perceptions: Is climate change dangerous? *Risk Analysis*, *25*(6), 1433–1442. doi:10.1111/j.1540-6261.2005.00690.x PMID:16506973

Leone, P. E., Krezmien, M., Mason, L., & Meisel, S. M. (2005). Organizing and delivering empirically based literacy instruction to incarcerated youth. *Exceptionality*, *13*(2), 89–102. doi:10.1207/s15327035ex1302_3

Leone, P. E., & Weinberg, L. A. (2012). *Addressing the unmet educational needs of children and youth in the juvenile justice and child welfare systems, 2012 Edition*. Georgetown Public Policy Institute, Center for Juvenile Justice Reform.

Levy, S. T., & Okal, Z. (2012). *Keep me Healthy! Participatory simulation for health education among students who are cognitively-impaired*. Retrieved from http://ccl.northwestern.edu/netlogo/models/HubNetDiseaseHubNet

Levy, S. T., Almog, H., Izkovich, A., Zlotnick, L., Eldar, S., & Ofeck, E. (2011; 2015). Dancing with Particles. Participatory simulation for social explorations of global warming. Developed at the Systems Learning & Development Lab (SLDL), University of Haifa.

Levy, S. T., Langbeheim, Roller, E., Peri, G., Grady Drozd, Y., & Azzam, Y., E. (2014). MeParticle WeMatter. Participatory simulation for role-playing particles in fluids. Developed at the Systems Learning & Development Lab (SLDL), University of Haifa.

Levy, S. T., Rotem, S., & Linyevsky, A. (2015). *TrafficJams. Participatory simulation for collaborative experiments in driving. Developed at the Systems Learning & Development Lab (SLDL)*. University of Haifa.

Lewis, J., & Kattmann, U. (2004). Traits, genes, particles and information: Re-visiting students understandings of genetics. *International Journal of Science Education*, *26*(2), 195–206. doi:10.1080/0950069032000072782

Lewis, J., & Wood-Robinson, C. (2000). Genes, chromosomes, cell division, and inheritance do students see any relationship? *International Journal of Science Education*, *22*(2), 177–195. doi:10.1080/095006900289949

Linn, M. C., Clark, D., & Slotta, J. D. (2003). WISE design for knowledge integration. *Science Education*, *87*(4), 517–538. doi:10.1002/sce.10086

Linn, M. c., Davis, E. A., & Bell, P. (Eds.). (2004). *Internet environments for science education*. Mahwah, NJ: Lawrence Erlbaum Associates.

Linn, M. C., & Eylon, B. S. (2011). *Science learning and instruction: Taking advantage of technology to promote knowledge integration*. New York: Routledge.

Linn, M. C., Eylon, B. S., & Davis, E. A. (2004). The knowledge integration perspective on learning. In M. C. Linn, E. A. Davis, & P. Bell (Eds.), *Internet environments for science education* (pp. 73–83). Mahwah, NJ: Erlbaum.

Linn, M. C., & Eylon, B.-S. (2006). Science education: Integrating views of learning and instruction. In P. A. Alexander & P. H. Winne (Eds.), *Handbook of Educational Psychology* (2nd ed.; pp. 511–544). Mahwah, NJ: Lawrence Erlbaum Associates.

Linn, M. C., & Hsi, S. (2000). *Computers, teachers, peers: Science learning partners*. Mahwah, NJ: Lawrence Erlbaum Associates, Inc.

Linn, M. C., Lee, H. S., Tinker, R., Husic, F., & Chiu, J. L. (2006). Teaching and assessing knowledge integration in science. *Science*, *313*(5790), 1049–1050. doi:10.1126/science.1131408 PMID:16931740

Lin, X., & Lehman, J. D. (1999). Supporting learning of variable control in a computer-based biology environment: Effects of prompting college students to reflect on their own thinking. *Journal of Research in Science Teaching*, *36*(7), 837–858. doi:10.1002/(SICI)1098-2736(199909)36:7<837::AID-TEA6>3.0.CO;2-U

Liu, L., Hao, J., von Davier, A., Kyllonen, P., & Zapata-Rivera, D. (2015). A tough nut to crack: Measuring collaborative problem solving. In Y. Rosen, S. Ferrara, & M. Mosharraf (Eds.), Handbook of Research on Technology Tools for Real-Life Skill Development (pp. 344-359). Hershey, PA: Information Science, IGI Global.

Looi, C. K., Zhang, B., Chen, W., Seow, P., Chia, G., Norris, C., & Soloway, E. (2011). 1:1 mobile inquiry learning experience for primary science students: a study of learning effectiveness. *Journal of Computer Assisted Learning*, *27*(3), 269–287. doi:10.1111/j.1365-2729.2010.00390.x

Loucks-Horsley, S., Stiles, K. E., Mundry, S., Love, N., & Hewson, P. W. (2009). *Designing professional development for teachers of science and mathematics*. Corwin Press.

Lyons, L., Silva, B. L., Moher, T., Pazmino, P. J., & Slattery, B. (2013). Feel the Burn: Exploring Design Parameters for Effortful Interaction for Educational Games. In *Proceedings of the 12th International Conference on Interaction Design and Children* (pp. 400–403). New York, NY: ACM. doi:10.1145/2485760.2485791

Maccini, P., Gagnon, J. C., Mulcahy, C. A., & Leone, P. E. (2006). Math instruction for committed youth within juvenile correctional schools. *Journal of Correctional Education, 57*(3), 210-229. Retrieved from http://www.jstor.org/stable/23282753

MacLeod, K. A. (2012). *Integrating science, technology, society and environment (STSE) into physics teacher education: Pre-service teachers' perceptions and challenges* (PhD Dissertation). University of Toronto, Toronto, Canada.

MacLeod, K.A. (2015). *Teaching and Learning in the Physical Sciences* [Syllabus]. Antigonish, NS: Physics Department and Chemistry Department, St. Francis Xavier University.

MacLeod, K. A. (2007). A practical example to understand momentum in 1D: The watergun experiment. *Physics Education*, *42*(5), 492–495. doi:10.1088/0031-9120/42/5/008

MacLeod, K. A. (2013). Physics education and STSE: Perspectives from the literature. *European Journal of Physics Education*, *4*(4), 1–12. doi:10.20308/ejpe.21926

MacLeod, K. A., & Kraglund-Gauthier, W. L. (2015). A case study of infusing technology into pre-service secondary science teacher learning: Conceptions and attitudes while navigating changing digital landscapes. *Journal of European Education*, *5*(2), 1–15. doi:10.18656/jee.24828

Maibach, E., Roser-Renouf, C., & Leiserowitz, A. (2009). *Global Warming's Six Americas 2009: An Audience Segmentation Analysis. Yale Project on Climate Change*. New Haven, CT: Yale University and George Mason University.

Makonin, S. (2013 -2016). *The Almanac of Minutely Power dataset.* Retrieved June 15th, 2016 from http://ampds.org/

Manyika, J., Chui, M., Brown, B., Bughin, J., Dobbs, R., Roxburgh, C., & Byers, A. H. (2011). Big data: The next frontier for innovation, competition, and productivity. *The McKinsey Global Institute*, *5*(33), 222.

Marais, F., & Combrinck, S. (2009). An Approach to Dealing with the Difficulties Undergraduate Chemistry Students Experience with Stoichiometry. *South African Journal of Chemistry. Suid-Afrikaanse Tydskrif vir Chemie*, *62*, 88–96.

Marx, R., Blumenfeld, P., Krajcik, J., & Soloway, E. (1997). Enacting project-based science. *The Elementary School Journal*, *97*(4), 341–358. doi:10.1086/461870

Mathieu, J. E., Tannenbaum, S. I., Donsbach, J. S., & Alliger, G. M. (2014). A review and integration of team composition models: Moving toward a dynamic and temporal framework. *Journal of Management*, *40*(1), 130–160. doi:10.1177/0149206313503014

Mayer, R. E., & Wittrock, M. C. (1996). Problem-solving transfer. In D. C. Berliner & R. C. Calfee (Eds.), *Handbook of educational psychology* (pp. 47–62). New York: Macmillan Library Reference USA, Simon & Schuster Macmillan.

McCright, A. M., & Dunlap, R. E. (2011). Cool dudes: The denial of climate change among conservative white males in the United States. *Global Environmental Change*, *21*(4), 1163–1172. doi:10.1016/j.gloenvcha.2011.06.003

McDermott, L. C. (1998). Research in physics education. *APS News*, *7*(1), 8.

McDermott, L. C. (2001). Oersted Medal Lecture 2001. Physics education research: The key to student learning. *American Journal of Physics*, *69*(11), 1127–1137. doi:10.1119/1.1389280

McDermott, L. C. (2006). Preparing K–12 teachers in physics: Insights from history, experience, and research. *American Journal of Physics*, *74*(9), 748–762. doi:10.1119/1.2209243

Merriam, S. (1998). *Qualitative research and case study applications in education*. San Francisco, CA: Jossey-Bass.

Meyer, A., & Rose, D. H. (2005). The future is in the margins: The role of technology and disability in educational reform. In D. H. Rose, A. Meyer, & C. Hitchcock (Eds.), *The universally designed classroom: Accessible curriculum and digital technologies* (pp. 13–35). Cambridge, MA: Harvard Education Press.

Meyer, A., Rose, D. H., & Gordon, D. (2014). Universal design for learning. *Theory into Practice*.

Michael, D. R., & Chen, S. (2006). *Serious games: games that educate, train, and inform*. Boston, MA: Thomson Course Technology PTR.

Michigan Department of Education. (2007). *Science grade level content expectations*. Retrieved July 9, 2014, from http://www.michigan.gov/mde

Minner, D. D., Jurist Levy, A., & Century, J. (2010). Inquiry-based science instruction – What is it and does it matter? Results from a research synthesis years 19842002. *Journal of Research in Science Teaching*, *47*(4), 474–496. doi:10.1002/tea.20347

Moher, T., Ching, C. C., Schaefer, S., Lee, V. R., Enyedy, N., Danish, J., … Fong, C. (2014). Becoming Reflective: Designing for Reflection on Physical Performances. *ICLS 2014 Proceedings*, 1273–1282.

Monaghan, J. M., & Clement, J. (1999). Use of computer simulation to develop mental simulations for understanding relative motion concepts. *International Journal of Science Education*, *21*(9), 921–944. doi:10.1080/095006999290237

Monaghan, J. M., & Clement, J. (2000). Algorithms, visualization, and mental models: High school students interactions with a relative motion simulation. *Journal of Science Education and Technology*, *9*(4), 311–325. doi:10.1023/A:1009480425377

Mueller, A., & Fleming, T. (2001). Cooperative learning: Listening to how children work at school. *The Journal of Educational Research*, *94*(5), 259–265. doi:10.1080/00220670109598761

Mulcahy, C. A., & Krezmien, M. P. (2009). Effects of a contextualized instructional package on the mathematics performance of secondary students with EBD. *Behavioral Disorders*, *34*(3), 136–150.

Mulcahy, C. A., Krezmien, M. P., Leone, P. E., Houchins, D. E., & Baltodano, H. (2008). Lessons learned: Barriers and solutions for conducting reading investigations in juvenile corrections settings. *Reading & Writing Quarterly*, *24*(2), 239–252. doi:10.1080/10573560701808635

Mulcahy, C. A., Krezmien, P. K., & Travers, J. (2015). Improving mathematics performance among secondary students with EBD: A methodological review. *Remedial and Special Education*, *37*(2), 113–128. doi:10.1177/0741932515579275

Mulford, D. R., & Robinson, W. R. (2002). An Inventory for Alternate Conceptions among First-Semester General Chemistry Students. *Journal of Chemical Education*, *79*(6), 739–744. doi:10.1021/ed079p739

Nation, M., Feldman, A., Smith, G. G., Wang, P., & Lou, Y. (2016). *Uncovering High School Students' Knowledge and Beliefs about Climate Change to Inform the Development of Embedded Curriculum Materials*. Paper presented at the Annual Meeting of the Association for Science Teacher Education, Reno, NV.

National Governors Association Center for Best Practices, Council of Chief State School Officers. (2010). *Common Core State Standards*. Washington, DC: Author.

National Research Council. (1996). *National science education standards*. Washington, DC: National Association Press.

National Research Council. (2000). *Inquiry and the National Science Education Standards*. Washington, DC: National Association Press.

National Research Council. (2007). *Rising above the gathering storm: Energizing and employing America for a brighter economic future*. Washington, DC: The National Academies Press.

National Research Council. (2011). *Learning science through computer games and simulations*. Washington, DC: The National Academies Press.

National Research Council. (2012). *A Framework for K-12 Science Education: Practices, Crosscutting Concepts, and Core Ideas*. Washington, DC: National Academies Press.

National Research Council. (2012). *A framework for K-12 science education: Practices, crosscutting concepts, and core ideas*. Washington, DC: The National Academies Press.

National Research Council. (2012). *A framework for K-12 science education: Practices, cross-cutting concepts, and core ideas*. Washington, DC: The National Academies Press.

National Research Council. (2012). *Education for life and work: Developing transferable knowledge and skills in the 21st century*. Washington, DC: The National Academies Press. doi:10.7226/13398

National Research Council. (2013). *New directions in assessing performance of individuals and groups: Workshop summary*. Washington, DC: National Academies Press.

Negroponte, N. (1995). *Being digital*. New York, NY: Alfred A. Knopf.

Nemirovsky, R., Tierney, C., & Wright, T. (1998). Body Motion and Graphing. *Cognition and Instruction, 16*(2), 119–172. doi:10.1207/s1532690xci1602_1

Neulight, N., Kafai, Y. B., Kao, L., Foley, B., & Galas, C. (2007). Childrens participation in a virtual epidemic in the science classroom: Making connections to natural infectious diseases. *Journal of Science Education and Technology, 16*(1), 47–58. doi:10.1007/s10956-006-9029-z

Newman, D., Finney, P. B., Bell, S., Turner, H., Jaciw, A. P., Zacamy, J. L., & Gould, L. F. (2012). Evaluation of the effectiveness of the Alabama Math, Science, and Technology Initiative (AMSTI). *Science, and Technology Initiative (AMSTI).*

NGSS Lead States. (2013). Next Generation Science Standards: For States by States. Author.

NGSS Lead States. (2013). *Next generation science standards: ForsStates, by states*. Washington, DC: The National Academies Press.

NGSS Lead States. (2013a). *Next Generation Science Standards: For States, by States*. Washington, DC: The National Academies Press.

NGSS Lead States. (2013b). Next Generation Science Standards: For States, by States: Vol. 2. *Appendixes*. Washington, DC: The National Academies Press.

Norooz, L., Mauriello, M. L., Jorgensen, A., McNally, B., & Froehlich, J. E. (2015). BodyVis: A New Approach to Body Learning Through Wearable Sensing and Visualization. In *Proceedings of the 33rd Annual ACM Conference on Human Factors in Computing Systems* (pp. 1025–1034). ACM. Retrieved from http://dl.acm.org/citation.cfm?id=2702299

Norris, C. A., & Soloway, E. (2011). Learning and schooling in the age of mobilism. *Educational Technology, 51*(6), 3–12.

Novodvorsky, I. (2006). Shifts in beliefs and thinking of beginning physics teacher. *Journal of Physics Teacher Education On-line, 3*(3), 11–17.

Nuffield Foundation. (2016). *A short history of curriculum development at the Nuffield Foundation.* Retrieved from http://www.nuffieldfoundation.org/short-history-curriculum-development-nuffield-foundation

Nurrenbern, S. C., & Pickering, M. (1987). Concept learning versus problem solving: Is there a difference? *Journal of Chemical Education, 64*(6), 508–510. doi:10.1021/ed064p508

O'Neil, H. F., Chung, G., & Brown, R. (1997). Use of networked simulations as a context to measure team competencies. In H. F. O'Neil Jr., (Ed.), *Workforce readiness: Competencies and assessment* (pp. 411–452). Mahwah, NJ: Lawrence Erlbaum Associates.

O'Neil, H. F. Jr, Chen, H. H., Wainess, R., & Shen, C. Y. (2008). Assessing problem solving in simulation games. In E. L. Baker, J. Dickieson, W. Wulfeck, & H. F. O'Neil (Eds.), *Assessment of problem solving using simulations* (pp. 157–176). Mahwah, NJ: Lawrence Erlbaum Associates.

O'Neil, H. F. Jr, & Chuang, S. H. (2008). Measuring collaborative problem solving in low-stakes tests. In E. L. Baker, J. Dickieson, W. Wulfeck, & H. F. O'Neil (Eds.), *Assessment of problem solving using simulations* (pp. 177–199). Mahwah, NJ: Lawrence Erlbaum Associates.

O'Neil, H. F. Jr, Chuang, S. H., & Baker, E. L. (2010). Computer-based feedback for computer-based collaborative problem solving. In D. Ifenthaler, P. Pirnay-Dummer, & N. M. Seel (Eds.), *Computer-based Diagnostics and Systematic Analysis of Knowledge* (pp. 261–279). New York: Springer-Verlag. doi:10.1007/978-1-4419-5662-0_14

OECD. (2013). *Draft PISA 2015 Science Framework.* Retrieved from http://www.oecd.org/pisa/pisaproducts/pisa2015draftframeworks.htm

OECD. (2013a). *PISA 2015 Science Framework.* OECD Publishing.

OECD. (2013b). *PISA 2015 Collaborative Problem Solving Framework.* OECD Publishing.

OECD. (2014). PISA 2012 Results: What Students Know and Can Do – Student Performance in Mathematics, Reading and Science (Volume I, Revised edition, February 2014). Pisa: OECD Publishing. doi:10.1787/9789264201118-en

Okal, Z. (2015). *Learning about contagion and health in human systems among students with retardation* (Unpublished master's thesis). Faculty of Education, University of Haifa.

Okeke, E. A. C., & Wood-Robinson, C. (1980). A study of Nigerian pupils understanding of selected biological concepts. *Journal of Biological Education, 14*(4), 329–338. doi:10.1080/00219266.1980.9654318

Olson, J. (1981). Teacher influence in the classroom: A context for understanding curriculum translation. *Instructional Science, 10*(3), 259–275. doi:10.1007/BF00139803

Ortiz Nieves, E. L., Barreto, R., & Medina, Z. (2012). JCE classroom activity #111: Redox reactions in three representations. *Journal of Chemical Education, 89*(5), 643–645. doi:10.1021/ed100694m

Pallant, A., & Tinker, R. F. (2004). Reasoning with atomic-scale molecular dynamic models. *Journal of Science Education and Technology, 13*(1), 51–66. doi:10.1023/B:JOST.0000019638.01800.d0

Papert, S. (1980). *Mindstorms: children, computers, and powerful ideas*. New York, NY: Basic Books.

Patton, M. Q. (2015). *Qualitative Research and Evaluation Methods* (4th ed.). Thousand Oaks, CA: SAGE Publications, Inc.

Pea, R. D. (2004). The social and technological dimensions of scaffolding and related theoretical concepts for learning, education, and human activity. *Journal of the Learning Sciences, 13*(3), 423–451. doi:10.1207/s15327809jls1303_6

Pedaste, M., Mäeots, M., Siiman, L. A., de Jong, T., van Riesen, S. A. N., Kamp, E. T., & Tsourlidaki, E. et al. (2015). Phases of inquiry-based learning: Definitions and the inquiry cycle. *Educational Research Review, 14*, 47–61. doi:10.1016/j.edurev.2015.02.003

Pedretti, E., & Little, C. (2008). *From engagement to empowerment*. Toronto, Canada: Pearson Education.

Penuel, W. R., Fishman, B., Yamaguchi, R., & Gallagher, L. P. (2007). What makes professional development effective? Strategies that foster curriculum implementation. *American Educational Research Journal, 44*(4), 921–958. doi:10.3102/0002831207308221

Peppler, K., Danish, J., Zaitlen, B., Glosson, D., Jacobs, A., & Phelps, D. (2010). BeeSim: Leveraging Wearable Computers in Participatory Simulations with Young Children. In *Proceedings of the 9th International Conference on Interaction Design and Children* (pp. 246–249). New York, NY: ACM. doi:10.1145/1810543.1810582

Peters, V. L., & Slotta, J. D. (2010). Scaffolding knowledge communities in the classroom: New opportunities in the Web 2.0 era. In M. J. Jacobson & P. Reimann (Eds.), *Designs for learning environments of the future: International perspectives from the learning sciences* (pp. 205–232). Secaucus, NJ: Springer. doi:10.1007/978-0-387-88279-6_8

Petrosino, A. J., Lehrer, R., & Schauble, L. (2003). Structuring Error and Experimental Variation as Distribution in the Fourth Grade. *Mathematical Thinking and Learning, 5*(2-3), 131–156. doi:10.1080/10986065.2003.9679997

Plass, J. L., Homer, B. D., & Hayward, E. O. (2009). Design factors for educationally effective animations and simulations. *Journal of Computing in Higher Education*, *21*(1), 31–61. doi:10.1007/s12528-009-9011-x

Pollock, E., Chandler, P., & Sweller, J. (2002). Assimilating complex information. *Learning and Instruction*, *12*(1), 61–86. doi:10.1016/S0959-4752(01)00016-0

Postman, N. (2009). *Teaching as a subversive activity*. Delta.

President's Council of Advisors on Science and Technology. (2010). *Prepare and inspire: K-12 educatioin in science, technology, engineering, and math (STEM) for America's future*. Retrieved from http://www.whitehouse.gov/administration/eop/ostp/pcast/docsreports

Project Tomorrow. (2010). *Creating our future: Students speak up about their vision for 21st century learning*. Irvine, CA: Project Tomorrow. Retrieved from http://www.tomorrow.org/speakup/pdfs/SU09NationalFindingsStudents&Parents.pdf

Pruneau, D., Liboiron, L., Vrain, E., Gravel, H., Bourque, W., & Langis, J. (2001). People's idea about climate change: A source of inspiration for the creation of educational programs. *Canadian Journal of Environmental Education*, *6*, 121–138.

Quinn, M. M., Rutherford, R. B., Leone, P. E., Osher, D. M., & Poirier, J. M. (2005). Youth with disabilities in juvenile corrections: A national survey. *Exceptional Children*, *71*(3), 339–345. doi:10.1177/001440290507100308

Ravitch, D. (2011). *The death and life of the great American school system: How testing and choice are undermining education* [Kindle version]. Retrieved from Amazon.com

Reiber, L. P. (2005). Multimedia learning in games, simulations, and microworlds. In R. E. Mayer (Ed.), *The Cambridge Handbook of Multimedia Learning* (pp. 549–567). Cambridge, UK: Cambridge University Press. doi:10.1017/CBO9780511816819.034

Reigeluth, C. M. (1994). Introduction: The imperative for systemic change. In C. M. Reigeluth & R. J. Garfinkle (Eds.), *Systemic change in education* (pp. 3–14). Englewood Cliffs, NJ: Educational Technology Publications, Inc.

Reigeluth, C. M., & Schwartz, E. (1989). An instructional theory for the design of computer-based simulations. *Journal of Computer-Based Instruction*, *16*(1), 1–10.

Reiser, B. J. (2004). Scaffolding complex learning: The mechanisms of structuring and problematizing student work. *Journal of the Learning Sciences*, *13*(3), 273–304. doi:10.1207/s15327809jls1303_2

Reiser, B. J., Tabak, I., Sandoval, W. A., Smith, B. K., Steinmuller, F., & Leone, A. J. (2001). BGuILE: Strategic and conceptual scaffolds for scientific inquiry in biology classrooms. In S. M. Carver & D. Klahr (Eds.), *Cognition and instruction: Twenty-five years of progress* (pp. 263–305). Mahwah, NJ: Lawrence Erlbaum Associates, Inc.

Reisslein, J., Atkinson, R. K., Seeling, P., & Reisslein, M. (2006). Encountering the expertise reversal effect with a computer-based environment on electrical circuit analysis. *Learning and Instruction, 16*(2), 92–103. doi:10.1016/j.learninstruc.2006.02.008

Reiter, S. (2003). Life skills versus the skill of living: The Circle of Internalization Method for the enhancement of the "Skill of Living". *Proceedings for the International Conference on Inclusive Education ICIE 2003 Inclusive Education: A Framework for Reform.*

Resnick, M., Berg, R., & Eisenberg, M. (2000). Beyond Black Boxes: Bringing Transparency and Aesthetics Back to Scientific Investigation. *Journal of the Learning Sciences, 9*(1), 7–30. doi:10.1207/s15327809jls0901_3

Richmand, E. (2014). New survey: Teachers say their voices aren't being heard. *Education Writers Association.* Retrieved June 16th 2015 from http://www.ewa.org/blog-educated-reporter/new-survey-teachers-say-their-voices-arent-being-heard

Rimor, R., Rosen, Y., & Naser, K. (2010). Complexity of social interactions in collaborative learning: The case of online database environment. *Interdisciplinary Journal of E-Learning and Learning Objects, 6*, 355–365.

Roberts, R., & Gott, R. (2003). Assessment of biology investigations. *Journal of Biological Education, 37*(3), 114–121. doi:10.1080/00219266.2003.9655865

Rodríguez-Triana, M. J., Holzer, A., Vozniuk, A., & Gillet, D. (2015). Orchestrating inquiry-based learning spaces: An analysis of teacher needs. In F. W. B. Li, R. Klamma, M. Laanpere, J. Zhang, B. F. Manjón, & R. W. H. Lau (Eds.), *Advances in web-based learning -- ICWL 2015, 14th International Conference, Guangzhou, China, November 5-8, 2015, Proceedings* (pp. 131-142). Cham, Switzerland: Springer. doi:10.1007/978-3-319-25515-6_12

Roschelle, J., Pea, R., Hoadley, C., Gordin, D., & Means, B. (2000). Changing how and what children learn in school with computer-based technologies. *The Future of Children, 10*(2), 76–101. doi:10.2307/1602690 PMID:11255710

Roschelle, J., & Teasley, S. D. (1995). The construction of shared knowledge in collaborative problem-solving. In C. E. O'Malley (Ed.), *Computer-supported collaborative learning* (pp. 69–97). Berlin: Springer-Verlag. doi:10.1007/978-3-642-85098-1_5

Rose, D. H., & Meyer, A. (2002). *Teaching every student in the digital age: Universal design for learning*. Alexandria, VA: Association for Supervision and Curriculum Development.

Roseman, J. E., Linn, M. C., & Koppal, M. (2008). Characterizing curriculum coherence. In Y. Kali, M. C. Linn, & J. E. Roseman (Eds.), *Designing coherent science education: Implications for curriculum, instruction, and policy* (pp. 13–36). New York: Teachers College Press.

Rosen, Y., & Mosharraf, M. (2015). Computer agent technologies in collaborative assessments. In Y. Rosen, S. Ferrara, & M. Mosharraf (Eds.), Handbook of Research on Technology Tools for Real-Life Skill Development (pp. 319-343). Hershey, PA: Information Science, IGI Global.

Rosen, Y., & Rimor, R. (2012). Teaching and assessing problem solving in online collaborative environment. In R. Hartshorne, T. Heafner, & T. Petty (Eds.), Teacher education programs and online learning tools: Innovations in teacher preparation (pp. 82-97). Hershey, PA: Information Science Reference, IGI Global.

Rosenbaum, E., Klopfer, E., & Perry, J. (2007). On location learning: Authentic applied science with networked augmented realities. *Journal of Science Education and Technology*, *16*(1), 31–45. doi:10.1007/s10956-006-9036-0

Rosen, Y. (2014). Comparability of conflict opportunities in human-to-human and human-to-agent online collaborative problem solving. *Technology, Knowledge and Learning*, *19*(1-2), 147–174. doi:10.1007/s10758-014-9229-1

Rosen, Y., & Foltz, P. (2014). Assessing collaborative problem solving through automated technologies. *Research and Practice in Technology Enhanced Learning*, *9*(3), 389–410.

Rosen, Y., & Rimor, R. (2009). Using collaborative database to enhance students' knowledge construction. *Interdisciplinary Journal of E-Learning and Learning Objects*, *5*, 187–195.

Roseth, C. J., Fang, F., Johnson, D. W., & Johnson, R. T. (2006, April). *Effects of cooperative learning on middle school students: A meta-analysis*. Paper presented at the Annual Meeting of the American Educational Research Association, San Francisco, CA.

Ryu, S., & Sandoval, W. A. (2012). Improvements to elementary childrens epistemic understanding from sustained argumentation. *Science Education*, *96*(3), 488–526. doi:10.1002/sce.21006

Saad, R., & BouJaoude, S. (2012). The relationship between teachers' knowledge and beliefs about science and inquiry and their classroom practices. *Eurasia Journal of Mathematics, Science & Technology Education*, *8*(2), 113–128.

Sandall, B. K. (2016). Wearable Technology and Schools: Where are We and Where Do We Go From Here? *Journal of Curriculum, Teaching, Learning, and Leadership in Education*, *1*(1). Retrieved from http://digitalcommons.unomaha.edu/ctlle/vol1/iss1/9

Schnotz, W., & Kurschner, C. (2008). External and internal representations in the acquisition and use of knowledge: Visualization effects on mental model construction. *Instructional Science*, *36*(3), 175–190. doi:10.1007/s11251-007-9029-2

Scholastic, Inc. (2016). America's teachers on teaching in an era of change. *Primary Sources* (3rd ed.). Retrieved June 16th 2016 from http://www.scholastic.com/primarysources/

Schwarz, C., & Passmore, C. (2012, July 14). *Preparing for NGSS: Developing and Using Models. NSTA Web Seminars*. Retrieved from http://learningcenter.nsta.org/products/symposia_seminars/ngss/files/PreparingforNGSS--DevelopingandUsingModels_9-25-2012.pdf

Seatter, C. S. (2003). Constructivist science teaching: Intellectual and strategic teaching acts. *Interchange*, *34*(1), 63–87. doi:10.1023/A:1024506417791

Seufert, T., & Brünken, R. (2006). Cognitive load and the format of instructional aids for coherence formation. *Applied Cognitive Psychology, 20*(3), 321–331. doi:10.1002/acp.1248

Shaffer, D. W., Hatfield, D., Svarovsky, G. N., Nash, P., Nulty, A., Bagley, E., & Mislevy, R. et al. (2009). Epistemic Network Analysis: A prototype for 21st century assessment of learning. *International Journal of Learning and Media, 1*(2), 1–21. doi:10.1162/ijlm.2009.0013

Shear, L., Bell, P., & Linn, M. C. (2004). Partnership models: The case of the deformed frogs. In M. C. Linn, E. A. Davis, & P. Bell (Eds.), *Internet environments for science education* (pp. 289–311). Mahwah, NJ: Erlbaum.

Sherin, M. G., Russ, R. S., Sherin, B. L., & Colestock, A. A. (2011). Accessing mathematics teachers' in-the-moment noticing. In M. G. Sherin, V. R. Jacobs, & R. A. Philipp (Eds.), *Mathematics teacher noticing: Seeing through teachers' eyes*. Routledge.

Shipman, H. L. (2006). Inquiry learning in college classrooms: For the times, they are, a changing. In L. B. Flick & N. G. Lederman (Eds.), Scientific inquiry and nature of science: Implications for teaching, learning, and teacher education (Vol. 25, pp. 357–389). Dordrecht, The Netherlands: Springer.

Simons, K. D., & Klein, J. D. (2007). The impact of scaffolding and student achievement levels in a problem-based learning environment. *Instructional Science, 35*(1), 41–72. doi:10.1007/s11251-006-9002-5

Slack, S. J., & Stewart, J. (1990). High school students problem-solving performance on realistic genetics problems. *Journal of Research in Science Teaching, 27*(1), 55–67. doi:10.1002/tea.3660270106

Slotta, J. D. (2004). The Web-based inquiry Science environment (WISE): Scaffolding knowledge integration in the science classroom. In M. C. Linn, P. Bell, & E. Davis (Eds.), *Internet environments for science education* (pp. 203–232). Mahwah, NJ: Lawrence Erlbaum & Associates.

Slotta, J. D., & Linn, M. C. (2009). *WISE science: Web-based inquiry in the classroom*. New York: Teachers College Press.

Smallen, D., & Leach, K. (1999). Making a NICER Transition to the Millennium: Five Keys to Successful Collaboration. *Educational Review, 34*(4).

Smith, G. A. (2007). Place-based education: Breaking through the constraining regularities of public school. *Environmental Education Research, 13*(2), 189–207. doi:10.1080/13504620701285180

Smith, G. G. (2012). Computer-Game Play as Imaginary Stage for Reading: Implicit Spatial Effects of Computer Games Embedded in Hard Copy Books. *Journal of Research in Reading, 35*(1), 1–19.

Smith, G. G., Li, M., Drobisz, J., Park, H.-R., Kim, D., & Smith, S. D. (2013). Play games or study? Computer games in eBooks to learn English vocabulary. *Computers & Education, 69*, 274–286. doi:10.1016/j.compedu.2013.07.015

Smith, T. M., Desimone, L. M., Zeidner, T., Dunn, A. C., Bhatt, M., & Rumyantseva, N. (2007). Inquiry-oriented instruction in science: Who teachers that way? *Educational Evaluation and Policy Analysis*, *9*(29), 169–199. doi:10.3102/0162373707306025

Solomonidou, C. (2009). Constructivist design and evaluation of interactive educational software: A research based approach and examples. *The Journal for Open and Distance Education and Educational Technology*, *5*(1), 6–24.

Songer, N. B. (2007). Digital resources versus cognitive tools: A discussion of learning science with technology. In S. Abell & N. Lederman (Eds.), *Handbook of research on science education* (pp. 471–491). Mahwah, NJ: Erlbaum.

Songer, N. B., Lee, H., & Kam, R. (2002). Technology-Rich Inquiry Science in Urban Classrooms: What are the barriers to inquiry pedagogy? *Journal of Research in Science Teaching*, *39*(2), 128–150. doi:10.1002/tea.10013

Spires, H. A., Lee, J. K., Turner, K. A., & Johnson, J. (2008). Having our say: Middle grade student perspectives on school, technologies, and academic engagement. *Journal of Research on Technology in Education*, *40*(4), 497–515. doi:10.1080/15391523.2008.10782518

Squire, K. (2005). Changing the game: What happens when video games enter the classroom? *Innovate: Journal of Online Education, 1*(6). Retrieved from http://www.innovateonline.info/index.php?view=article&id=82

Squire, K. (2010). From information to experience: Place-based augmented reality games as a model for learning in a globally networked society. *Teachers College Record*, *112*(10), 2565–2602.

Squire, K., & Jan, M. (2007). Mad City Mystery: Developing scientific argumentation skills with a place-based augmented reality game on handheld computers. *Journal of Science Education and Technology*, *16*(1), 5–29. doi:10.1007/s10956-006-9037-z

Squire, K., & Klopfer, E. (2007). Augmented reality simulations on handheld computers. *Journal of the Learning Sciences*, *16*(3), 371–413. doi:10.1080/10508400701413435

Stake, R., & Easley, J. (1978). *Case studies in science education* (Vols. 1-2). Center for Instructional Research and Curriculum Evaluation, University of Illinois at Urbana-Champaign.

Stark, L. A., & Pompei, K. (2010). Making genetics easy to understand. *Science*, *327*(5965), 538–539. doi:10.1126/science.1183029 PMID:20110495

Steinkuehler, C., & Duncan, S. (2008). Scientific habits of mind in virtual worlds. *Journal of Science Education and Technology*, *17*(6), 530–543. doi:10.1007/s10956-008-9120-8

Steinkuehler, C., & Squire, K. (2014). Video games and learning. In R. K. Sawyer (Ed.), *The Cambridge handbook of the learning sciences* (2nd ed.; pp. 377–396). New York, NY: Cambridge University Press. doi:10.1017/CBO9781139519526.023

Stieff, M., & Wilensky, U. (2003). Connected Chemistry-Incorporating interactive simulations in chemistry classroom. *Journal of Science Education and Technology*, *12*(3), 285–302. doi:10.1023/A:1025085023936

Stieff, W. (2011). When is a molecule three dimensional? A task-specific role for imagistic reasoning in advanced chemistry. *Science Education*, *95*(2), 310–336. doi:10.1002/sce.20427

Stroup, W. M., & Wilensky, U. (2014). On the embedded complementarity of agent-based and aggregate reasoning in students developing understanding of dynamic systems. *Technology. Knowledge and Learning*, *19*(1-2), 19–52. doi:10.1007/s10758-014-9218-4

Subramaniam, K. (2012). How WebQuests Can Enhance Science Learning Principles in the Classroom. *The Clearing House: A Journal of Educational Strategies, Issues and Ideas*, *85*(6), 237–242. doi:10.1080/00098655.2012.698323

Suits, J. P., & Sanger, M. J. (2013). Dynamic visualization in chemistry courses. In J. P. Suits, & M. J. Sanger (Eds.), *Pedagogic Roles of Animations and Simulations in Chemistry Courses. ACS Symposium Series*, (*vol. 1142*, pp, 1-13). Washington, DC: American Chemical Society. doi:10.1021/bk-2013-1142.ch001

Suits, J. P., & Diack, M. (2002). Instructional design of scientific simulations and modeling software to support student construction of perceptual to conceptual bridges. *World Conference on Educational Multimedia Hypermedia and Telecommunication*, (vol. 1, pp. 1904-1909). Denver, CO: AACE.

Sweller, J., van Merrienboer, J., & Paas, F. (1998). Cognitive architecture and instructional design. *Educational Psychology Review*, *10*(3), 251–296. doi:10.1023/A:1022193728205

Takacs, J., Pollock, C. L., Guenther, J. R., Bahar, M., Napier, C., & Hunt, M. A. (2014). Validation of the Fitbit One activity monitor device during treadmill walking. *Journal of Science and Medicine in Sport / Sports Medicine Australia, 17*(5), 496–500. https://doi.org/10.1016/j.jsams.2013.10.241

Tamir, P., & Jungwirth, E. (1975). Students growth as a result of studying BSCS biology for several years. *Journal of Research in Science Teaching*, *12*(3), 263–279. doi:10.1002/tea.3660120309

Tannis, L. 2014. Educating incarcerated youth: Exploring the impact of relationships, expectations, resources and accountability. Palgrave Macmillan UK. doi:10.1057/9781137451026

Tapscott, D., & Williams, A. D. (2006). *Wikinomics: How mass collaboration changes everything*. New York, NY: Penguin Group.

Taylor, K. H., & Hall, R. (2013). Counter-mapping the neighborhood on bicycles: Mobilizing youth to reimagine the city. *Technology, Knowledge and Learning, 18*(1–2), 1–29. https://doi.org/10.1007/s10758-013-9201-5

Terwogt, M. M., Stegge, H., & Rieffe, C. (2003). Childrens understanding of inherited resemblance: The case of two parents. *International Journal of Behavioral Development*, *27*(4), 366–374. doi:10.1080/01650250344000037

Texas Education Agency. (2010). *Texas essential knowledge and skills*. Retrieved July 9, 2014, from http://tea.state.tx.us

Thayne, J., & Lee, V. R. (2016). *Making statistics matter: Connecting statistical inquiry to the life of the students.* Presented at the Association for Educational Communications and Technology Conference, Indianapolis, IN.

Therrien, W. J., Taylor, J. C., Watt, S., & Kaldenberg, E. R. (2013). Science instruction for students with emotional and behavioral disorders. *Remedial and Special Education*, *35*(1), 15–27. doi:10.1177/0741932513503557

Thomas, D., & Brown, J. S. (2011). *A new culture of learning: Cultivating the imagination for a world of constant change*. Lexington, KY: CreateSpace Independent Publishing Platform.

Tinker, R., & Wilensky, U. (2007). *NetLogo Climate Change model*. Retrieved from http://ccl.northwestern.edu/netlogo/models/ClimateChange

Tissenbaum, M., Lui, M., & Slotta, J. D. (2012). Co-Designing Collaborative Smart Classroom Curriculum for Secondary School Science. *Journal of Universal Computer Science*, *18*(3), 327–352.

Tsirgi, J. E. (1980). Sensible reasoning: A hypothesis about hypotheses. *Child Development*, *51*(1), 1–10. doi:10.2307/1129583

Tsui, C., & Treagust, D. F. (2007). Understanding genetics: Analysis of secondary students conceptual Status. *Journal of Research in Science Teaching*, *44*(2), 205–235. doi:10.1002/tea.20116

Turner, V., Gantz, J., Reinsel, D., & Minton, S. (2014). *The Digital Universe of Opportunities: Rich Data and the Increasing Value of the Internet of Things*. Retrieved June 15th, 2016 from http://www.emc.com/leadership/digital-universe/index.htm

University Corporation for Atmospheric Research (UCAR). (2015). *How much has the global temperature risen in the last 100 years?* Boulder, CO: University Corporation for Atmospheric Research.

van Aalst, J. (2000). An introduction to physics education research. *Canadian Journal of Physics*, *78*(1), 57–71. doi:10.1139/p00-005

van Aalst, J. (2005). Critical details and physics teaching. *Canadian Journal of Science. Mathematics and Technology Education*, *5*(3), 417–422. doi:10.1080/14926150509556670

van Joolingen, W. R. (1999). Cognitive tools for discovery learning. *International Journal of Artificial Intelligence in Education*, *10*, 385–397.

van Joolingen, W. R., & de Jong, T. (1991). Supporting hypothesis generation by learners exploring an interactive computer simulation. *Instructional Science*, *20*(5-6), 389–404. doi:10.1007/BF00116355 van Joolingen, W. R., & Zacharia, Z. C. (2009). Developments in inquiry learning. In N. Balacheff, S. Ludvigsen, T. de Jong, & S. Barnes (Eds.), *Technology-enhanced learning* (pp. 21–37). Dordrecht, The Netherlands: Springer. doi:10.1007/978-1-4020-9827-7_2

Varma, K., & Linn, M. C. (2012). Using interactive technology to support students understanding of the greenhouse effect and global warming. *Journal of Science Education and Technology, 21*(4), 453–464. doi:10.1007/s10956-011-9337-9

Veermans, K., van Joolingen, W. R., & de Jong, T. (2006). Use of heuristics to facilitate scientific discovery learning in a simulation learning environment in a physics domain. *International Journal of Science Education, 28*(4), 341–361. doi:10.1080/09500690500277615

Venville, G., Gribble, S. J., & Donovan, J. (2005). An exploration of young childrens understandings of genetics concepts from ontological and epistemological perspectives. *Science Education, 89*(4), 614–633. doi:10.1002/sce.20061

Verlet, L. (1967). Computer experiments on classical fluids I. Thermodynamical properties of Lennard-Jones molecules. *Physical Review, 159*(1), 98–102. doi:10.1103/PhysRev.159.98

Vicsek, T. (2002). Complexity: The bigger picture. *Nature, 418*(6894), 131–131. doi:10.1038/418131a PMID:12110869

Vogel, J. J., Vogel, D. S., Cannon-Bowers, J. A. N., Bowers, C. A., Muse, K., & Wright, M. (2006). Computer gaming and interactive simulations for learning: A meta-analysis. *Journal of Educational Computing Research, 34*(3), 229–243. doi:10.2190/FLHV-K4WA-WPVQ-H0YM

Wagler, M., & Mathews, J. (2012). Up River: Place, ethnography, and design in the St. Louis River estuary. In S. Dikkers, J. Martin, & B. Coulter (Eds.), *Mobile media learning: Amazing uses of mobile devices for teaching and learning* (pp. 23–40). Pittsburg, PA: ETC Press.

Webb, N. M. (1995). Group collaboration in assessment: Multiple objectives, processes, and outcomes. *Educational Evaluation and Policy Analysis, 17*(2), 239–261. doi:10.3102/01623737017002239

Webb, N. M., Nemer, K. M., Chizhik, A. W., & Sugrue, B. (1998). Equity issues in collaborative group assessment: Group composition and performance. *American Educational Research Journal, 35*(4), 607–651. doi:10.3102/00028312035004607

Webb, N. M., Nemer, K. M., & Zuniga, S. (2002). Short circuits or superconductors? Effects of group composition on high-achieving students science performance. *American Educational Research Journal, 39*(4), 943–989. doi:10.3102/00028312039004943

Weintrop, D., Beheshti, E., Horn, M., Orton, K., Jona, K., Trouille, L., & Wilensky, U. (2016). Defining computational thinking for mathematics and science classrooms. *Journal of Science Education and Technology, 25*(1), 127–147. doi:10.1007/s10956-015-9581-5

Wieman, C. E., Adams, W. K., & Perkins, K. K. (2008). PhET: Simulations that enhance learning. *Science, 322*(5902), 682–683. doi:10.1126/science.1161948 PMID:18974334

Wildman, J. L., Shuffler, M. L., Lazzara, E. H., Fiore, S. M., Burke, C. S., Salas, E., & Garven, S. (2012). Trust development in swift starting action teams: A multilevel framework. *Group & Organization Management, 37*(2), 138–170. doi:10.1177/1059601111434202

Wilensky, U. (1999). *NetLogo*. Retrieved from http://ccl.northwestern.edu/netlogo/

Wilensky, U., & Stroup, W. (1999). Learning through participatory simulations: Network-based design for systems learning in classrooms. *Proceedings of Computer Supported Collaborative Learning (CSCL'99)*.

Wilkerson-Jerde, M. H., Gravel, B. E., & Macrander, C. A. (2015). Exploring shifts in middle school learners modeling activity while generating drawings, animations, and simulations of molecular diffusion. *Journal of Science Education and Technology, 24*(2-3), 396–415. doi:10.1007/s10956-014-9497-5

Williams, M., Linn, M. C., Ammon, P., & Gearhart, M. (2004). Learning to teach inquiry science in a technology-based environment: A case study. *Journal of Science Education and Technology, 13*(2), 189–206. doi:10.1023/B:JOST.0000031258.17257.48

Willingham, D. T. (2010). *Why don't students like school: A cognitive scientist answers questions about how the mind works and what it means for the classroom*. San Francisco, CA: Jossey-Bass. doi:10.1002/9781118269527

Wolfer, A. J. (2000). *Introductory College Chemistry Students' Understanding of Stoichiometry: Connections between Conceptual and Computational Understandings and Instruction* (Unpublished doctoral dissertation). Oregon State University, Corvallis, OR.

Wolfer, A., & Lederman, N. (2000). *Introducing college chemistry students understanding of stoichiometry: Connections between conceptual and computational understandings and instructions*. Paper presented at the meeting of the National Association for Research in Science Teaching, New Orleans, LA.

Wood-Robinson, C. (1994). Young peoples ideas about inheritance and evolution. *Studies in Science Education, 24*(1), 29–47. doi:10.1080/03057269408560038

Yaron, D., Karabinos, M., Lange, D., Greeno, J. G., & Leinhardt, G. (2010). The ChemCollective: Virtual labs for introductory chemistry courses. *Science, 328*(5978), 584–585. doi:10.1126/science.1182435 PMID:20431007

Yarroch, W. L. (1985). Student understanding of chemical equation balancing. *Journal of Research in Science Teaching, 22*(5), 449–459. doi:10.1002/tea.3660220507

Yerrick, R. (2010). *How notebook computers, digital media, and probeware can transform science learning in the classroom.* Available from http://www.apple.com/education/ resources/

Young, M. F., Slota, S., Cutter, A. B., Jalette, G., Mullin, G., Lai, B., & Yukhymenko, M. et al. (2012). Our princess is in another castle: A review of trends in serious gaming for education. *Review of Educational Research, 82*(1), 61–89. doi:10.3102/0034654312436980

Zacharia, Z. C., Manoli, C., Xenofontos, N., de Jong, T., Pedaste, M., van Riesen, S. A., & Tsourlidaki, E. (2015). Identifying potential types of guidance for supporting student inquiry when using virtual and remote labs in science: A literature review. *Educational Technology Research and Development*, *63*(2), 257–302. doi:10.1007/s11423-015-9370-0

Zervas, P. (Ed.). (2013). *The Go-Lab inventory and integration of online labs – labs offered by large scientific organisations. D2.1. Go-Lab Project (Global Online Science Labs for Inquiry Learning at School).* Retrieved from http://www.go-lab-project.eu/sites/default/files/files/deliverable/file/Go-Lab-D2.1.pdf

Zhang, J. (1998). A distributed representation approach to group problem solving. *Journal of the American Society for Information Science*, *49*(9), 801–809. doi:10.1002/(SICI)1097-4571(199807)49:9<801::AID-ASI5>3.0.CO;2-Q

Zoubi, H. (2015). *Studying complex systems by participation in a collaborative simulation study unit on the topic of global warming* (Unpublished master's thesis). Faculty of Education, University of Haifa.

About the Contributors

Ilya Levin received his Ph.D. degree in Computer Engineering in 1987 from the Institute of Computer Technologies, Latvian Academy of Science. From 1987 he was the Head of the Computer Science Department in the Leningrad Institute of New Technologies (in the former USSR). He moved to Israel 1990. Between 1993 and 1997, Ilya Levin was the Head of the Computer Systems Department in Holon Institute of Technology. In 1997, he worked as a Research fellow in the Computer Science Department of University of Massachusetts. During four years between 2003-2006, Dr. Ilya Levin was an Associate Professor of the School of Engineering in the Bar Ilan University. Presently, Prof. Ilya Levin is a Full Professor in the School of Education of Tel Aviv University. His recent research interests include: Computer Design, Cultural Studies of Information Society. Ilya Levin is the author of around 160 research papers both in Computer Engineering and in Humanities.

Dina Tsybulsky received her PhD degree in Science Education in 2014 from the Hebrew University, Jerusalem. Her research interests include: inquiry-based learning biology and chemistry, nature of science and science education in digital age. Currently, Dina is a lecturer and a research fellow of the School of Education of Tel Aviv University. Her postdoctoral research deals with transformations of science teachers' worldviews and their implementations in teachers' "digitalized" educational practices.

* * *

Sergei Abramovich, PhD (mathematics), is Professor in the School of Education and Professional Studies at State University of New York at Potsdam. He is the author/editor of seven books and numerous articles on the use of computers in mathematics education. His service to educational community includes membership on editorial boards of nine professional journals published in the United States and elsewhere.

Metin Besalti is a PhD student in the Instructional Technology program at University of South Florida. He is specifically interested in game-based learning environment related to science learning.

Christina Bosch is a doctoral student in special education at the University of Massachusetts, Amherst. She is an ARISE fellow, and a research assistant on Project RAISE, an NSF funded project to support inquiry science in juvenile corrections settings.

Denise M. Bressler is a Research Associate with the Center for Mathematics, Science, and Computer Education at Rutgers, The State University of New Jersey. Denise is passionate about the potential for learning with mobile technologies. Her research revolves around mobile game-based learning and mobile augmented reality (AR) experiences that promote flow and science learning. Denise created a series of mobile AR games called School Scene Investigators which can be played with WiFi-enabled iOS devices and quick response (QR) codes posted in a school environment. Denise received her Ph.D. in Learning Sciences and Technology from Lehigh University.

Ryan F Cain is a doctoral student in the Department of Instructional Technology and Learning Sciences. Prior to moving to Utah State University, Ryan was an elementary school science teacher in New York City and a photographer. Ryan is largely interested in the maker movement and its implications for elementary science classrooms, and he has also been involved in leading professional development and research on maker education.

Joel R. Drake is a doctoral candidate in the Department of Instructional Technology and Learning Sciences at Utah State University. An avid cyclist and former professional technical writer, Joel holds degrees in Physics, English, and Instructional Technology. In addition to supporting curriculum development projects for elementary classrooms, Joel is studying informal physics knowledge among amateur road cyclists.

Allan Feldman is a professor of science education in the Teaching and Learning Department of the College of Education at the University of South Florida. His scholarship focuses on science teacher education, and in particular how practicing science teachers learn from their practice in a variety of subjects including physics, environmental education, and education for sustainability in formal and informal settings. He is co-editor of the book "Educating Science Teachers for Sustainability", which was published in 2015 by Springer. He has written a number of papers on action research, and self-study as well as on science teacher education. Much of his scholarship has been on teacher learning through action research and self-study, and through participation in scientific research groups. He is currently co-editor of Educational Action Research, and has published theoretical and empirical studies of action research and self-study.

Tanya Gupta is an Assistant Professor of Chemistry at South Dakota State University (SDSU). Her research interests focus on development and integration of simulations and games for teaching and learning; guided-inquiry based pedagogy and curriculum development for inquiry-based interdisciplinary science instruction, and Eye-tracking methodology to study student problem solving behavior. She maintains an active research group that includes undergraduate and graduate students working on these projects in chemical education and has made several research presentations at the ACS National Meetings, the Biennial Conference in Chemical Education, and Gordon Research Conference in Chemistry Education. Dr. Gupta is an active member of the American Chemical Society's Committee on Computers in Chemical Education. In this role Dr. Gupta serves ACS-CCCE by organizing & presiding symposia and conferences centered on use of technology in Chemical Education, and as an internal editor for ConfChem papers & presentations. Dr. Gupta serves as a panelist for the National Science Foundation. Her research career involves experience of working on several NSF and HHMI funded projects in chemical education and interdisciplinary areas. She also has an extensive experience in science teacher professional development. She has extensive teaching experience that includes various chemistry courses for undergraduate, graduate students in STEM and non-STEM majors, pre-service, and in-service teachers.

Tracey Hall conducts research on instructional interventions in reading, writing, literacy in the content areas, and assessment. Her work includes the design and evaluation of instructional approaches and assessment tools.

Merle Heidemann is a Senior Specialist, Emerita College of Natural Science, Michigan State University.

Tasos Hovardas has served as a Senior Teaching Fellow at the Department of Education, University of Cyprus. His research interests involve a primary focus on environmental education and science education, with a concentration on designing and assessing computer-supported inquiry learning.

Michael Krezmien is an associate professor and the director of the Center for Youth Engagement.

Victor R. Lee is an associate professor of Instructional Technology and Learning Sciences at Utah State University. His academic research focuses on new uses for wearable technologies to support learning and research, the Quantified Self movement, and the Maker Movements as they apply to formal and informal educational settings.

Sharona T. Levy is a faculty member at the University of Haifa, part of two divisions: Preschool Education and Development and Technologies in Education at the Faculty of Education. Her prime interest is in facilitating and studying people's learning about complex systems. To this goal, she work with a wide span of age-groups, conducting research into people's reasoning about systems they encounter in everyday life and about systems which they construct and explore in the domain of science and technology; and develop computer-based and physical learning environments. She also focus on science and technology education for young children. Sharona T. Levy received her PhD from Tel-Aviv University in 2002, at the Knowledge Technology Lab and was a postdoctoral fellow at the Center for Connected Learning and Computer-based Modeling.

Anna Lewis is an Assistant Professor in Science Education. She teaches both pre-service and in-service K-12 formal and informal educators. She conducts and evaluates professional development programs for both state and nationally funded grants.

Katarin MacLeod, Ph.D., is an associate professor at St. Francis Xavier University in Antigonish, Nova Scotia, Canada. Her background includes graduate work in experimental superconductivity and science (physics) education. She currently teaches a range of science education courses from undergraduate to graduate level and her research includes physical science education from K-20.

George Matsumoto is the senior education and research specialist at the Monterey Bay Aquarium Research Institute (MBARI). In addition to conducting research, he coordinates educational outreach. He and his team developed the EARTH professional development summer program and have been successfully offering this program each summer since 2006.

Akash Mehta is materials chemist by background, he has his Ph.D. in Inorganic Materials research from Max Plank Institute of Chemical Physics for Solids, Germany. He pursued his first post doctoral research at Ames Laboratory, USA in materials chemistry with Late Dr. John D. Corbett. His second post-doctoral work was in the area of chemical education research with Dr. Thomas Greenbowe at Iowa State University. Dr. Mehta is passionate about teaching and learning and his research efforts focus in the area of chemical education research particularly for various organic chemistry concepts and laboratories. After his post-doc Dr. Mehta taught chemistry at Des Moines Area Community. college. University of Northern Iowa and at South Dakota State University. Now he is an independent researcher and consultant for various science/ chemical education research projects.

Molly Nation is a PhD student in the Science Education program at University of South Florida. She is specifically interested in how teachers' beliefs influence their practice and the learning of their students.

Martina Nieswandt is an associate professor in science education and the associate dean for research and engagement at the College of Education at the University of Massachusetts, Amherst.

Wardell Powell is a postdoctoral fellow on the Project RAISE team, an NSF funded project designed to support inquiry science in juvenile corrections settings.

Yigal Rosen is a Senior Research Scientist in The Vice Provost for Advances in Learning (VPAL) at Harvard University. He leads the Harvard VPAL Research Team that advances big data research initiatives in higher education, such as the use of data from HarvardX Massively Open Online Courses and Harvard's learning platform Canvas. The interdisciplinary team is comprised of researchers from computer science, statistics, data science and educational psychology fields. Prior to joining Harvard he was a Senior Research Scientist at Pearson, leading research and development of innovative learning and assessment technologies of higher-order skills, including PISA, NAEP and PARCC programs. He obtained his Ph.D. degree in Education from the University of Haifa, being the youngest recipient of a doctoral degree in the University. Yigal was a post-doctoral fellow at Harvard University Graduate School of Education and at Tel Aviv University School of Education. Yigal is a leading editor of the recently published book, "Handbook of Research on Technology Tools for Real-World Skill Development".

Glenn Smith combines web-based eBooks with computer games for stealth assessment of comprehension and content learning. Glenn is an Associate Professor in Educational & Psychological Studies, College of Education, at University of South Florida. Glenn researches: (a) computer games in eBooks, (b) online learning, and (c) improving spatial skills with computer games. Glenn holds a Ph.D. in Educational Media & Computers, and masters and bachelor degrees in Computer Science. He worked as a computer programmer: a) in Hollywood, California developing television commercials and opening segments for television shows, including Steven Spielberg's Amazing Stories, b) for Boeing Computer Services and c) for Microsoft Corporation.

Michelle Williams is CEO of Williams Learning Solutions, Inc.

Nikoletta Xenofontos holds a Bachelor's degree in Education and a Master's degree in Science Education. She is currently a PhD candidate in the Department of Educational Sciences at the University of Cyprus, in the Learning in Natural Sciences program. Her research interests focus on enhancing students' conceptual understanding and acquisition of inquiry skills through the use of computer supported inquiry learning.

Zacharias C. Zacharia is the director of the Research in Science and Technology Education Group at the University of Cyprus and an Associate Professor of Science Education in the Department of Educational Sciences at the University of Cyprus. His research interests focus on the implementation of computer supported inquiry learning in science education.

Index

21st Century Teaching 247, 265

A

activity data 2, 4, 21, 26, 28
Activity Tracker 2, 9-10, 15, 26, 28
Agent 85, 89, 91-93, 95, 101, 103, 105-106, 146, 185
Agent-Based Models 159, 185
Algorithms 5, 188-189, 192, 203, 216, 218
ARIS 144-145, 149-151, 157
Assessment 9, 53, 67, 72, 79, 84-86, 88-92, 94-95, 99-106, 109, 111, 117, 120, 129, 152-153, 156, 189, 192, 195-196, 203, 205-206, 211, 223, 247, 250, 259, 277
Augmented Reality (AR) 136, 157

B

Big Data 219-223, 225, 229, 234, 236, 239-244

C

Chemistry Education Research 247, 262, 265
Client 163-164, 178, 180, 185
Climate 32-38, 43-46, 48, 50, 52-57, 135, 141, 165, 184-185, 199, 222
Climate Change 32-38, 43-46, 48, 50, 52-57, 135, 141, 165, 184-185, 199
Climate Change Science 30, 32-34, 37-38, 46
Co-Design 224, 286-287, 295
Coherence 79-80, 131, 223, 233, 235, 237, 244
Coherence in an Inquiry-Based Learning Activity Sequence 80
Collaboration 84-86, 88-91, 95-96, 101, 103, 105-106, 134-136, 140-141, 144, 147-149, 156, 164, 177-178, 181, 234, 236, 241, 243, 246, 265, 286
Collaborative learning 84, 90, 101-103, 105, 134-135, 143, 153-154, 184, 194
Collaborative Problem Solving 84-85, 87-88, 90-91, 95-96, 102-106, 135, 141, 143
collaborative skills 101
Combined Effect of Software Scaffolds 58, 80
Complex Systems 158-161, 171, 177, 179, 181-182, 184-185
Computer Agent 85, 89, 92-93, 95, 101, 105-106
computer-based assessment 88, 91, 100
Computer Model 185
computer-supported learning environments 59-61, 73, 81
Concept Introduction 194, 196, 199-200
Constructivism 25, 201, 203, 218

D

Design Principle 164-165, 167-168, 174, 185
Dynamic Visualizations 107, 111, 113, 121, 126-127, 133, 194

E

eBooks 30, 37, 56
elementary school 21, 23, 109-110, 127, 130, 293
Emergence 161, 185
Engagement 2-3, 23, 45, 60, 90-91, 94, 97-99, 134, 136, 139, 144-145, 149-151, 153, 156-157, 163-164, 170, 178, 180, 194, 223, 233, 238, 246, 252, 255, 264, 268-269, 275-277, 280, 285-287, 289-290, 295

F

Fitbit 4, 6, 9-12, 14, 17, 20-21, 25-26, 28
Flipped Classroom 265
Flow 65, 145-147, 152-154, 157, 160, 164, 192, 194

G

Game-Based Learning 136-137, 140, 148-149, 151, 157
Games 18, 27, 30, 32, 34, 36-37, 48, 50-51, 53-57, 104, 134, 136-144, 146-151, 153-157, 163, 217
GPS 25, 138, 140, 143
group composition 88, 90, 99, 101, 106
groups 9, 13, 46, 49, 84, 86, 88, 90-91, 96, 98-101, 104, 116, 122, 126, 137-139, 141, 143-145, 148, 162, 173, 178, 224-225, 236-237, 257-258

H

Harmful Algal Blooms (HABs) 43, 56
Human Variability 280, 295

I

IBL-STEM 265
Immersive 143, 149, 151, 153, 182
Information and Communications Technology (ICT) 30-32, 56
Inheritance of Traits 107, 109-110, 112-114, 117, 121, 123, 125-127
INPLACE 134, 136, 140-152, 157
Inquiry 1, 3, 7-11, 21-24, 26, 28, 30-32, 34-40, 42-43, 45, 49-54, 56-62, 64-68, 70, 72-76, 78-81, 85, 107-109, 111, 127, 129-140, 142, 146-147, 150-152, 155, 190, 194-196, 211, 213, 219, 223-224, 227, 231-232, 236, 239-241, 243-246, 253, 255, 261-262, 265, 267-270, 272, 275-278, 290-291, 295
Inquiry Learning 3, 22, 60, 62, 68, 72, 74-76, 79-80, 155, 219, 244-245, 265
Inter-Contextuality 80
Interdependent 136, 140-142, 148, 157, 163, 229
Introducing and Removing Scaffolding (Fading In 81

J

Juvenile Justice 267-269, 272-273, 275, 287, 292-293

K

Knowledge Integration 54, 111, 117, 125, 130, 132-133
Knowledge Integration Framework 133

L

Laboratory 6, 38-39, 43, 52-53, 62, 68, 70, 77, 187-190, 194-195, 204-211, 214, 220, 247, 249, 257, 263, 268, 277, 290
Learning 3, 7-8, 13, 22-28, 30-31, 33-34, 36-38, 47-53, 56, 58-63, 65-81, 84, 90-91, 95-96, 99-109, 111-113, 115, 117, 122, 126-140, 142-144, 146-148, 150-159, 161, 163-167, 170-172, 174-185, 192-195, 199, 201-204, 206, 211-217, 219-221, 223, 226, 229, 231-241, 244-248, 250, 252-265, 267-272, 274-278, 280-281, 285-286, 289-295
Learning Cycle 194-195, 199, 211-212, 216
Life Cycle of Plants 107, 126

M

middle school 1, 26, 75, 105, 110, 130, 145, 150, 152, 184, 220, 250, 257
Mobile 4, 56, 134-136, 138-145, 147-155, 157, 182, 220, 249-250, 256, 269, 273, 286-287, 290
Model-based Learning 158, 165, 201
Models 30-32, 34-38, 40-41, 45, 50, 52-53, 55, 57, 103-104, 131, 159, 161-162, 165, 168, 178, 181, 183-185, 187-190, 192, 194, 198, 202-203, 210, 214, 216, 218, 227, 232, 236, 240, 242, 295

N

NGSS 3, 8-11, 13, 15, 17-18, 20-21, 24, 33, 35, 43, 55, 108, 110, 117, 122, 131, 227, 271, 276, 281, 294

O

Organic Chemistry 187, 201, 204

P

Participatory Simulations 25, 27, 158-159, 179-180, 182, 184
Pedagogical Content Knowledge (PCK) 244
Pedagogy 31, 59, 111, 132, 141, 227, 234, 250, 252, 257, 260, 262-263
Personally-Relevant Data 2, 28
Physical Sciences 245-247, 252-254, 256, 260, 264, 266
Physics Education Research 247, 251, 262-266
Pollination Animation 113-114, 116, 119, 121-122, 125-126
Post-Activity Application 208
Prior Knowledge Activation 196
Problematizing Student Inquiry 58, 60-61, 74, 81
Problem Solving 76, 84-96, 99, 101-106, 135, 141, 143, 176, 186-189, 199, 201, 206-210, 213-214, 217-218, 248, 265
Problem-Solving 85-86, 93-96, 104-105, 131, 175, 186, 206, 211, 214, 219
Professional Development (PD) 219-220, 222, 224, 244
Professional Learning Communities (PLC) 244
Project Based Inquiry Science 267, 275-277, 290
Project-Based Learning 277, 291, 295
Prototyping 270, 286, 295

Q

QR Codes 143-144

R

Role-Play 159, 162, 185
Roles 11, 88, 90, 94, 96, 99, 137-139, 141-142, 144, 148, 181, 212, 214-217, 225, 239, 241, 255

S

Science Education 25-26, 28, 31, 34-35, 51-56, 59, 75-78, 80, 84, 104, 107-108, 112, 128-132, 143, 151-153, 155-156, 182-184, 212-217, 219-220, 242, 247-248, 251, 260-263, 268-269, 275-276, 287, 289, 291
Science Practices 24, 35, 43, 51
Scientific Inquiry 31-32, 35-36, 52-53, 76, 78, 131, 134, 137, 139, 142, 146-147, 151, 262, 265, 295
Scientific Models 30, 34-36, 38, 45, 52, 192, 202
Sea Level Rise 32, 37-38, 40, 57
Serious Educational Games (SEGs) 32, 37, 57
Simulation 50, 79-80, 92, 104, 111-113, 158-160, 162-181, 183-188, 192-201, 204-208, 210-212, 216, 218, 265, 289
software scaffolds 58-62, 68, 73-74, 80-81
Special Education 165, 175, 181, 271, 273, 291, 293-295
statistics education 1, 3
Stoichiometry 186-189, 195, 199-201, 203-216, 218

structuring student inquiry 73, 81
Structuring Student Inquiry in Computer-Supported Learning Environments 81
STSE Education 266
Subject Matter Knowledge (SMK) 244

T

Teaching 4, 25-26, 28, 30-34, 36, 49-54, 56, 75-78, 89, 91, 101, 103, 105, 109, 115-116, 119, 128-132, 152-153, 157, 165, 182-183, 194, 201-203, 213-216, 218, 220, 223, 225, 227, 230-231, 234, 237-239, 242-244, 246-247, 252-266, 274, 287, 291-292, 294
Technology 1, 3, 26-32, 37, 41, 51, 54, 56, 59, 75-76, 80, 85, 91, 95, 101, 103, 105, 108-109, 115, 125, 127, 129-130, 132-133, 135-136, 138, 144, 151-153, 155-156, 164, 182-184, 212-214, 216-217, 225, 243, 245-247, 250-251, 253, 255-256, 258-261, 264-266, 268-269, 272, 275-277, 280, 286-287, 289-291, 293
Technology Education 182, 243, 265
Tinkerplots 6, 8, 16, 18, 25, 28
Transfer of Knowledge or Skills 81
Typicality 6, 29

U

Universal Design for Learning (UDL) 268-269, 275, 295
Upper Elementary 107-110, 117, 127

V

Variability 3, 6-7, 29, 280, 295
Variation of Traits 107, 109-110, 112-113, 117, 121, 125-127

W

Wearable Technology 1, 28-29
Web-Based Inquiry Science Environment (WISE) 107, 132-133

www.ingramcontent.com/pod-product-compliance
Lightning Source LLC
LaVergne TN
LVHW062112200926
848484LV00007B/1652